# Imperfect Markets and Imperfect Regulation

# Imperfect Markets and Imperfect Regulation

## An Introduction to the Microeconomics and Political Economy of Power Markets

Thomas-Olivier Léautier

The MIT Press
Cambridge, Massachusetts
London, England

This book was set in Times by Westchester Publishing Services. Printed and bound in the United States of America.

Library of Congress Cataloging-in-Publication Data

Names: Léautier, Thomas-Olivier, author.
Title: Imperfect markets and imperfect regulation : an introduction to the microeconomics and political economy of power markets / Thomas-Olivier Léautier ; foreword by Jean Tirole.
Description: Cambridge, MA : MIT Press, [2019] | Includes bibliographical references and index.
Identifiers: LCCN 2018017791 | ISBN 9780262039284 (hardcover : alk. paper)
Subjects: LCSH: Power resources—Economic aspects. | Electric utilities.
Classification: LCC HD9502.A2 L4185 2019 | DDC 381/.101—dc23 LC record available at https://lccn.loc.gov/2018017791

10 9 8 7 6 5 4 3 2

*A Oriane-Zénaïde et François-René.*

*SGANARELLE:* *(...) Qu'est-ce que vous croyez?*
*DOM JUAN:* *Ce que je crois?*
*SGANARELLE:* *Oui.*
*DOM JUAN:* *Je crois que deux et deux sont quatre, Sganarelle, et que quatre et quatre sont huit.*
*(Molière, Dom Juan, Act 3, Scene 1)*

Contents

# Foreword

The power industry has a bright future and is also facing new challenges. The demand for electricity will grow with the demise of fuel-powered mobility and the advent of the electric car and with related developments in other human activities. Yet it itself has been a carbon-intensive industry and has yet to complete its transition to cleaner sources of energy. This revolution will affect the way each segment of the industry operates: generation and transmission as well as local consumption. Smarter grids will also reinforce the impact of price signals.

Guiding the power industry and its regulators through this complex transition requires powerful tools. Making the various agents (residential consumers, industrial users, transmission operators, grid owners, and producers) accountable for the consequences of their choices?—on the environment, on shortages, on cost?—has always been a leitmotiv of electricity economics and of economics more generally. But new ideas are needed to meet the new challenges: How should we provide incentives to install peak capacity that will be used only a few hours per year? How do we connect the short-term incentives provided by the spot prices with the long-term investment required to guarantee network stability? How do we design and monitor power markets to limit exercise of market power? Which retail contracts provide the best incentives to consumers? Which consumer classes should be equipped with smart meters first? How do we create incentives to build and adequately maintain the transmission infrastructure? How will renewable energy sources transform the generation mix?

Thomas-Olivier Léautier's book is valuable to anyone interested in these issues, for three reasons. First, it is the first book to lay out answers to all these questions, grounded in economic analysis and consistent with experience in power markets. It provides a clear and structured synthesis of dozens of academic articles and uses economic theory to shed light on industry-shaping events, such as the California energy crisis in the 2000–2001 winter and the collapse in carbon prices in Europe in the 2010s. Second, it presents the economic analysis that underlies policy issues, starting from the simplest situation and progressively adding complexity and different features. This enables the reader to see how issues relate to one another. Finally, it carefully separates the recommendation from economic theory from the constraints of real-life policy making.

Léautier was trained as an engineer at the French elite engineering school Polytechnique and received a PhD in economics from MIT. He later acquired much academic and nonacademic experience with the electricity industry, and he is knowledgeable with its state-of-the-art economics of energy; indeed, he regularly contributes to it himself.

I cannot think of anyone better qualified to write this book. Léautier is heir to the tradition of French "ingénieurs-économistes" (engineer-economist), which started in the nineteenth century and in the twentieth century counted among its ranks many top civil servants and private sector leaders, whose mantra was to apply rigorous analysis to improve efficiency and find innovative ways of allocating scarce resources.

Among them is Marcel Boiteux, one of the main characters staged at the beginning of this book, who coinvented peak-load pricing in the 1940s. He applied it early on at the French utility company, Électricité de France, before becoming the designer of the French nuclear program when he was chairman of the utility in the 1970s. Boiteux is, to the best of my knowledge, the only nonacademic to have presided over the prestigious Econometric Society, a learned society created in 1930 by Frisch and Joseph Schumpeter and developed in the 1930s by Hotelling, Pigou, Divisia, Roy, Keynes, Irving Fischer, and other luminaries. Pierre Massé is another case in point. In the 1940s, while at Électricité de France, he made important contributions to dynamic stochastic programming to optimize the use of electric generation as well to a number of economic problems; he later became head of the French planning board.

Like their nineteenth-century forebears (such as Jules Dupuit, who, in 1844, invented the notions of consumer surplus and consumer segmentation?—in economics jargon, second-degree price discrimination?—while working as an engineer on roads and train services), the members of the Econometric Society responded to the problems they actually faced in their work environment. Similarly, Léautier efficiently uses his dual training as engineer and economist to bring relevant answers to today's questions. The industry has now abandoned its planning logic and in most countries has turned to the private sector for enhanced efficiency and to a regulator to constrain market power. But while new challenges emerged, the basic challenges remain.

Though familiarity with basic economic concepts is an obvious requirement, the book is a satisfying read. It is self-contained and remarkably luminous. The first part of each chapter presents the main arguments, illustrated with graphs. It is aimed at interested but nontechnical readers. The second part of each chapter presents the equations underlying the arguments, which will be useful for students and researchers. It leaves aside the ideological debates that typically divide proregulation and antiregulation advocates and focuses on how good economics can help contribute to the common good?—without prejudice. Readers of this book will find themselves at the frontier of present knowledge on the economics of the power industry. Whether the reader is a student, an academic, a regulator, or an industry participant, this book is a must-read.

Jean Tirole
Chairman, Toulouse School of Economics
2014 Nobel laureate in economics

# Acknowledgments

This textbook on the microeconomics of power markets is almost an autobiography. I have been fortunate enough to study, comment on, and participate in the electric power industry for more than twenty years. Along the way, I have met and worked with truly great minds, who have inspired me and taught me all I know about power markets. I am grateful for their comments, insights, and encouragements, which have led to this text. Of course, the opinions expressed and the policy recommended in this text, and all errors and omissions are my responsibility, and can be attributed to my employers, the Toulouse School of Economics, and EDF.

## Academic Influence

Paul Joskow, who was then head of the economics department at MIT, introduced me to the wonders of the power industry one sunny afternoon in May 1995. Paul handed me an article by Bill Hogan and asked me to read it and tell [him] what I thought about it. This was simultaneously a perfect introduction to power economics, since Bill's article is probably the most influential academic article on the power industry published in the 1990s, and a true example of the MIT spirit: trust the graduate students to come up with ideas. Under Paul's guidance, my thoughts on Bill's article formed the backbone of my PhD dissertation and led to three academic publications. Throughout pursuit of my PhD, Paul proved a thoughtful and present adviser.

My MIT luck did not stop here. A few weeks later, Jean Tirole, who was visiting MIT for the summer, asked me what I was working on. He was intrigued by the restructuring of the power industry, which had started in Great Britain in 1990 and was under way in the United States, and offered to cosupervise my dissertation. As a PhD adviser, Jean was simply perfect, exceedingly insightful and wonderfully dedicated. As a testimony to his immense talent, he received a Nobel Prize in economics, a rare distinction. As a testimony to his wonderful personality, the entire economics profession was truly happy for him—an even rarer feat.

Bengt Holmstrom was my third MIT PhD adviser. Bengt is known for his pathbreaking work on the theory of the firm, for which he received a Nobel Prize in economics. He provided extremely valuable insights in the theory part of the dissertation and advised me throughout the process.

I met Claude Crampes for the first time in Buenos Aires in January 1995, as we were both on a mission for the World Bank. Almost twenty years later, in the fall of 2006, Claude welcomed me to the Toulouse School of Economics, and we have worked closely together over the past ten years: we wrote research articles, developed and taught a graduate course on the economics of electricity markets and networks, developed and taught an executive education seminar, and wrote a blog on the same topic. My debt and gratitude to Claude is boundless. This book would not exist without him: he tested every idea exposed in these pages and patiently read countless previous versions. In addition, Claude shepherded me through the perils of academic life. His advice was always spot on, and his support unwavering.

I have had the luck and the privilege to work with Richard Green, professor at Imperial College in London, over the past five years. The output of this collaboration is presented in chapter 9. Richard is both an intuitive economist (he elegantly draws results where I painstakingly derive them) and a prolific researcher, who has explored most facets of the electric power industry. In particular, he produces superb work at the intersection of economics and engineering.

I have recently started working with Estelle Cantillon, who teaches economics at the Université Libre de Bruxelles. Estelle's interest in power markets arises from her path-breaking research in market design. I am confident she will uncover many of the remaining mysteries of price formation in power markets.

I have been inspired by Bill Hogan, who teaches economics at the Kennedy School of Government at Harvard University. Bill is probably the academic who has most influenced the actual design of power markets. Bill possesses simultaneously unshakable convictions and unlimited energy to share these convictions. I believe he serves as a standard for the engaged academic. This book is my modest attempt to follow in his footsteps.

For ten years, the Toulouse School of Economics has provided a highly stimulating and supportive environment for my research, leading to the writing of this book. I am deeply indebted to my fellow researchers: Jean-Charles Rochet, my first academic co-author; Christian Gollier, who tirelessly and passionately advocates and supports energy and climate change research at TSE; Jacques Crémer, who recruited me at TSE; Bruno Jullien and Patrick Rey, infinite sources of wisdom and knowledge about industrial organization and microeconomic theory; Michel Lebreton, who opened my eyes to the beauty of cooperative game theory; David Salant, a leading expert of auction theory who read and provided insightful comments on my final draft; Bert Willems (visiting TSE from Tilburg University), who shares many of my research interests; and Pierre Dubois, who introduced me to applied econometrics. All were incredibly generous with their time and advice, answered my questions, and supported me throughout the process. In particular, I have had the privilege to work with two outstanding PhD students: Nicolas Astier and

Xavier Lambin. Nicolas and I developed a formal analysis of the adoption of demand response, which is presented in chapter 5. Xavier and I have formalized the economic impact of setting up a capacity market in one market on an adjacent market. As all professors know, witnessing students learn and grow, and sometimes contributing to the process, is one of the great joys of academic life.

Finally, Jean-Michel Glachant has welcomed me to the Florence School of Regulation, which has become my second academic home. Jean-Michel and his team are bridging the gap between academic research and policy making by conducting applied research, holding mixed academics/practitioners conferences, and organizing training courses. In particular, my colleague Fabien Roques and I developed a course on advanced topics in power market design, from which this book is inspired. I am grateful for this opportunity.

**Practitioners' Influence**

I spent seven years in the power practice of McKinsey and Company, working mostly in North America. It was my privilege to advise companies during the restructuring of the power industry in the late 1990s and to be an immediate witness to the collapse of Enron and other merchant power producers. During my tenure, I worked with outstanding thought leaders, in particular, Thomas Seitz, Eric Lamarre, Claude Généreux, Raoul Oberman, and Les Silverman. They and other McKinsey colleagues firmly and relentlessly questioned my thinking, leading me away from easy shortcuts, on to the path of clear and rigorous exposition of complex issues.

This book grew out of an executive education course I have developed for Électricité de France (EDF) over the past ten years, jointly with Fabien Roques and Damien Heddebaut. Fabien is not only a great economist (like Richard Green, he received his doctorate at Cambridge in Great Britain, working with David Newbery) but also an outstanding economic consultant. He knows the European policy debates extremely well and shapes many an argument. Damien knows the power industry like the back of his hand. He is a bottomless source of insights on new rules and emerging technologies. Working with Fabien and Damien greatly enhanced my understanding of the current evolutions in the industry. I am also grateful to David Jestaz, who ran EDF's Corporate University for Management as I developed the course, and Nicole Verdier-Naves, who was (and still is) overseeing EDF's overall leadership and development. Both gave Fabien, Damien, and me entire academic freedom to develop the course as we saw fit, irrespective of EDF's strategy or stated positions.

Nicolas Couderc, who currently runs the French portfolio of EDF's renewables division, has also contributed greatly to this book. First, he introduced me to the hard truth of climate change. He is one of the few persons I know who anticipated the sharp decline in the cost of producing electricity using renewable energy sources. He has read and corrected drafts of this text and raised numerous questions. Answering them has enriched this text.

Finally, I have been blessed to work with market designers Gordon van Welie, whom I met for the first time on September 11, 2001, as he started his tenure as CEO of ISO New England, and Brent Layton and Carl Hansen, who are respectively the chairman and CEO of New Zealand Electricity Authority. All three have shown me the challenges of reconciling economic principles with the realities of electricity markets, in particular, the governance of the industry. We have had countless and always enriching conversations about the subtle art of market design, which attempts to turn economic convictions into practical rules that can be enacted and implemented. This text owes much to these conversations.

# 1 Introduction

"Communism is Soviet power plus the electrification of the whole country."
—Vladimir Ilyich Ulyanov, Lenin (1920)

The combustion engine and the electric power grid have arguably shaped the infrastructure and the economy of the member countries of the Organization for Economic Cooperation and Development (OECD) over the course of the twentieth century. In the twenty-first century, the imperative to reduce—and even reverse—net emissions of carbon dioxide ($CO_2$) to limit the impact of climate change will most likely drive the combustion engine off the road and lead electricity—provided it is produced without $CO_2$ emission—to occupy a greater share of the energy mix.[1] This energy transition will require a massive investment in new technology, hence sound energy policies and economic interventions. This textbook aims to provide policy and decision makers with a well-grounded understanding of the microeconomics of the power industry to facilitate this energy transition and reduce its cost.

## 1.1 Why Electricity?

Comrade Lenin was right: the electricity industry has truly powered the twentieth century. The electricity fairy has reached towns, factories, and countrysides, increasing human

1. Precisely, the share of primary energy used for power generation will grow from 42 percent in 2015 to 47 percent in 2035. While this evolution appears minor, it is in fact significant, given the inertia of our energy production system. Source: British Petroleum (2017).

productivity and well-being in its wake. Numerous statistical studies have estimated that throughout the twentieth century a 1 percent increase in electricity consumption has been associated with a 1 percent increase in GDP. Per capita electricity consumption is a reliable indicator of a country's level of development and is routinely reported by the World Bank and other development agencies. As a result, the U.S. Academy of Engineers found in 2000 that electrification, not space travel nor the conquest of the atom, was the greatest engineering achievement of the twentieth century.

The electricity fairy still has a long way to go before she reaches all of mankind and delivers on her promise of global growth. Today, 1.2 billion people, mostly living in Africa, still have no access to electricity, while many more have only limited access; for example, Indians consume only about 20 percent of the world's average. The electric power industry must therefore continue to provide reliable access to affordable electricity in OECD countries, where demand is expected to grow at a modest 1.2 percent per year between 2012 and 2040 and to expand coverage to emerging countries, where demand is expected to grow at a brisk 2.5 percent per year over the same period (US Energy Information Administration 2016; British Petroleum 2017).

The electricity fairy now faces a significant obstacle in its expansion: the carbon constraint.

Most observers agree that climate change is a physical reality, caused by increased concentration of greenhouse gases, chiefly carbon dioxide ($CO_2$). Empirical evidence suggests that cumulative $CO_2$ emissions and average surface temperature are approximately linearly related. Thus controlling $CO_2$ emissions is necessary to contain climate change. Specifically, recent scientific analysis suggests that limiting the average temperature increase to less than 2°C with more than 50 percent probability, which is the agreement negotiated at the United Nations Climate Change Conference held in Paris in December 2015 (known as the Conference of Parties [COP 21]), will require world $CO_2$ emissions from fossil fuel and industry to become negative sometime soon after midcentury (Intergovernmental Panel on Climate Change 2014). A tall order indeed!

To achieve this goal, a revolutionary transformation of our energy system is required, which includes three dimensions: first, we need to reduce the economy's energy intensity by breaking the historical correlation between economic growth and energy-usage growth. Recent evidence suggests this is already occurring, at least in OECD countries (see, e.g., US Energy Information Administration 2016).

Second, since electricity is the only non-$CO_2$-emitting energy we currently know how to produce, we need to electrify transportation and heating: fuel-powered cars will gradually cede the roads to electric cars, and heat pumps will warm residential, commercial, and industrial buildings. In OECD countries, these two effects will roughly balance each other, leading to the slow growth in electricity demand mentioned above.

Third, we need to decarbonize our entire electricity generation fleet: coal- and gas-burning power plants will be replaced by low-carbon technologies, such as wind, solar, hydro, and nuclear. The magnitude of this challenge cannot be overstated: coal-fired power plants are the highest $CO_2$ emitters, generating almost a ton of $CO_2$ per Megawatt-hour produced. Today, coal-fired plants contribute around 40 percent of the electricity produced worldwide. More problematic, they are dominant in China (75 percent of electricity produced in 2012, expected to decrease to 63 percent in 2020) and India (72 percent in 2012, expected to decrease to 68 percent in 2020), where half of the growth in electricity consumption is expected to take place over the next decades.

A radical transformation of the power industry will entail three major components. First, with *decarbonization*, renewable and carbon-free sources will produce a significant share of consumed electricity. Second, with *decentralization*, small, decentralized units will be located near or at the consumption centers, reducing the need to transport electricity; electricity storage will be economical. The final element is *demand response*: since wind and solar production are not controllable (e.g., solar panels never produce at night, and wind turbines' production is subject to the whims of Eole, the God of wind), customers' ability to adjust their demand in times of scarcity and storage will become an essential part of the power industry.

This represents a formidable scientific, technical, and financial challenge. Engineers will develop technologies to harness the energy from the sun and the wind, to capture and store carbon, and to manage the spatial and temporal volatility of energy production and usage. This will require tremendous investment, both in R&D to develop new technologies and in capital to deploy these new technologies. In its Roadmap to 2050, the European Union estimates the required increase in capital investment at €270 billion annually, or 1.5 percent of GDP, in line with other estimates.

Predictions on the speed of the upcoming revolution in the power industry range from one extreme to the other: Jeremy Rifkin (2011) envisions a third industrial revolution, driven by data, that will radically transform the industry in a few years, while the International Energy Agency (IEA 2015), stressing the long period required to replace installed generation assets, suggests a more conservative forecast. The future probably reconciles both views: the transformation will be as profound as predicted by Jeremy Rifkin, but it may take longer to come about. Technology and policy will ultimately determine the speed of transformation.

One key driver of change is the speed at which low-carbon-production technologies and storage technologies become competitive. In the fall of 2017, developers offered to produce solar power in Saudi Arabia for less than $20/MWh, which is less than the variable cost of a gas-fired power plant. While this price has surprised investors, it is at the time of this writing established as the standard. Should prices for solar electricity, in particular for concentrated solar power, which reduces the variability of output, be durably established below $30/MWh, the transformation will come faster than expected by the IEA.

Electricity storage constitutes another technological uncertainty. Today, most electricity storage technologies are not economical, and many observers do not foresee a significant change in the near future. On the other hand, numerous firms are investing significant amounts in developing batteries, for example Tesla and Panasonic are building a Gigafactory to capture economies of scale in battery production, and investors are financing numerous other storage solutions. Should this second group be correct, the energy transition will proceed more quickly than expected.

On the policy front, the main source of uncertainty is the speed of adoption of a carbon price in key economies. At the time of this writing, only Europe and a few U.S. states have functioning carbon markets, and the $CO_2$ price in the European market is too low to impact investment decisions. Should China and India set a carbon price, and should it rise to \$30/ton, the decarbonization of the power industry could occur much faster.

Other policies will play a crucial role in speeding up (or slowing down) this energy transition: for example, the rules defining demand-response mechanisms or the level and structure of subsidies to renewable energy sources (RES).

## 1.2 The Challenge

Minimizing the cost of the energy transition requires sound policies and investment decisions. Unfortunately, recent experiences suggest that policy and decision makers have sometimes been led astray.

### 1.2.1 A Brief History of the Power Industry

The first power companies were local, small power plants serving customers in their immediate vicinity. However, very rapidly electricity was produced, transported, distributed, and sold to final customers by vertically integrated monopolies, serving large territories. In the United States, these were publicly held (i.e., owned by private investors and publicly listed) and regulated. In the rest of the world, they were state owned. This industry structure has enabled massive development of the industry, hence massive productivity and well-being increase throughout the twentieth century.

Vertically integrated regional monopolies were the product of the technological features of the industry. Economies of scale in production favored large centralized production units serving a large number of customers over smaller decentralized units serving one or a small number of customers. Similarly, the transmission and distribution grids are natural monopolies: it is always cheaper to have a single grid in a country or a city. Finally, integration of production and grid assets, at both the planning and operation stages, was efficient. These three technological features led to regional vertically integrated monopolies.

Vertically integrated regional monopolies, either regulated or state owned, were also consistent with the prevailing economic wisdom of the time, which gave governments an essential role in providing infrastructure and basic goods such as electricity.

Starting in 1990, developed countries embarked on a restructuring process and have opened parts of the industry to competition: while the transmission and distribution networks remain regulated monopolies, generation and retail are deemed competitive. Wholesale markets replace hierarchies as coordination mechanisms for multiple producers and retailers in the short term. Anticipation of future wholesale prices replaces central planning to guide investment decisions in the middle to long term. The technological structure of the industry, however, is unchanged: electricity is generated in large centralized units and transported to customers using transmission and distribution grids. Restructuring started in Great Britain in 1990s, then reached the United States in the mid-1990s and Europe in the following decade.

This restructuring was made possible by two technological advances. The first data revolution, the massive increase in computing power, enabled the development of spot markets. A new power generation technology, the combined-cycle gas turbine, facilitated entry into the (now-competitive) power generation business: it is simple (compared with coal or nuclear) to build and operate and requires much lower capital investment.

Restructuring was also consistent with a change in the prevailing view on how best to organize the production and delivery of essential goods, such as electricity. Economists and policy makers had learned from the Great Depression that markets could not be trusted and that government intervention was needed to hold what Daniel Yergin and Joseph Stanislaw call the "commanding heights" of the economy (Yergin and Stanislaw 1998). The electric power industry was thus state owned or heavily regulated.

In the 1960s, the liberal revolution was led in academic economic circles by Friedrich von Hayek, an Austria-born economist, and Milton Friedman. Both were faculty members of the University of Chicago, and both were recipients of the Nobel Prize in Economics: in 1974 for Hayek and 1976 for Friedman. By the late 1970s, their ideas had reached policy makers. Margaret Thatcher was an avowed disciple of Hayek's, and Ronald Reagan shared Friedman's liberal views. Thus liberalization of the power industry, along with liberalization of multiple other industries, aimed to transfer investment and operations decisions from bureaucrats to firms competing in markets.

The industry's performance since restructuring has been uneven at best. Restructuring has been great for utilities CEOs and senior executives, who have seen their pay rise to match that of other corporate chieftains, for management consultants who have advised utilities to embark on bold strategic moves, for investment bankers and lawyers who have executed the resulting transactions, and for economic professors who have advised on market reforms and market design.

Restructuring has been less uniformly favorable for shareholders and customers. While there certainly are success stories, many failures are well publicized. The California debacle (reviewed in chapter 3) is probably the best known: during the winter of 2000–2001, electricity traders exploiting incoherent market rules were able to game the market, thus forcing rolling blackouts on the sixth-largest economy on the planet. On the other side of the Atlantic, European utilities have collectively lost more than half a trillion euros between 2008 and 2013, which strongly suggests that they are no longer a safe stock to hold. This story is vividly recounted by Jean Pierre Hansen and Jacques Percebois (2017).

Over the next few years, the emergence of economic decentralized power production technologies, coupled with the availability of local optimization techniques, has the potential to radically transform the technological structure of the industry. Large users, for example, factories but also office buildings and shopping malls, will become producers and will at times produce locally and sell their electricity surplus to the wholesale market and at other times purchase from the wholesale market.

This profound transformation is made possible by two technological advances. First, the second digitization, the widespread availability of measuring devices and optimization algorithms, makes local optimization possible. Second, the rapid decrease in cost of decentralized production makes it competitive against centralized production.

This transformation is also consistent with evolving social norms, which favor local production and consumption.

### 1.2.2 A Very Brief History of Electricity Policies

This massive technological leap to occur will require a sound regulatory framework. Policy makers and regulators must design rules that protect consumers from abuse of market power but also encourage and facilitate innovation. Sadly, the experience of the past twenty-five years is less than encouraging.

Electricity policies instituted during the twentieth century have been reasonably successful. Electrification proceeded in most OECD countries at a reasonable pace. The main challenge was the regulation of natural monopolies, the theory and practice of which were gradually developed over the years.

However, since restructuring, electricity policies have been sometimes egregious mistakes leading to disaster (e.g., California), sometimes overly cautious, hence impeding change (e.g., the insistence by U.S. regulators on constant retail prices as a benchmark), and often inconsistent. Early mistakes have led to reregulation of the industry. For example, in Great Britain, where electricity sector reform was pioneered in 1990, new market rules are being implemented that de facto reregulate the electricity generation mix, and the government is imposing caps on retail prices. These examples (and others) are detailed later in this book.

Thus as the power industry faces its most profound technological transformation, policy makers and citizens are unsure which industry structure and policies are most appropriate to usher in these changes.

## 1.3 The Answer: The Toolbox

Three reasons contribute to policy makers' poor performance over the past twenty-five years. The first is their lack of familiarity with the specific microeconomics of the power industry. This is probably best captured in the following quote, from Paul Joskow and Richard Schmalensee's (1983, page 9) classic textbook: "Currently electric power is supplied by complex and highly developed systems with unusual technical characteristics. These make it likely that reliance on an economist's instinct, developed through countless examples drawn from agriculture and manufacturing, will produce incorrect conclusions." Second, like most stakeholders, policy makers believe that electricity is different and cannot be treated like any other economic good.

Third, markets are by nature imperfect, and imperfect regulation or policies can make them more so. The only certainty is that market participants will take advantage of every loophole and every inconsistency to increase their profits. The California saga (reviewed in chapter 3) is a textbook example of a suite of well-intended rules that led to a disaster.

The objective of this book is to modestly contribute to the quality of policies and investment. Of course, no economics textbook is sufficient to deliver sound decisions. Yet I believe this book can contribute in two ways. First, by describing the economics underlying power markets. While they are different from those underlying agriculture and manufacturing, they are fairly straightforward. There are no magic intuitions nor fundamental microeconomic differences between the electricity industry and, for example, the airline or the hotel industries.

An electric power system is a truly awesome engineering feat. Yet its operations follow the rules of physics (electromagnetism, mechanics, chemistry, etc.) and those of economics. As we do for all other economic activities, we can compute the cost of producing and delivering power; we can decompose this cost in a variable part, which increases with the volume of electricity produced and transported, and a fixed part, which does not; and we can determine whether the variable part increases more or less rapidly than the volume of electricity produced and transported.

The electricity fairy truly brings light into the darkness. Billions of human lives have been bettered, millions have been saved by electricity. Yet as they do for all other economic activities, users put values to the electricity they consume, which vary across users, usages, and circumstances. Aluminum smelters are located where electric power is cheap. Factories and office buildings invest in light-emitting diode (LED) lamps to reduce their electric power consumption. Over the years, dozens of academic studies have confirmed that households are willing to reduce their consumption when the power price increases significantly.

Combining these values for electricity with the cost of production, leads to prices, that in turn, guide production, consumption, and investment decisions. This is the story this text tells.

This text's second contribution is to point out the challenges from turning microeconomics results into policies. Policy makers have to consider and trade-off multiple objectives, economic efficiency being only one of them. By construction, most policies are political compromises and will deviate from economic orthodoxies. The European Union is simultaneously pushing for the creation of a wholesale power market, thus making the spot price the cornerstone of decision making, and the growth of renewable energy sources.

To achieve the second objective, RES receive subsidies, making the spot price almost irrelevant (the underlying mechanism is reviewed in chapter 9). By pointing at specific junctures where economics and political economy diverge, that is, where economic results are politically challenging, this text will hopefully support policy makers' choices and the implications of different policies.

### 1.3.1 The Central Intuition of Power Economics

The main insight of the microeconomics of power markets is that average prices and costs, which are used in industries for which storage is possible, are not sufficient to describe the electricity industry. When discussing the economics of a car factory, an oil refinery, or a steel mill, executives (and policy makers) compare the average price of the product with its average production cost. If the former exceeds the later, the factory (or the refinery or the mill) is built (or continues to operate). If this is not the case, corrective action is required.

This average price versus average cost logic does not apply to the power industry, where on-peak prices can be 1,000 times larger than off-peak prices. A baseload power plant may profitably produce for more than 8,000 hours every year (out of 8,760 hours per year) and yet fail to cover its fixed costs. On the other hand, a peaking plant may cover its fixed costs in less than 20 hours per year on average. Chapter 3 demonstrates that, at the long-term equilibrium, all production technologies but one have average costs higher than the average price of electricity.

This most important result is called peak-load pricing. It structures the entire economics of the power industry, and is almost seventy years old: in 1949, a young researcher named Marcel Boiteux, freshly back from World War II, developed the theory of peak-load pricing (Boiteux 1949). A simplified version of the result is that the price of electric power should be equal to its variable cost of production, for example, €30/MWh for 8,700 hours per year, and spike to around €1,000/MWh on average for the remaining 60 hours of the year. Fixed costs of production are thus entirely recovered during these 60 hours.

Peak-load pricing applies to other industries, for which demand varies during the year and the good provided is not storable. Two often-used examples are plane tickets and hotel rooms. As all travelers know, flying is often cheaper in the spring than the summer, and hotel rooms at a ski resort are more expensive in the winter than in the summer.

The true specificity of the power industry lies in the numbers, not the peak-load pricing logic. Hotel room prices rarely exceed ten times the lowest prices. In the power industry, highest prices may be a hundred times the lowest ones. To understand this, remember that capital cost is recovered in around 60 hours per year. For a hotel, this would be the equivalent of having three nights a year to recover capacity costs and making no profit all other nights. Most hotels are full more often than three nights a year, hence capital is recovered during more often than three nights, hence on-peak prices need not rise so high. Furthermore, hotels and airlines are able to charge prices higher than their variable costs during off-peak periods, which contribute to fixed cost recovery. This is not the case for electric power, which is a commodity: power produced by plant A cannot be distinguished from power produced by plant B (when they are connected to the same uncongested grid).

Peak-load pricing in its simplest form is introduced in chapter 3. Additional features are progressively added, as different issues are examined. For example, peak-load pricing explains why the introduction of demand response and RES do not modify the time-weighted electricity price in the long run but have different impacts in Great Britain and in the Southeast of the United States. Peak-load pricing also explains why RES entry leads to negative prices: in some hours, producers have to pay to generate electricity and inject it into the grid, while users receive a payment to consume power and take it out of the grid. Peak-load pricing also explains why transporting a Megawatt-hour from Warsaw to Lisbon (or from Boston, Massachussetts, to Orlando, Florida) should be (almost) free most of the time and how it should be priced otherwise.

### 1.3.2 The Main Policy Challenge

The practical implications of peak-load pricing are the main cause of divergence between economics and policy making, which largely explains the complexity of power market design.

Economists argue that setting a very high price, for example €3,000 per Megawatt-hour during on-peak hours is efficient, as it leads consumers who value a Megawatt-hour at less than €3,000 to reduce their consumption. Electricity is therefore allocated to users who value it the most.

When demand does not sufficiently respond to price, administrative rationing may be required: for a few on-peak hours, a fraction of customers are not served. This practice is known by the self-explanatory term rolling blackout.

Economists argue that rolling blackouts, while unpleasant, are likely to be transitory. When customers are rationed, the price of power goes to a extremely high level, for example €20,000 per Megawatt-hour. Expecting such a price, most consumers will find a way to voluntarily reduce their consumption. Thus, in the long term, demand will respond to price, rolling blackouts will never occur, and prices will rise to balance supply and demand, to a few thousands euros per Megawatt-hour for a few peak hours.

The appeal of this cold logic is lost to policy makers—and voters, who are more concerned about fairness than efficiency: is it fair that people pay dearly for electricity precisely when they need it more? How can we advocate a market design that leaves open the possibility that a fraction of the population is not served precisely when they need electricity the most?

This tension is not limited to peak-load pricing in the electricity industry. Many other recommendations from economists, while perfectly legitimate in classrooms and seminar rooms, are exceedingly difficult to implement, for they lack political appeal. Optimal taxation constitutes a simple example of this tension. Suppose a community has built a bridge, and hence needs to charge a toll to cover the fixed cost. Two persons present themselves simultaneously. Mrs. *A* absolutely needs to cross to bring medicine to her very sick child. Mr. *B* would like to cross to go and watch a movie. Economic efficiency suggests that, since Mrs. *A* clearly values crossing the bridge much more, she should be charged much more than Mr. *B*. This insight was formalized in 1920 by F. Ramsey. However, most of us would find it morally unsound (not to say repugnant) to charge a grieving mother. The story does not stop here. If Mrs. *A* is, in fact, the wealthiest person in the country, our objections to levying a heavy charge weakens.[2]

Policy makers therefore develop market rules to mitigate the undesirable political impact of peak-load pricing. As will be discussed in this text, these rules lead to a very complex market architecture—with no guarantee to achieve the objective.

### 1.3.3 Outline of This Book

This book examines analytically the various segments of the power industry: generation, retail, and the impact of the transmission grid.

Each chapter addresses one specific issue and is structured in two parts: the first part presents the main results using pictures and verbal arguments. A few analytical tools are introduced. It is aimed at readers who are interested in the electricity industry but are not analytically inclined; for example, practitioners (including regulators and policy makers) and students taking a graduate course in management or public policy.

The second part presents an analytical exposition of these results. It is aimed at graduate students and researchers specializing (or at least strongly interested) in the economics of electricity markets who want to understand, apply, and expand on the underlying models.

This book is structured in four parts. Part 1 examines electricity generation, which represents the largest share of the costs of the industry. Today, electricity generation, while open to competition, remains heavily regulated. Generation has been the subject of significant academic interest. Chapter 2 presents the fundamental peak-load pricing model, which, as previously argued, is the cornerstone of the economics of the power industry. It starts from the simplest model, involving a single production technology and perfectly price-reactive customers, then progressively incorporates more realistic features.

Chapter 3 discusses market power in electricity generation, that is, the ability for producers to manipulate wholesale markets to increase their profits above the perfectly competitive level. This has long been a concern for economists, and, as illustrated by the California crisis, is a serious risk in electricity markets. Chapter 3 illustrates various strategies producers can choose from to exert their market power and discusses possible remedies. Its main policy recommendations are that, markets be designed to be as robust as possible to exercise their market power, and that market monitors be endowed with significant analytical capabilities and investigative authority to effectively police electricity markets.

2. I am grateful to my colleague Michel Lebreton for this illuminating example.

Part 2 examines electricity retailing. Historically, retailing was a very simple technical activity, which involved mostly billing and collections. Retailing was often part of the distribution activity. With the advent of restructuring, retailing has become an autonomous business. There were high hopes of intense competition and innovation. Alas, as chapter 4 discusses, retail competition is not as strong as initially envisioned, because customers are sticky. Thus retail competition is better represented by models of imperfect competition, where customers can be swayed by a good offer but have a preference for their historical supplier. Chapter 4 also discusses vertical integration and forward contracting.

The next innovation in power retailing is demand response, the subject of chapter 5. Today, most customers pay a constant price for every Megawatt-hour, disconnected from its true value as expressed in the wholesale spot market. This is, of course, economically inefficient. Demand response to price is essential as noncontrollable RES enter the market. Chapter 5 presents various approaches to implementing demand response. Its main conclusion is that demand response will start with large users (or grouping of small users).

Part 3 discusses the transmission and distribution grids. They remain natural monopolies, hence subject to regulation. Chapter 6 discusses the standard trade-off between rent extraction / and cost minimization present in all regulatory situations (Laffont and Tirole 1993), and presents three particular issues to be addressed. First, the grid operator must provide nondiscriminatory access to all market participants. This has led to structural separation for the transmission grid and a code of conduct for the distribution grids. Second, the pricing of network usage is complex, since electricity flows according to the laws of physics, not contracts, and lines are sometimes congested. Fortunately, a simple solution exists, called nodal pricing. Third, regulation must provide incentives for optimal investment in the grid and in generation.

Chapter 7 discusses how imperfectly competitive generators can take advantage of the limits on lines' transfer capabilities to exercise market power. For example, it presents different models where producers artificially create congestion on the transmission grid to increase their profits. These additional opportunities for market manipulation created by the transmission grid reinforce chapter 3's policy recommendations.

Part 4 discusses two current policy issues: the dynamics of renewables subsidies (chapter 8) and capacity mechanisms (chapter 9). This part's contribution is threefold. First, it illustrates the interplay between imperfect markets and imperfect regulation in two real and important cases. In both instances, an imperfection was detected, and policies were designed to correct this imperfection. In both instances, the effectiveness of the policies and their side effects are widely debated by academics and practitioners.

Second, it presents important analytical results that shed economic light on the policy debate. Chapter 8 determines analytically how the generation mix and the value of renewable energy evolve as the share of RES increases. It also illustrates that renewables subsidies could never stop, if the value of renewable energy were to fall faster than its production cost. Chapter 9 argues that capacity mechanisms are mainly motivated by political considerations, not economic ones; it then proceeds to compare the performance of different proposed mechanisms.

Finally, these two chapters illustrate how the basic peak-load pricing model introduced in chapter 2 can be extended to examine analytically critical policy issues.

Concluding observations and future avenues of research are presented in chapter 10. In particular, it revisits the main divergences between microeconomics and political economy.

### 1.3.4 What Is Covered, What Is Not Covered

The objective of this book is to provide interested but nontechnical readers with a compendium of all important results in the microeconomics of power markets. The model for this text is a corporate finance textbook for MBA students.

This book mostly presents economic results. These results have been "proved" by models. Contrary to other sciences, for example, physics, there is more than one way to represent an economic phenomenon. Most of the models I use are simple enough that their solution can be determined analytically. I believe this is sufficient for readers to build their intuition for the main results. Researchers build more sophisticated models to better represent the complexity reality. Solving these richer models often requires numerical analysis. While these models provide more precise answers, I do not believe using them profoundly modifies the economic intuition.

This book also presents facts, for example, cost figures and description of markets. These facts are used to illustrate the main results. Most examples come from Europe, with which I am most familiar, although I try to provide examples from other countries. Facts presented in this textbook will soon be outdated. For example, the cost of RES has decreased dramatically over the past five years and is likely to continue to do so in the next five years. I do not believe this limits the usefulness of this text: its objective is to provide readers with a toolbox—results they can use to interpret and understand the facts—not with an up-to-date factbase.

To simplify the exposition and keep the text accessible to non economists, I chose to leave out two important aspects of power markets.

First, I do not open the black box of price formation. Most of this text assumes that a black box generates a price that balances supply and demand at every hour. This is simultaneously true and untrue for power markets.

Power markets are different from most other markets: they are ultimately centralized, that is, one entity, called the system operator (SO), is responsible for administering real-time prices and quantities. Thus a price-generating black box does exist: the system operator's dispatch algorithm. Few other markets are organized this way. Another example is the ride-hailing application Uber, for which an algorithm matches supply and demand in real time and produces time- and demand-varying prices.

On the other hand, dispatch algorithms do not produce a single price. In fact, as discussed in chapter 2, they produce multiple prices (day-ahead, hour-ahead, five-minute real-time increments), which then gives market participants opportunities to design sophisticated bidding strategies. Understanding this black box is a fertile field of economic research.

Second, most results assume the industry is at its long-term equilibrium, that is, that the generation mix is selected to exactly match demand. This long-term equilibrium can be perfectly competitive (chapters 2, 5, 6, and 8) or imperfectly competitive (chapters 3, 7, and 9). Of course, no industry is ever at the long-term equilibrium. It has either too much or too little capacity installed, since input prices and demand conditions change every month while it takes years to build or retire generation assets. Imbalances from the long-term equilibrium correct themselves over time, albeit sometimes slowly. Understanding out-of-equilibrium dynamics is also a fertile field of research.

However, the long-term equilibrium provides a sound analytical benchmark, hence is used in this text.

# I WHOLESALE MARKETS

# 2 Marcel Boiteux Forever: Peak-Load Pricing

Price is a signal, not a punishment.
—Marcel Boiteux, testimony to the French National Assembly

The theory of peak-load pricing, developed in 1949 by Marcel Boiteux, provides a simple yet robust description of the economics of power markets. This chapter presents its main elements.

## 2.1 The Simplest Peak-Load Pricing Story

### 2.1.1 Set-Up

The simplest situation is characterized by a perfectly elastic demand and a single production technology. These two assumptions make peak-load pricing results easy to derive and understand. As discussed later in this chapter, they are not essential: the economic intuition is unchanged if these assumptions are relaxed.

#### 2.1.1.1 Demand

**Units** First, a word on units. The main unit of measure for electric energy used in this text is the Megawatt-hour (MWh). Kilowatt-hours (kWh) and Terawatt-hours (TWh) are sometimes used. A Megawatt-hour is 1,000 kWh, a Terawatt-hour a million MWh. To provide orders of magnitude, average residential consumption worldwide is 5 MWh/year, and France consumes 500 TWh/year. Wholesale electricity prices are usually expressed in €/MWh, while retail prices are expressed in Euro cents per kilowatt-hour—ct/kWh. One ct/kWh is equal to € ten/MWh. This text sometimes uses $/MWh and £/MWh. Given current exchange rates, all three units are roughly equivalent and are used indifferently.

The rate at which energy is produced or consumed is called power. Throughout this text, we consider the hourly rate, hence it is measured in Megawatts, that is, Megawatt-hours per hour. This is the appropriate unit for wholesale market transactions. Kilowatts (kW) and Gigawatts (gW) are sometimes used. A Gigawatt is 1,000 Megawatts, and 1,000 kilowatts is a Megawatt. As illustrated in figure 2.1, peak demand for a country such as France is 92,400 Megawatts, or 92.4 Gigawatts. Kilowatts (kW) are used to measure residential customers' maximum demand, which is typically lower than 10 kW per household in OECD countries.

**Load duration curve** Electricity demand varies greatly across hours within a year and across years. Electricity demand is higher during the day than at night and, higher on weekdays than on weekends. In northern Europe and Canada, electric heating leads to higher demand in the winter than in the summer. In the South of the United States, air conditioning leads to higher demand in the summer than in the winter. Power engineers represent this variation using a load duration curve, which displays demand for every half hour, ordered from the highest to the lowest. The load duration curve for France in 2009 is presented in figure 2.1. The peak demand, reached for only one half hour on February 8 at 7:00 p.m., was 92,400 MW. The minimum demand that year was 31,526 MW.

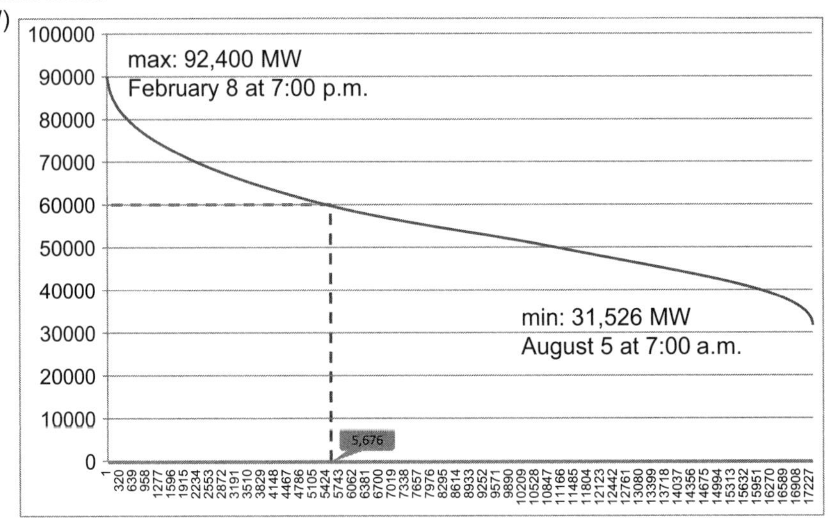


**Figure 2.1**
Load duration curve for France in 2009. All half hours of a given year are represented on the horizontal axis, demand is represented on the vertical axis. The load duration curve represents demand for every half hour, ordered from the highest to the lowest.

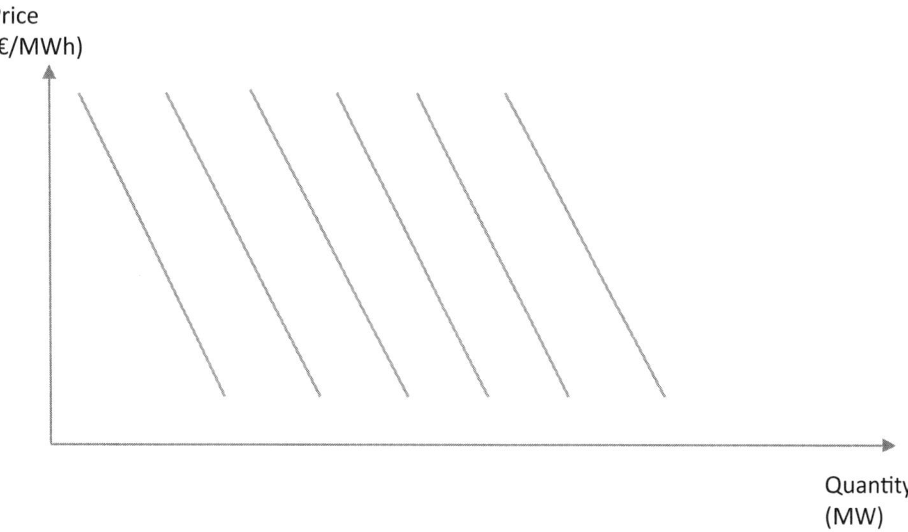


**Figure 2.2**
Inverse demand curves for different half hours of the year. In Europe, the left-most curve corresponds to a summer morning, the right-most to a winter evening.

Load duration curves are used to determine the number of half hours demand exceeds a given level. By construction, demand exceeds the minimum of 31,526 MW for all half hours of the curve. The curve shows that demand exceeds 60,000 MW for 5,676 half hours of the year. These are not consecutive half hours, some might be February evening, others might be January midday. This nonchronological representation of electricity demand is extremely powerful, as shown throughout this chapter.

**Inverse demand curves** In this section, we assume that consumers adjust their consumption to respond to variations in the spot price (i.e., the real-time price) of electric power. When the spot price increases, consumers reduce their demand. The natural representation of demand would be to have price on the horizontal axis and quantity demanded at that price on the vertical axis. However, for reasons that will soon become clear, economists prefer to represent inverse demand, that is, a diagram with quantity on the horizontal axis and price on the vertical axis. For a given price (on the vertical axis), the quantity consumed is measured on the horizontal axis. Demand decreases as the price increases, hence inverse demand curves are sloping downward, as seen in figure 2.2.

Since demand varies across hours of the year, we have numerous downward sloping inverse demand curves, one for each hour, as illustrated in figure 2.2.

The basic model also assumes all customers are identical. Thus, they all have the same load duration curve and the same sensitivity to prices.

#### 2.1.1.2 Supply

**Variable and fixed cost of production** The second assumption is that only one technology is used. It is characterized by a variable cost of production per unit, expressed in €/MWh, and a an hourly fixed cost of production per unit, also expressed in €/MWh. The variable cost is essentially the cost of fuel burned to generate electricity. It is assumed constant and is denoted $c$. The variable cost depends on the technology used and on fuel prices and usually ranges between 20 and 100 €/MWh for nuclear and thermal plants. The variable cost is essentially zero for RES such as wind turbines and solar panels.

The fixed cost of production includes the annual capital charge but also fixed operations and maintenance costs. It is also assumed to be constant per unit of capacity, hence is called capacity cost in this text, and is denoted $r$. It also depends on technology. As shown later, the relevant fixed cost for the story is that of a gas turbine, around €60,000/MW/year. The fixed cost used is the hourly fixed cost, that is, the annual fixed cost is divided by 8,760 hours per year.[1] To fix ideas in our example, we choose $c = €50/\text{MWh}$ and $r = \frac{60{,}000}{8{,}760} = €6.85/\text{MWh}$.

**Constant returns to scale in power generation** Consider a power plant of capacity $K$ Megawatt, which means it is impossible to produce output $Q > K$. As discussed above, the hourly capacity cost is assumed to be proportional to installed capacity and equal to $rK$. It must be paid every hour of the year. If the plant produces $Q \leq K$ Megawatt-hour during a given hour, the total cost of production for this hour is $rK + cQ$.

This representation of the cost of producing electricity is of course an approximation. In reality, the variable cost increases as production gets closer to the maximum feasible capacity, and the capacity cost per unit often decreases as capacity increases, as it includes a portion that is independent of capacity. For example, a power developer needs to pay lawyers to write up the contracts with the building contractors and equipment manufacturers. Legal fees are not proportional to the size of the asset, hence the power developer will pay an amount independent of the size of the asset. However, this approximation is close enough to reality that we can safely use it.

Under this approximation, electric power generation for a given technology exhibits constant returns to scale: producing 200 Megawatt-hours using a 200 Megawatt power plant costs exactly the same as producing them using two 100 Megawatt plants of the same technology.

### 2.1.2 The Problem

Inverse demand is downward sloping and time dependent. Furthermore, electricity cannot be stored economically on a large scale. This raises two questions: How should we price electricity? How much capacity should we build?

1. Numbers are expressed following the North American convention: a comma "," separates thousands, and a dot "." separates the integer from the decimal parts, e.g., 8,760.00.

The term *should* is ambiguous. The problem is first solved from the perspective of a benevolent central planner, hence the questions can be rephrased as What are the optimal electricity prices and capacity? As is often the case in economics, if competition is perfect, which is assumed in this chapter, the equilibrium reached by industry participants decentralizes the optimum, hence the questions can be rephrased as: What electricity prices and capacity arise in equilibrium? Chapter 3 shows that if competition is imperfect, the equilibrium differs from the optimum.

Electrical engineers and economists have attempted to find a rigorous answer to these two questions since the early days of the power industry in the 1890s. The formal answer was provided by a young French economist, Marcel Boiteux, upon his return from the World War II.

Before describing the solution, observe that other goods share these two features, for example, hotel rooms and plane tickets. Neither can be stored: a seat on the 8 p.m. flight from New York to Paris must be "consumed" at 8 p.m. Demand for both varies over time: more seaside hotel rooms are requested in the summer than in the winter. The solution to the electricity-pricing problem has been applied to these other industries where it is known as yield management.

### 2.1.3 The Solution

#### 2.1.3.1 Optimal prices

A general result in economics is that to maximize the net surplus from consumption, price should be equal to the marginal cost of production and to the marginal surplus from consumption. Understanding this result requires a few definitions.

**Consumer surplus** The consumer surplus, sometimes called the gross surplus or the surplus from consumption, is the surplus (or the utility, or the pleasure, or the benefit) that a consumer derives from consuming a given quantity of a good. To compute his gross surplus, economists estimate the surplus that a consumer derives from consuming each unit of the good, then add these surpluses.

Suppose that, for a given hour in a winter evening, a family consumes 5 kWh of electricity. One kWh goes to heating, hence is valued at 30 cts, that is, generates a surplus of 30 cts. Another kWh goes to lighting, which matters slightly less, 15 cts. Another kWh goes to the various screens (television, computers, etc.), and is valued at 10 cts. Finally, two other kWh go to domestic appliances (dish washer, washer, drier) that could be run later; these valued at 5 cts. The gross surplus from these 5 kWh is the sum of the surplus from each kWh: $30 + 15 + 10 + 2 \times 5 = 65$ cts.

Suppose now that many infinitesimally small units are consumed. They can be ordered by decreasing per unit surplus: the first unit generates the highest per unit surplus, the next very slightly less, and soon. If we plot the surplus per unit as a function of the number of units consumed, we obtain a downward sloping curve. This is the inverse demand curve, such as the ones presented in figure 2.2.

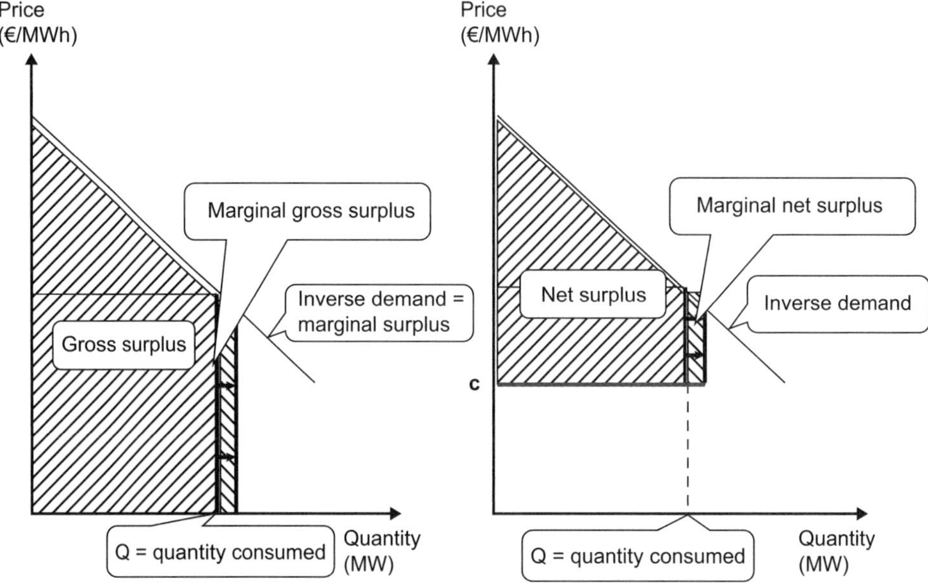


**Figure 2.3**
Gross consumer surplus and marginal gross consumer surplus (left panel) and net surplus and marginal net surplus (right panel).

The gross surplus derived from consuming quantity $Q$ is represented by the hatched surface under the inverse demand curve on the left panel of figure figure 2.3: it is the surface under the inverse demand curve up to the vertical line at quantity $Q$.

**Short-term net surplus** The net surplus from consumption is the surplus from consuming a given quantity minus the cost of producing this quantity. It measures the economic value generated by production and consumption. In the short term, the net surplus is the consumers' surplus minus the variable production cost. For example, if the variable cost of producing electricity is 3 cts/kWh, the short-term net surplus is $65 - (3 \times 5) = 50$ cts.

The net surplus from producing and consuming quantity $Q$ is represented on the right panel of figure 2.3 as the surface under the inverse demand curve and above the production cost $c$, up to the vertical line at $Q$.

**Marginal surplus** The word *marginal* is often used in this text. It refers to the last unit produced or consumed, called the marginal unit, or to an attribute of the marginal unit. For example, the marginal surplus (sometimes called the marginal value) is the surplus of the last unit consumed. In the above example, the marginal surplus is equal to 5 cts/kWh. Similarly, the marginal cost is the cost of producing the last unit produced.

The marginal gross surplus is represented on the left panel of figure 2.3, the marginal net surplus on the right panel. For every infinitesimally small quantity, the inverse demand is the marginal gross surplus.

**Optimal production and consumption** The objective of a benevolent central planner is to maximize the net surplus. Production and consumption are short-term decisions, hence the short-term optimum is to maximize the short-term net surplus. The optimal production and consumption plan is: every unit that generates a positive short-term net surplus is produced and consumed, while no unit that generates negative short-term net surplus is produced. The optimal quantity produced and consumed therefore sets the marginal short-term net surplus to zero. Under reasonable conditions, this optimal quantity exists and is unique.[2]

**Equilibrium price** Which price leads to the optimum? Consider first the consumers. If the marginal surplus was higher than the price, consumers will consume more, since this would increase their surplus. Thus they consume exactly all units whose surplus is higher than the price they pay. Consumption for any given price is such that the marginal surplus is equal to the price.

Consider now producers. If the price is higher than the cost, producers will produce more to capture positive profits. Production for any given price is such that the cost of the last unit produced (also called the marginal cost) is equal to the price.

Under reasonable conditions, there exists a unique equilibrium price such that supply is exactly equal to demand, that is, quantity produced is exactly equal to quantity consumed. At this price, the marginal surplus is equal to the price, which is also equal to the marginal cost. The marginal net surplus is equal to zero: the equilibrium leads to the optimum.

**Off-peak and on-peak prices** How does that insight apply to peak-load pricing? The key is to separate two different periods. In the off-peak period, production is lower than installed capacity, hence producing a marginal Megawatt-hour requires essentially fuel costs. Thus the off-peak price is equal to the variable cost, which we have denoted $c$. Consumption then adjusts to this variable cost, that is, consumption is such that the marginal surplus is exactly equal to the variable cost of production $c$. This situation is observed on the left of figure 2.4.

In the on-peak period production and consumption are precisely equal to installed capacity. The price is equal to the value of the last unit consumed, the value of the marginal Megawatt-hour. This situation is observed on the right of figure 2.4.

As a short cut, I use "off-peak" on "on-peak" to refer respectively to "off-peak period" and "on-peak period."

2. Sufficient conditions for the existence and unicity of the optimal quantity produced and consumed are presented in section 2.3.

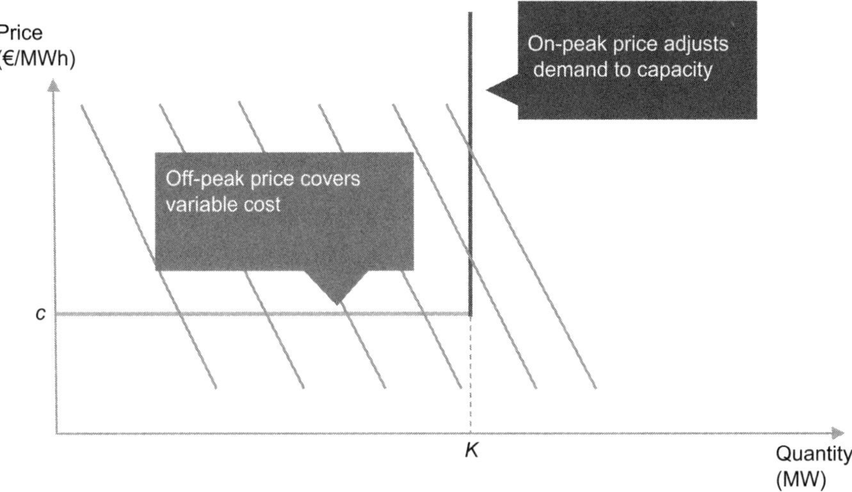


**Figure 2.4**
Demand curves and supply curve for a single-generation technology.

Observe the duality between off- and on-peak: off-peak, price is set by the variable cost and determines consumption. The dual situation occurs on-peak: price is set by the value of the marginal Megawatt-hour such that cumulative consumption equals capacity.

**Supply curve** A useful concept for examining markets and price is the supply curve, which traces the short-run marginal cost, that is, the cost of producing a marginal unit of a good for various quantities of this good, when capacity is already built. In our example, the supply curve is L-shaped.

The cost of off-peak production of one additional Megawatt-hour is the variable production cost (essentially the fuel cost), denoted $c$. The off-peak supply curve is the horizontal segment on the left of figure 2.4. On-peak production equals installed capacity, and the cost of producing one additional Megawatt-hour exceeds the variable production cost, since this would require additional capacity. The on-peak supply curve is the vertical segment on the right of figure 2.4.

### 2.1.3.2 Optimal and long-term equilibrium capacity

**Optimal capacity** Capacity choice is a long-term decision, hence the long-term optimum is to maximize the average hourly long-term net surplus, which is the average hourly short-term net surplus minus the hourly capacity cost.

Consider adding a (marginal) Megawatt of generation capacity. Installed capacity has no impact off-peak, hence the analysis is limited to on-peak hours. For every on-peak hour, since consumption is exactly equal to installed capacity, adding one Megawatt of generation capacity leads to the consumption of one additional Megawatt-hour, which

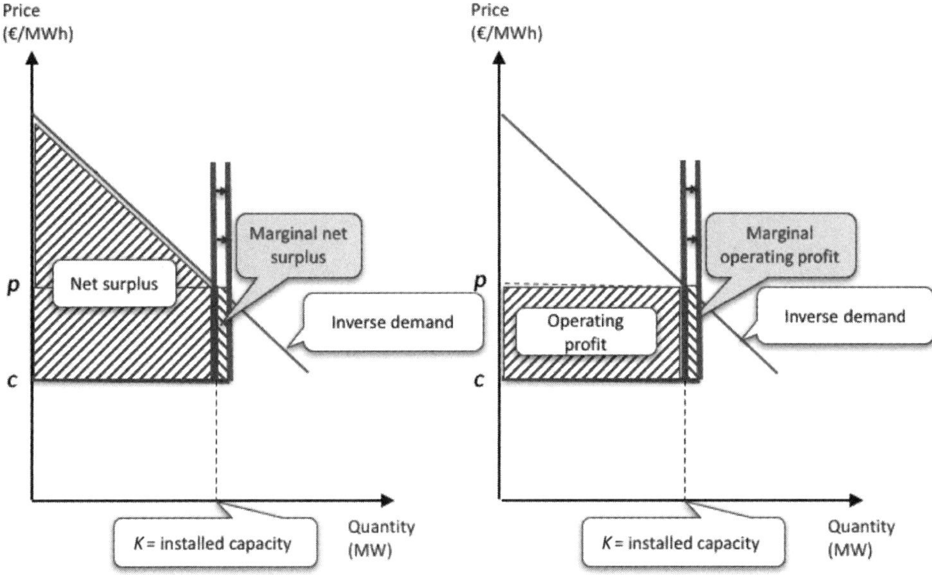


**Figure 2.5**
Short-term net surplus and marginal short-term net surplus (left panel) and operating profit and marginal operating profit (right panel) for any on-peak hour.

generates hourly net surplus equal to marginal surplus minus the variable cost of production $c$. For any on-peak hour, the left panel of figure 2.5 presents the short-term net surplus, which is the surface below the inverse demand curve and above the variable production cost $c$, and the marginal short-term net surplus from a marginal increment in generation capacity, which is the rectangle whose base represents the capacity increment and whose height represents the marginal net surplus minus the variable production cost $c$.

The average hourly net surplus generated by a marginal Megawatt of generation capacity is thus the average overall on-peak hours of these hourly net surpluses. It is represented by the shaded triangle in figure 2.6.

On the other hand, adding one Megawatt of generation capacity costs the hourly capacity cost $r$. Under reasonable conditions, there exists a unique optimal capacity that precisely equates the average hourly marginal net surplus and the hourly capacity cost.[3]

**Long-term equilibrium capacity** This condition is both an optimality condition, that is, it maximizes the net surplus, and also an equilibrium condition. Consider a producer adding a (marginal) Megawatt of generation capacity. She realizes no operating profit off-peak, since she sells at a price equal to her variable cost of production $c$, hence her analysis is limited to on-peak hours. For any on-peak hour, the right panel of figure 2.5 presents her

3. Sufficient conditions for the existence and unicity of the optimal capacity are presented in section 2.3.

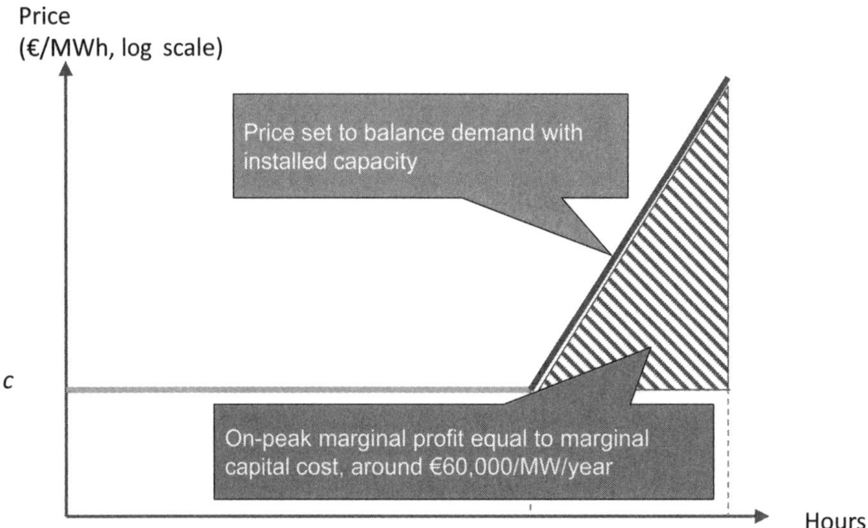


**Figure 2.6**
Pricing structure. Hours are represented on the horizontal axis, prices on the vertical axis. For most hours, price is equal to the variable cost $c$. On-peak, price rises above the variable cost. The operating margin is equal to the fixed cost.

operating profit, which is a rectangle whose base represents the generation capacity and whose height represents the price $p$ minus the variable production cost $c$, and the marginal operating profit from a marginal increment in generation capacity, which is the rectangle with base the capacity increment and of height the price $p$ minus the variable production cost $c$. Comparing the left and right panels of figure 2.5 illustrates that the marginal operating profit is exactly equal to the marginal net surplus, since the price is equal to the marginal surplus, even though the net surplus exceeds the operating profit.

If competition is perfect, producers build capacity until the last unit precisely breaks even, that is, until the average hourly marginal operating profit is precisely equal to the hourly fixed cost $r$. Otherwise, if the average hourly marginal operating profit exceeds (respectively is lower than) the hourly fixed cost, producers would increase (respectively decrease) installed capacity. This is known as a free-entry condition.

Since marginal operating profit is equal to the marginal net surplus, the long-term equilibrium is also the optimum.

**Resulting price structure** The above story seems simple enough. However, it has profound implications. The price in off-peak hours is the variable cost, which we estimated around €50/MWh. Meanwhile, the on-peak margin has to cover the capacity cost, which we have estimated at €60,000/MW/year. The pricing structure is illustrated in figure 2.6.

Different combinations of on-peak hours and prices could yield the same on-peak margin: many on-peak hours at low on-peak prices or few on-peak hours at high on-peak prices. Which combination occurs depends on the sensitivity of power demand to prices. If electric power demand is not very sensitive to prices, which is currently the case, on-peak price will rise very high, hence on-peak hours will be few. On the other hand, if electric power demand is very sensitive to power prices, the equilibrium price would not rise very high, hence the number of on-peak hours would be large.

Demand sensitivity to price is determined by two factors: the share of consumers adapting their consumption in response to wholesale spot prices and each consumer's sensitivity to wholesale spot price.

In this section, I assume that all consumers respond to wholesale spot prices. Using a simple specification for demand presented later in this chapter, I estimate that, if individual customers' sensitivity is at the high end of empirical estimates, on-peak would last around 600 hours per year and price would rise to around €350/MWh, seven times off-peak price. Hence, if electric power demand is very sensitive to prices, the power industry will be close to the hotel or the airlines industries.

On the other hand, if all customers respond to price but their individual sensitivity is at the low end of empirical estimates, on-peak would last less than 150 hours per year and price would rise to around €1,200/MWh, more than twenty times the off-peak price.

In reality, not all customers respond to prices. Suppose, for example, 20 percent of demand responds to price, which is a higher share than most markets experience today but will be a reasonable target in a few years, once demand response policies (presented in chapter 5) are implemented. Using again the simple specification for demand presented later in this chapter, I estimate that, if individual customers' sensitivity is high, on-peak lasts around 200 hours per year, and price would rise to around €800/MWh, more than fifteen times the off-peak price. If individual customers' sensitivity is low, on-peak lasts around fifty hours per year and price would rise to around €3,300/MWh, more than sixty times the off-peak price.

Basic microeconomics suggests that, owing to low demand sensitivity to prices, on-peak prices twenty or sixty times higher than the off-peak prices are perfectly legitimate.

**Long-run marginal cost** In the long run, before capacity is built, producing one Megawatt-hour requires building one Megawatt of capacity and burning fuel for one hour. The hourly long-run marginal cost is thus $(c+r)$.

The free-entry condition implies that the average hourly price is equal to the hourly long-run marginal cost. An additional Megawatt of capacity produces at every hour, hence its average hourly revenue is the average hourly price. If the later exceeded (respectively was below) the hourly long-run marginal cost, producers would profitably increase (respectively decrease) installed capacity.

#### 2.1.3.3 Imperfect competition

Chapter 3 discusses imperfect competition in detail. However, since this issue is extremely important, it is briefly discussed here.

**Perfect competition assumptions** As established above, the industry equilibrium is optimal, in both the short and long terms. In the short term, I have implicitly assumed that producers offer their energy at their variable cost. They could instead collude and offer a price higher than their variable cost. Most readers will find this unlikely off peak but would agree it is a strong possibility on-peak. Alternatively, producers could reduce their production on peak, which would reduce demand, hence increase the equilibrium price.

Similarly, the free-entry condition is a perfectly competitive outcome. It assumes that producers install every Megawatt of capacity that is profitable. They could instead install less capacity, which would increase the on-peak prices. These strategies (and others) are discussed in chapter 3. In this chapter, I assume producers do not to use them.

**Characterization of imperfect competition** On-peak prices are equal to the value of the marginal Megawatt-hour consumed, hence exceed the variable cost of production. This result is sometimes wrongly interpreted as exercise of market power. The faulty argument runs as follows:

> Under perfect competition, price is equal to the variable cost of production. Price strictly exceeding variable cost of production is evidence of market power.

On-peak price exceeding variable cost of production does not constitute evidence of market power. Consider, for example, a market in which producers offer their entire capacity at their variable cost of production. Surely, this would be considered a perfectly competitive market. Yet the on-peak price is the value consumers put on the marginal Megawatt-hour, which exceeds the variable cost of production. The correct argument is as follows:

> Under perfect competition (1) price is equal to short-run marginal cost, which exceeds variable cost of production on-peak, and (2) capacity is selected such that average price is equal to the hourly long-run marginal cost.

Therefore, evidence of imperfect competition could be either price exceeding the variable cost of production off-peak, or average price exceeding the hourly long-run marginal cost.

### 2.1.4 Policy Implications

This pricing structure creates three challenges for policy makers. First, they often do not fully appreciate the consequences of peak-load pricing. Second, even when they do, they find these consequences unpalatable and difficult to explain to voters. Third, they realize that ensuring fairness in power markets is challenging.

Before proceeding, a word on governance of market design. In most countries or states, policy makers (i.e., the executive and legislative branches of government) set the broad policy directions, for example, decide to open generation to competition, and entrust the specific design of market rules to regulatory agencies. The exact split of responsibilities vary by countries, some legislations are more prescriptive than others. In this text, I endow "policy makers" with political concerns, that is, the need to get reelected, while I assume regulator have mostly technical concerns.

**Appreciation** In my experience, most policy makers and even many observers of the power industry fail to appreciate peak-load pricing and its consequences.

While peak-load pricing theory was well known in academic circles by the early 1990s, its application to the restructured power industry shocked many observers. For example, power prices in the Midwest of the United States skyrocketed to $10,000/MWh in late July and early August of 1999. The *Wall Street Journal* was so shocked it put that information on its front page.

This episode was a textbook example of peak-load pricing: at the time, the Midwest had a mostly coal-based power system. Power prices were hovering around the variable cost of coal production, around $14–16/MWh. At night, power price sometimes dipped below $10/MWh, as producers preferred to lose operating margins for a few hours rather than shut down at night and restart in the morning and pay the attendant costs. When temperature rose in late July, demand grew, driven by air-conditioning consumption, but also supply was reduced: high river temperature reduced cooling water available for power plants, hence allowed production. The system was on-peak for a few afternoons, hence the price duly rose to very high levels.

The California crisis offers another fascinating example of peak-load pricing. The California market opened on April 1998. At the time, the main concern was that wholesale prices were too low. Observers were puzzled by the numerous instances of equilibrium prices equal to zero. Again, peak-load pricing provides an explanation: when demand is lower than inflexible production capacity (nuclear and run-of-river hydro), the price is set at the variable cost of the marginal production unit, which is zero. (The case of inflexible nuclear is formally derived in chapter 8.)

By the summer of 2000, two short years afterward, the situation was radically different: demand was close to available capacity, and price rose to very high levels.

A segment of the excellent movie *The Smartest Guys in the Room*, illustrates market participants' puzzlement (Gibney 2005). A former Enron trader remembers: "power trades at around $50/MWh. When it gets to €100/MWh, something is happening. Can you imagine $1,000/MWh?"

Had this trader studied peak-load pricing, he would have known that prices at $1,000/MWh are to be expected for a few hours in a well-functioning power market. I am not suggesting prices in California were the outcome of a well-functioning market. There is, in fact, evidence that the market was manipulated, that is, that there were too many hours with high prices (examples of academic analyses include Joskow and Kahn

2002 and Wolak 2003). I am merely observing that if even a trader fails to understand the basic microeconomics of the power industry, it is highly unlikely that policy makers and citizens do, which, of course, makes it extremely difficult to design well-functioning power markets.

**Acceptability** Once they fully appreciate that power prices can legitimately rise to very high levels for a few hours, policy makers need to make the fact politically acceptable. Cold economic logic dictates that prices spike to extremely high levels, which generates profits for producers and leads consumers to reduce their consumption (and sometimes be subject to involuntary curtailment, as discussed in section 2.2.1) precisely when consumers' demand is the highest. A trained economist finds this the most efficient approach to cover the fixed cost of generation assets. In fact, this point can even be proved as was done informally earlier in this chapter (and done formally in sections 2.3 and later).

This beautiful logic is lost on most voters, who simply find that producers shamelessly profit from consumers' misery. Two factors make this unacceptable to voters: electricity is perceived to be an essential good, and, historically, it was sold at a regulated fixed price, hence very high on-peak prices were never observed.

Experience proves that a modicum of public support is required to design markets. Explaining why power prices can legitimately rise to very high levels for a few hours is difficult, but it is also necessary.

**Fairness** The California story also illustrates the final challenge for policy makers: How do we make sure that on-peak prices are fair, that is, that producers do not overcharge during on-peak hours to reap operating profits higher than their fixed cost? This is a legitimate public policy concern. We (policy makers, academics, and consumers) would not want to pay more for electric power than we need to. Furthermore, given the specificities of the power industry, we have limited confidence that we can rapidly and effectively detect exercise of market power. This point is discussed in detail in chapter 3.

## 2.2 A More Realistic Story

Reality appears to be much more complex than the above example suggests. Surely, the economics of these "complex and highly developed systems with unusual technical characteristics" (Joskow and Schmalen 1983) cannot boil down to such a simple story. Well, in fact it (almost) can, even though multiple new elements must be added: demand does not respond to wholesale prices; there exist more than one technology to produce power; running a power system requires energy but also operating reserves and other ancillary services; and electric power is transported across continents. Adding these does not significantly alter the story. The first three points are covered here, the last one in chapter 7. Finally, this section discusses "security of supply," a term loosely used in policy debates, which covers, in fact, three distinct time horizons.

The main message of this section is that the standard model does surprisingly well at explaining the main economic intuitions of power markets. It misses key ingredients, such as intertemporal linkages and geographical differentiation, hence some numbers may not be completely accurate, but its logic is robust.

### 2.2.1 Non-Price-Responsive Demand

The basic model assumes that consumers are identical and adjust their consumption to respond to variations in the spot price of electric power. The first assumption is clearly not realistic: customers have different usages for power, hence different needs. At best, we can group customers by classes (e.g., industrial, commercial, residential). This richness in usage does not modify the structure of the analysis: all we need in the previous section is a downward sloping aggregate demand curve, which can be built by the aggregation of different customers' demand curves.

**Retail and wholesale prices** The second assumption is not met, either. Most customers purchase electricity from a retailer (or supplier), usually through long-term contracts, while retailers purchase power from producers on wholesale markets.

This general description covers multiple situations. In Europe, wholesale markets are decentralized, that is, buyers and sellers enter into bilateral transactions that the market operator aggregates. In North America, wholesale markets are centralized, that is, the market operator runs an auction to collect all offers from producers and demand from retailers and consumers and than determines the equilibrium production and price. This distinction is ignored in this text, since both market organizations lead to the same outcome under perfect competition. As discussed in box 2.2.1, this theoretical result is confirmed experimentally.

Wholesale markets exist for multiple dates. The most important is the spot market, since the hourly wholesale spot price is the value of energy at every hour. In most markets, the spot market is, in fact, a day-ahead market, not a true real-time market. Since technological constraints imply that most power plants must decide today whether or not to be online tomorrow, buyers and seller agree today on the quantities each will buy and sell and on the price for electricity for tomorrow between 4:00 and 5:00 p.m. Since demand and supply conditions may vary between 4:00 p.m. today and 4:00 p.m. tomorrow, adjustment markets also exist, which also produce prices for electricity for tomorrow between 4:00 and 5:00 p.m. Thus most electric power markets are two-settlement markets, in which the price for electricity for a given hour is determined twice. This text does not open the black box of price formation and assumes a single wholesale spot price exists.

Forward wholesale markets also exist, where producers and customers can exchange power for the next weeks, months, and years. In addition, financial instruments are available. For example, a producer can sell a future contract, which pays the difference between the spot price at a give date and the forward price. The interplay between spot and forward markets is discussed in section 4.3.

**Box 2.2.1**
Law of one price

When I teach this material, some students wonder why the price is not equal to the average cost of producing a Megawatt-hour. This result is a very powerful insight, called the marginalist revolution, which economists understood and formalized in the late 1800s. The optimal decision is to produce and consume a unit if and only the value derived from its consumption exceeds the cost of its production. Therefore, production and consumption optimally occur for all units whose marginal value exceeds their marginal cost: we are concerned only with the values of the marginal unit.

To decentralize the optimum to competitive buyers and sellers, the price is set at the marginal cost, which is also the marginal value in equilibrium.

Our students appreciate this somehow theoretical insight through a game we have them play: each student can either be a consumer/buyer or a producer/seller of four units of a commodity. Producers have different cost structures, and consumers have different values structures. Buyers and sellers negotiate the sale price directly, that is, no central market-clearing institution sets the price.

When the students start playing the game, each producer is willing to sell at a price slightly above its production cost, and each consumer is willing to buy at a price slightly below its valuation. After a few rounds of the same game, students learn what the equilibrium price is. Even a producer whose cost is €50/unit sells at the equilibrium price (say €100/unit), thus capturing a positive profit. Symmetrically, even a buyer whose value is €140/unit buys at the equilibrium price, thus capturing positive surplus.

As it turns out, the equilibrium price is the production cost of the last unit produced, and the value of the last unit consumed. I have had students play this game more than a hundred times. The result is always the same, with graduate students in economics, executives, engineers, judges, and so on: the law of one price applies.

The structure described above applies to most commodities, for example, oil, metals, and agricultural products. In most of these industries, customers face the wholesale spot prices, sometimes with a lag. For example, drivers pay the wholesale spot price for gasoline at the pump (plus a variety of taxes), and wheat retail prices follow the wholesale spot prices.

In the power industry, by contrast, most customers pay a constant retail price, sometimes called a flat rate. Historically, meters could only record consumption between two readings, hence customers were paying the same price for every Megawatt-hour they consumed, irrespective of the true value of energy at the hour of consumption.

If all customers face a constant retail price, the inverse demand is a vertical line: demand is perfectly independent of price, as illustrated on the left panel of figure 2.7. This may have been a reasonable representation of demand twenty or thirty years ago, but it is unrealistic in most markets. Today, most electro-intensive customers purchase directly from the wholesale markets. Smart meters and enhanced communications technology

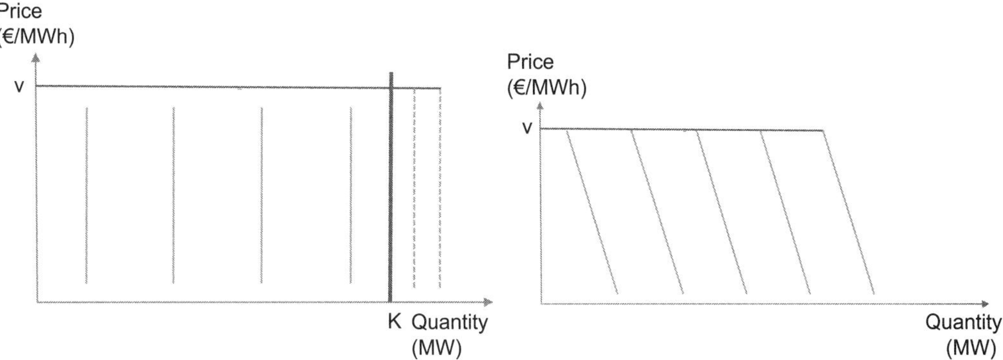


**Figure 2.7**
Demand curves if demand is perfectly inelastic (left panel) and partially elastic (right panel). On the left panel, in every hour, demand does not vary with price. When demand is lower than installed capacity $K$, it can be entirely served. When demand exceeds installed capacity $K$, it must be reduced through involuntary curtailment. The price is then set at the VoLL: inverse demand is a horizontal line at the VoLL. On the right panel, demand is slightly reduced as price increases, up to the VoLL.

enable retailers to record hourly demand and to charge a different price for every hour, even for residential customers, although most retail contracts, in particular for residential customers, still offer a flat rate. Demand response is discussed in details in chapter 5.

As long as a positive fraction of customers respond to the spot price, the aggregate demand remains downward sloping. The inverse demand is more vertical than would be the ease if all customers respond to price, that is, a given change in price yields a lower change in demand, as illustrated in the right panel of figure 2.7.

**Administrative curtailment and the value of lost load** When demand is vertical, the peak-load pricing logic applies differently: demand can no longer be adjusted to capacity through an increase in price. When demand (at any price) exceeds capacity, administrative curtailment is required. The SO implements rolling blackouts, that is, selectively shuts down regions for a few hours. The curtailment plan is usually approved by the government.

When electricity demand slightly exceeds available electricity generation, the frequency of electric power is reduced. Sometimes SOs slightly allow a temporary reduction before resorting to selectively curtailing customers. These episodes are known as brownouts. These are costly, as electric equipment is designed to be operated at the standard frequency and deteriorates when used at another frequency.

When demand is curtailed, the SO values electricity at the value of lost load (VoLL). Unless rationing is efficient, the marginal VoLL, that is, the amount a user is willing to pay for a marginal Megawatt-hour when rationing occurs, is higher than the marginal value of power when it does not. The formal definition of the VoLL and the proof of this property are presented in section 2.4. An example illustrates the argument.

Consider two customers: customer A uses electricity for heating with value €200/MWh, while customer B uses electricity for lighting with value €100/MWh. When there is no rationing, each consumes one kWh, and the marginal value is €100/MWh.

Suppose now rationing must be implemented, and only 1.9 kWh is available. If rationing is efficient, the lowest value usage is curtailed: customer A uses 1 kWh to heat her house, and customer B is rationed and uses 0.9 kWh to light her house. How much would the SO value the marginal 0.1 kWh? He would deliver it to user B, hence value it at €100/MWh. Thus the marginal values with and without rationing are equal.

In practice, however, rationing is often inefficient, that is, the SO cannot curtail consumer B alone and instead must curtail both. Each will have 0.95 kWh available for heating and 0.95 kWh available for lighting. A marginal 0.1 kWh would be used for heating and lighting, hence would be valued at €150/MWh, which is higher than would be the case without rationing.

Another argument presented in box 2.2.2. for why the (marginal) VoLL is higher than the (marginal) value of electricity can be found in an unexpected place: the French novel, *Du côté de chez Swann*, by Marcel Proust (1919). It is less rigorous, but I like its poetry.

**What is the VoLL?** In the above two-usage example, the SO is able to compute the VoLL. This is not the case in reality, since a customer's VoLL depends on a number of factors. First, it depends on the usage. Students in a classroom hit by a power outage are probably not willing to pay much to get the light back and resume the course. In fact, most would be willing to pay (at least a small sum) to enjoy a break in the sunny courtyard. On the other hand, a patient receiving open-heart surgery would be willing to pay a significant sum to avoid a curtailment.

Second, the VoLL depends on the duration of the outage. Supermarkets have deep-frost fridges, which keep perishable products at extremely low temperatures. They are not willing to pay much to avoid an outage that lasts only a few minutes. In fact, in some instances, they are willing to reduce their fridge's consumption and resell part of their energy into the market for a short duration. On the other hand, they would be willing to pay a large amount to avoid an outage lasting a few hours, which would destroy all their stocks.

Third, the VoLL depends on the information given to customers. If you know you will be curtailed tomorrow at 10:00 a.m., you do not step in an elevator at 9:59 a.m. On the other hand, if the curtailment catches you unaware, and you end up stuck in a cramped elevator with foul-smelling colleagues, you will be willing to pay a significant sum to get power back; hence terminate your ordeal.

It is therefore not surprising that estimates of the VoLL vary in an extremely wide range, from $2,000/MWh in the British pool in the 1990s to $200,000/MWh (Cramton and Lien 2000). For some reason, many countries choose 20,000 as their estimate of the VoLL–irrespective of the currency: for example, the security of supply standard used in France is consistent with VoLL around €20,000/MWh, and the New Zealand Electricity Authority estimates the VoLL around NZ$20,000/MWh.

**Box 2.2.2**
Literary digression

Charles Swan, the main character of the first volume, has an affair with a courtesan, Odette de Crécy, accurately portrayed by Ornella Mutti in Volker Schlöndorff's *Swan in Love*, an excellent cinematographic rendition of this complex and rich story. (Schlöndorff 1984) Swan knows Odette to be a courtesan who has sexual intercourse with other men. In France at the turn of the twentieth century, such relations were perfectly acceptable for a man of excellent social standing, such as Swan.

As long as Odette is available to Swan, he does not miss her. However, when, one evening Odette is no longer available to him (and, in fact, is plying her trade with another man), her value to him increases so much that he decides to marry her, knowing full well that neither she nor their children will ever be able to join him in his elite circle of friends. This decision is all the more remarkable given that Swan knows (even if he does not articulate it clearly, which is the charm of the novel) that Odette, married or not, will always remain a courtesan and will never fully belong to him. Indeed, a few year later, Swan acknowledges that fact:

"To think that I've wasted years of my life, that I've longed to die, that I've experienced my greatest love, for a woman who didn't appeal to me, who wasn't even my type!"

Despite his future regrets, that fateful evening when he cannot find her, Swan's value for Odette increases a thousandfold. So does the value of power when rationing occurs.

In some way, the intuition could be made identical (one has to be careful over interpreting Swan's psyche). Swan puts different values to one hour with Odette. When he decides not to see her, the value Swan places on an hour with Odette is lower than his value for the another activity he chooses. When she disappears, the value Swan puts on this lost hour with Odette is the average value of all the hours he could have spent with Odette, including the highest valued ones. Hence, the value is higher with than it would be without rationing.

**Resulting demand curves** When all customers face a constant retail price, the inverse demand curve is vertical for prices up until the VoLL, horizontal afterward (left panel of figure 2.7). When only a small fraction of customers respond to spot prices, there may also be instances when administrative curtailment is required. In that case, inverse demand is a steeply sloping line up until the price equals VoLL, and then a horizontal line when the price is equal to the VoLL (right panel of figure 2.7).

**Resulting price structure** The peak load pricing logic still applies to that inverse demand curve. The only difference is that the visible hand of an administrative intervention replaces the invisible hand of market forces to adjust demand to available capacity through curtailment and to set the price at VoLL.

When demand does not respond to price, the latter is equal to the variable cost (around €50/MWh for almost all hours) and is set by the SO at the VoLL for the remaining hours, during which power is curtailed. This is illustrated on the left panel of figure 2.8.

When demand partially responds to price, the latter is equal to the variable cost for most of the hours. When demand (for price equals variable cost) is equal to installed capacity,

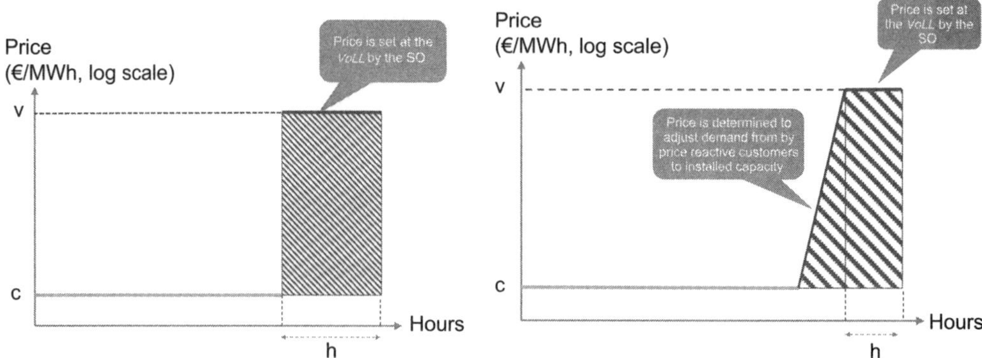


**Figure 2.8**
Price structure when no customer is price responsive (left panel) and when a fraction of customers is price responsive, and curtailment occurs (right panel). When no customer responds to price, the SO starts curtailing customers when demand (at the variable cost of production) is equal to installed capacity and sets the price at the VoLL. When a fraction of customers respond to prices, the price rises on peak, as in section 2.1. If the price rises to the VoLL, the SO starts curtailing constant-price customers and sets the price at the VoLL.

price increases, and price-responsive customers reduce their demand accordingly. As was the case in section 2.1, this is voluntary demand reduction and not involuntary curtailment. As long as price is lower than the VoLL, the SO does not curtail any customers.

If price rises up to the VoLL, the SO starts curtailing non-price-responsive customers and sets the price at the VoLL, which represents the value of an additional Megawatt-hour. Price-responsive customers adapt their demand to this price. This price structure is illustrated on the right panel of figure 2.8.

**When administrative curtailment is not required** The left panel of figure 2.8 represents the pricing structure of the past, when no customer was price responsive. The right panel of figure 2.8 represents today's and tomorrow's pricing structure, when a small but growing fraction of customers are price responsive. In a few years, the fraction of price-responsive customers will be large enough that the pricing structure will be that of section 2.1. This naturally raises the question: How much demand response is required for curtailment never to be necessary?

Intuitively, the higher the share of demand responds to the wholesale spot price, the lower the probability administrative curtailment is required. Numerical simulations, presented later in this chapter, put a number on this intuition. I find that rationing is not required for the optimal generation mix when more that 3.9 percent of demand is price responsive if the price elasticity of demand is low. If demand elasticity is high, rationing is no longer required when more than 13.9 percent of demand is price responsive. This result may seem counterintuitive: a less elastic demand results in less curtailment. The intuition is that, for a given share of price-responsive customers, optimal capacity is higher when demand is less elastic, hence curtailment is less frequent.

As discussed later, these results are obtained under specific assumptions on the shape of the demand function and the uncertainty. Further research is required to confirm these values. Still, they are good news. If less than 20 percent of demand responding to price is indeed sufficient for curtailment never to occur at the optimal capacity, rationing will soon be a practice of the past.

**Optimal capacity** As in section 2.1, the optimal capacity is such that the average hourly marginal operating profit of a Megawatt of capacity is equal to its hourly fixed cost.

If no customer responds to price, the marginal operating profit is VoLL minus production cost—usually approximated by VoLL-times the number of curtailment hours. This corresponds to the hatched rectangle on the left panel of figure 2.8. For example, many analysts suggest a value of €20,000/MWh to be a reasonable estimate for the VoLL. Thus if the marginal fixed cost of capacity is taken at €60,000/MW/year, the optimal capacity is such that power is curtailed three hours per year on average, since €60,000/MW/year = €20,000/MW/hour times three hours/year.

If a fraction of customers respond to price, the marginal operating profit is the price minus the production cost, where the price is set to balance demand from price-responsive customers with installed capacity (hatched triangle on the right panel of figure 2.8) and then set at the VoLL by the SO (hatched rectangle on the right panel of figure 2.8).

**Engineering generation adequacy criterion** Power engineers do not design power systems using the VoLL. Rather, they use a physical generation-adequacy criterion, for example, demand should be curtailed for all but three hours per year on average. Or equivalently. The criterion is determined administratively and may or may not coincide with the economic optimum.

To determine how much installed generation capacity is required to meet this criterion, engineers and statisticians first compute the distribution of possible future peak demand, considering different scenarios for weather and economic growth (and other relevant variables). Then, they determine the relevant percentile of the distribution. For example, since (8760-3)/8760 = 99.966 percent, the relevant percentile corresponding to 3 hours per year adequacy criterion is 99.966 percent. Suppose the 99.966 percent of the distribution of peak demand is 108 GW; the probability that peak demand exceeds 108 GW is only 0.034 percent, which is equivalent to saying that, on average, peak demand will be lower than 108 GW for 99.966 percent of the time, or all but three hours per year.

Then, system planners decide how much generation capacity to build, so that, on average, generation assets produce 108 GW for a few peak hours. In general, planners take into account forced (or unplanned outages) in power generation units. For example, if they assume a forced outage rate of 7 percent, they will determine that the adequate generation capacity, which guarantees that, on average, demand is met for all but three hours per year, is 116 GW (since $116 \times 0.93 \simeq 108$).

This does not mean that three hours of rolling blackouts will occur every single year. If the weather is mild, and plants' operating conditions are good, no rolling blackout may occur for one or more years. On the other hand, if the weather is unfavorable (a winter colder than average in Europe, or a summer hotter than average in the United States), and if plants' operating conditions are poor, three or more hours of rolling blackouts may occur. On average, however, if the engineers and statisticians computations are correct, rolling blackouts should occur about three hours per year.

The outcome of the generation adequacy computation is often expressed as a capacity margin, that is, the generation capacity minus the expected peak demand, as a fraction of expected peak demand. Suppose for example expected peak demand is 100 GW. The capacity margin is $\frac{116-100}{100} = 16\%$. Adequate generation capacity is 16 percent higher than expected peak demand: 8 percent results from demand being higher than its expectation, and another 8 percent from forced outages in production. This number is actually representative of capacity margins used by engineers up until renewable energy sources were introduced. If the capacity margin exceeds 20 percent, too few on-peak hours occur, wholesale spot price remains close to marginal cost, and capital cost cannot be recovered. If the capacity margin falls below 10 percent, rolling blackouts are likely to exceed three hours per year on average.

The capacity margin loses its meaning as renewable energy sources enter electricity markets, since they produce on average 30 percent of the time, and their production cannot be controlled.

While still in use today, generation adequacy is a concept of the past. As demand becomes progressively more price responsive, it will adjust to existing supply through an increase in price, not through administrative curtailment. Anticipating this trend, some countries such as New Zealand have abolished the engineering reliability criterion altogether.[4] Other countries are holding on to the criterion.

**Formal equivalence between the VoLL and the generation-adequacy criterion** Both approaches are formally equivalent. Since the product of the (expected) number of hours of curtailment times the VoLL is equal to the marginal fixed cost of capacity, choosing a generation-adequacy criterion is mathematically equivalent to choosing a VoLL, and vice versa. In our example, if the capital cost of a peaking turbine is €60,000/MW/year, a VoLL set at €20,000/MWh is mathematically equivalent to a generation-adequacy criterion that available generation should exceed demand for all but three hours per year, on average. The higher the generation-adequacy criterion (i.e., the lower the expected number of curtailment hours), the higher the VoLL.

However, setting a VoLL or setting a generation-adequacy criterion lead to dramatically different market designs. In the first approach, known as the energy-only market design, regulatory intervention is limited to setting the price for electricity when rolling blackouts are required. In the second approach, policy makers require the SO to set up additional an

4. New Zealand is largely hydro based. The supply risk is a dry year during which rolling blackouts occur for a few winter days.

capacity mechanism to guarantee the generation-adequacy criterion is met, which is a much more complex market design. The various capacity mechanisms and their implications are discussed in chapter 9.

**Political economy of non-price-responsive demand** The foregoing analysis is not surprising to economists: price should be high when demand is curtailed. The value of €20 000/MWh is very high, but it appears to be the value policy makers and systems operators place on lost load. Risk-neutral investors should include these extremely rare but extremely profitable events in their analysis and invest optimally (given the estimate of VoLL); hence, absent market imperfections, curtailment should reach three hours per year on average.

Even assuming no market imperfection, the implications of this situation are difficult for policy makers to accept. If they find prices at €1,000/MWh difficult to explain, how can they explain to consumers–who are also voters—that producers pocket extremely high profits from prices at €20,000/MWh, precisely when a fraction of customers are curtailed? This is an extreme version of the high-price problem previously discussed.

The capacity adequacy criterion, that is, the expected number of hours a fraction of demand will be curtailed, poses another challenge to policy makers. No elected official in an OECD country is willing to stand in front of his constituents and explain, "You were curtailed for two hours yesterday. Do not worry, it is part of the plan. In fact, it is economically optimal." Remembering that Gray Davis, while governor of California, was recalled in the middle of his term following rolling blackouts, elected officials would prefer no curtailment to occur, ever. They implicitly place an infinite value on lost load, which is, of course, not economical.

Policy makers like neither curtailment nor high prices and would prefer a market design that guarantees no curtailment (or very, very low probability of curtailment) and prices at a low level. These preferences are clearly not compatible with the underlying micro-economics. In particular it explains the drive for capacity mechanisms, which is discussed in details in chapter 9. This is the most important divergence between the micro-economics and the political economy of power markets, that structures most market design choices.

### 2.2.2 Multiple Technologies

Peak-load pricing extends naturally to multiple technologies. When only one technology is present, the supply curve is $L$-shaped: horizontal at the variable production cost until capacity, then vertical when capacity is reached. When multiple technologies are present, the supply curve is a staircase, that is, a succession of $L$s.

**Characteristics of multiple technologies** To make things concrete, suppose three technologies are available: nuclear, combined-cycle gas turbine (CCGT), and open-cycle gas turbine (OCGT). Table 2.1 presents estimates of the variable and fixed costs of each technology, based on data from the International Energy Agency.

**Table 2.1**
Cost of different production technologies

| Technology | Fixed cost per year (€/MW-year) | Fixed cost per hour (€/MWh) | Variable cost per hour (€/MWh) |
|---|---|---|---|
| Nuclear | 299,000 | 34 | 10 |
| Combined cycle (CCGT) | 72,000 | 8 | 90 |
| Gas turbine (OCGT) | 60,000 | 7 | 130 |

Source: International Energy Agency (2010) median case with two modifications: gas price €40 MW/hr and $CO_2$ price €50/ton.

In table 2.1, the technologies are ordered by increasing operating cost: nuclear is cheaper than CCGT, which is cheaper than OCGT. The technologies are also ordered by decreasing fixed costs: if one technology has lower short-term marginal cost than another, it has higher fixed costs. This makes sense: if a technology is cheaper to build and to run than all others, it will be the only one installed. Similarly, no one would install a technology more expensive to run and build than the others.

**Screening curves and optimal usage of different technologies** The trade-off of fixed versus variable costs of generation is illustrated on the screening curves presented in figure 2.9. Hours are represented on the x-axis, and the y-axis presents the total annual cost of producing one MW during a strip of any number of hours. For example, producing one MW for one hour using nuclear technology costs €299,000 in fixed cost, plus €10 per hour of production. The total cost of serving a strip of hours using a nuclear power plant is therefore a straight line of intercept €299,000 and slope €10 per hour. Similarly, the total cost of serving a strip of hours using a CCGT is a straight line of intercept €92,000 and slope €70 per hour.

Figure 2.9 illustrates that hourly long-run marginal cost are increasing as operating cost increases. Consider the OCGT. Its fixed costs are lower than those of the CCGT. Its variable cost has to be high enough so that its total cost crosses the total cost of the CCGT for less than 8,760 hours, otherwise the CCGT will never be installed: the long-run marginal cost of the OCGT exceeds that of the CCGT. Similarly, the long run marginal cost of the CCGT exceeds that of the nuclear.

This particular cost structure translates into a differentiated usage pattern. Remember that demand varies significantly across months, weeks, and days. The issue is this: Under which circumstances should a specific technology be turned on? Since nuclear is the most expensive to build, and the cheapest to run, it should run all the time. It is the baseload technology. In markets where nuclear is not present, coal is often the baseload technology.

At the other extreme, since the OCGT is cheap to build and expensive to run, it should be turned on for high-demand situations (winter evenings in Europe, summer afternoons in the

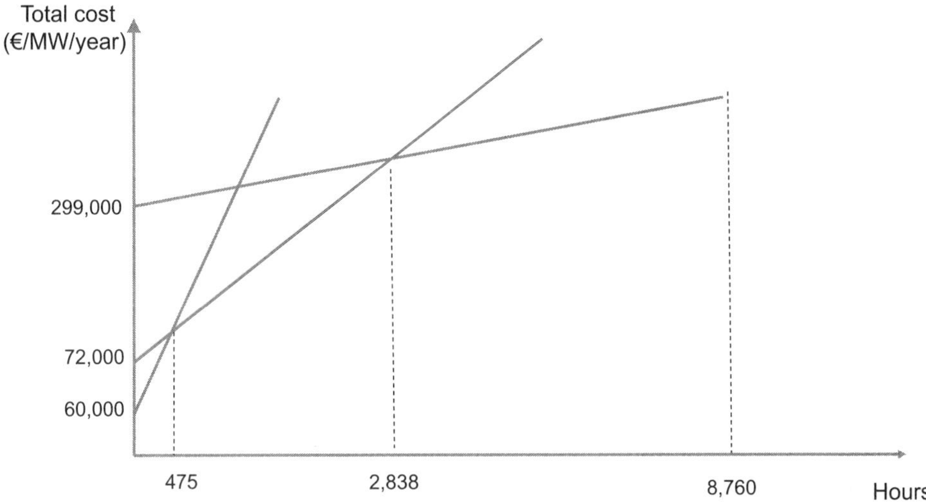


**Figure 2.9**
Screening curves for three generation technologies: nuclear, CCGT, and OCGT. Hours are represented on the *x*-axis, and the *y*-axis presents the total annual cost of producing 1 MW during a strip of any number of hours.

South of the United States). This corresponds to a peaking usage. Finally, CCGT being the intermediary technology (sometimes called semibase), it starts running for intermediary demands, for example, 2,000–5,000 hours per year.

These results can be illustrated using the screening curves presented in figure 2.10. The three lines cross: the total cost of CCGT crosses the total cost of OCGT for 475 hours and crosses the total cost of nuclear for 2,838 hours. As presented on the left panel, to produce a strip of hours lasting more than 2,838 hours, nuclear is the cheapest technology, measured in €/MW/ year. The right panel shows that, to produce a strip of hours lasting less than 475 hours, OCGT is the cheapest technology. For intermediate number of hours, CCGT is the cheapest technology. This analysis constitutes an excellent and highly illustrative first approximation. It is rigorously exact if and only if no customer responds to prices.

**Resulting supply curve** The supply curve is presented in figure 2.11. The first flat portion corresponds to the hours when nuclear is the only technology producing. The price is thus equal to the variable production cost of nuclear, which determines demand. Then comes the first vertical portion of the supply curve: demand is equal to nuclear production, which is equal to nuclear capacity, and the price rises to precisely balance demand and nuclear capacity. If nuclear were the only technology present, this would be the end of the story.

However, since another technology present, the story continues. The CCGT is turned on as soon as demand is high enough that the price exceeds its variable production cost, that

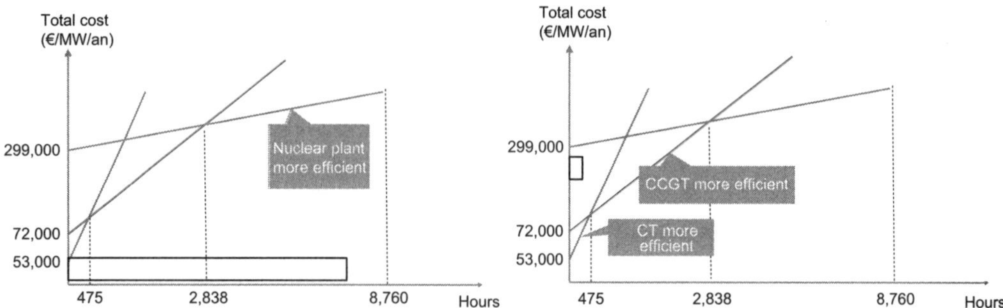


**Figure 2.10**
Optimal usage of a nuclear power plant (left panel) and an open-cycle gas turbine (right panel).

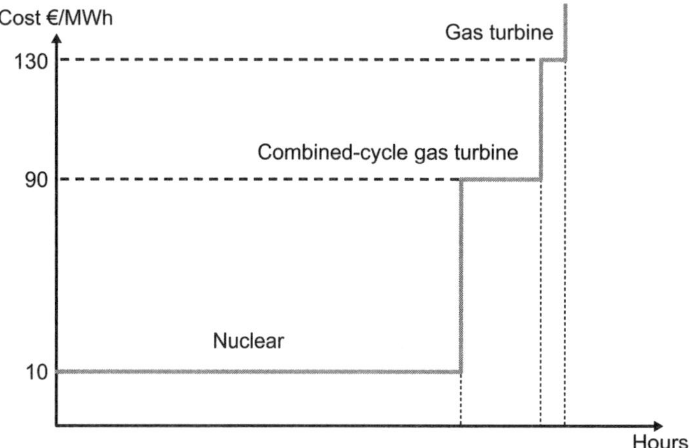


**Figure 2.11**
Supply curve for three technologies.

is, when users are willing to pay more for a Megawatt-hour than the cost of gas to produce it. Then, we look at the second flat portion of the supply curve: the price is equal to the variable production cost of CCGT, which determines demand.

Then comes the second vertical portion of the demand curve: the price increases so that demand is equal to the cumulative nuclear plus CCGT capacity.

Finally, we travel the third *L* of the supply curve: a flat portion where price is the variable production cost of OCGT, then a vertical portion where price is such that demand is equal to the cumulative nuclear plus CCGT plus OCGT capacity.

**Optimal cumulative capacity and generation mix** If demand responds to price, the logic presented in section 2.1 also applies here to determine the optimal capacity and generation mix. The OCGT captures positive operating margin only when it produces at capacity, that is, when demand is equal to the cumulative capacity of all three technologies. The optimal cumulative capacity is therefore such that the OCGT's average hourly operating margin is exactly equal to its hourly capacity cost. It is solely determined by the cost of the marginal technology, in this case the OCGT, and the demand function. Thus the optimal cumulative installed capacity does not depend on the entire generation mix, only on the marginal technology.

Once the cumulative capacity has been determined, we need to determine the optimal generation mix. By construction, each technology produces if and only if its operating margin is nonnegative. The free-entry condition for each technology is that average hourly operating margin, generated during the fraction of hours it produces at capacity, is exactly equal to hourly capacity cost.

Derivations presented later in this chapter prove these conditions can be interpreted as follows: the equilibrium mix is such that substituting one Megawatt of one technology by one Megawatt of the next technology generates no gain nor loss.

**Optimal cumulative capacity and generation mix if no customers respond to price** Power engineers, who historically assumed that demand does not respond to prices, determined the optimal capacity and generation mix by combining the load duration curve (figure 2.1) and the screening curves (figure 2.9). This analysis is presented in figure 2.12.

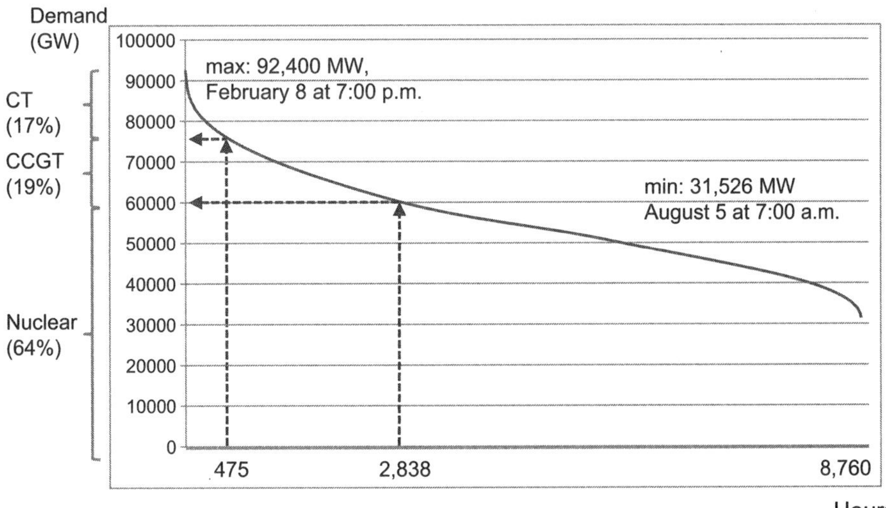


**Figure 2.12**
Determination of the optimal generation mix combining the screening curves and the load-duration curve.

First, cumulative capacity is determined to meet peak demand, up to the generation adequacy criterion, for example, 116 percent of expected peak demand.[5]

Second, analysis of the screening curves shows that nuclear is the most efficient technology to serve a strip of demand lasting more than 2,838 hours. As discussed in section 2.1, the load duration curve shows that demand exceeds 60,000 MW for 5,676 half hours of the year, which is exactly 2,838 hours. An equivalent formulation is that the size of demand strips lasting more than 2,838 hours is 60,000 MW. Therefore, optimal nuclear capacity is 60,000 MW, or 64 percent of peak demand.

Similarly, the load duration curve shows that demand exceeds 77,000 MW for 475 hours per year, or, equivalently, the size of demand strips lasting more than 77,000 MW is 475 hours. Therefore, the optimal CCGT capacity is 77,000–60,000 = 17,000 MW, or 19 percent of peak demand.

Finally, OCGT constitutes the remaining generation capacity, 17 percent of peak demand.

Finally, figure 2.12 illustrates that nuclear capacity (for example) exceeds nuclear generation by 5,676 nonconsecutive half hours per year, which corresponds to the first flat portion of the supply curve (figure 2.11).

**Invariance of the long-run average price** We have seen that the baseload technology runs all hours. Thus a baseload producer receives the (time-weighted) average price over the year. If this average is higher than the hourly long-run marginal cost, an additional unit is added into the market, hence the average price is reduced. This process is repeated until the average price is exactly equal to the hourly long-run marginal cost. Conversely, if the average price is lower than the hourly long-run marginal cost, baseload installed capacity is retired until the average price is exactly equal to the hourly long-run marginal cost. This is the free- entry condition. The average price is entirely determined by the hourly long-run marginal cost of the baseload technology.

This property holds exactly for the long-run optimum. As we have just discussed, industry equilibrium at any time differs from this long-run optimum. Still, this property is directionally correct: if time-weighted average price is expected to remain for the next few years below the hourly long-run marginal cost of the baseload technology, investors in baseload assets will tend to retire them.

This property is very important, and produces counterintuitive results. For example, increasing the share of price-responsive customers while increasing the net surplus does not modify the time-weighted average price in the long term. This issue is examined in detail in chapter 5.

5. To simplify the example, I choose installed capacity equal to peak demand and round up the numbers.

Similarly, since the mid-2000s, RES producers (e.g., wind and solar) have been subsidized, hence have massively entered into electricity markets. Since their marginal cost is essentially zero (wind is free), and their fixed cost is subsidized, one would expect the average price to be significantly reduced. And indeed, this is what we observe in Europe. However, this average price reduction is a short-term effect. Since the price is lower than the fixed cost, capacity is being retired, hence the average price increases. This process will stop precisely when the average price equals the hourly long-run marginal cost of the baseload technology, in this case nuclear. Thus renewables entry has no impact on the average price in the long-term equilibrium, which is determined by the cost of the baseload technology. This property holds until baseload technology is pushed out of the market, a situation we examine in chapter 8.

**Long-run marginal costs** We have seen that long-run marginal costs are higher for a CCGT and an OCGT than for the baseload nuclear plant. We have also seen that the time-weighted average price is exactly equal to the long-run marginal cost of baseload technology. Therefore, the time-weighted average price is strictly lower than the long-run marginal costs of a CCGT and an OCGT. If they were producing all the time, a CCGT and an OCGT would not cover their cost. How can they cover their cost by operating less than full time? The answer is that these technologies operate for less but higher-priced hours. The number of hours a plant is operating is not sufficient to determine its profitability. What matters is the number of hours a plant is producing at capacity, which are the only hours when it captures positive operating profit, and the price during these hours.

**Is the generation mix ever optimal?** The analysis above appears simple enough. In reality, the generation mix is, of course, never optimal, for three main reasons, which are not specific to the power industry.

First, generation assets last more than twenty years. Investment is decided today, based on assumptions of future screening curves (i.e., fixed and variable costs of generation technologies) and future load-duration curves. These assumptions are almost always wrong: fuel prices change, taxes or subsidies are decided that impact costs, demand grows more or less than expected, and so on.

Second, adjustment is not easy. When excess capacity has been installed, it is not immediately shut down, as has been observed in the Northeast of the the United States in the early 2000s and in Europe since 2010. Investors may keep assets running even if they do not fully recover their cost of capital, as long as they cover their variable costs. This decision could be rational: investors could be betting that future prices will rise high enough to exceed tomorrow's long-run marginal cost and cover today's losses. Or this decision could be less rational, for example, attributable to institutional reluctance to restructuring.

Third, I have assumed throughout this chapter that industry equilibrium is optimal. In reality, this is not true. Investors do not necessarily coordinate to reach the optimal generation mix. They have to tendency to overinvest when prices are high and underinvest when prices are low. This boom-bust investment cycle is discussed in chapter 3.

### 2.2.3 Noncontrollable and Storage Technologies

**Noncontrollable technologies** The previous story has focused on controllable technologies: operators can decide to produce or not, given the economic conditions. In practice, not all technologies are controllable. For example, if an operator were to close the valves of a dam on a river without a reservoir, the water would overflow. Thus a dam without a reservoir, known as run-of-river hydro, is noncontrollable.

Renewable energy sources, such as wind turbines and photovoltaic panels, are technically controllable (at least large installations). However, in most jurisdictions at the time of this writing, they are given physical priority, that is, system operators are required to dispatch as much energy as they produce, hence are in practice noncontrollable.

To accommodate noncontrollable technologies, economists have developed the concept of residual demand, that is, total demand net of noncontrollable production. The above story still applies by simply replacing total demand by residual demand. This is illustrated in chapter 8, which examines the impact of RES entry on power markets.

**Storage technologies** We have so far ignored the possibility of storing electricity. In fact, storage technologies, such as dams with reservoirs, known as storage hydro, have existed for a long time. Such reservoirs are an important part of the supply in Brazil, Quebec, British Columbia, and Scandinavia.

A slightly different story applies in this case. Consider first the problem of the operator of the reservoir. To simplify the argument, suppose her planning horizon is one year. She can produce of finite volume of Megawatt-hour. Her problem is then to decide at which hours she will release water, hence produce her target Megawatt-hour.

Suppose first the operator knew exactly the quantity of water she can use and the prices for every hour of the coming year. She would then order the hours by decreasing price and produce at the maximum in the highest hour, then the next, and so on, until her water is exhausted. Thus she would maximize the value of her water.

However, future water inflows and prices are unknown. Thus the operator problem is one of probability. Fortunately, the solution to this class of problems, known as optimal control problems, was developed in the 1940s and 1950s. It is used for controlling dams but also inventories, satellites, and so on.

For the reservoir, the optimal rule turns out to be very simple: the operator determines a usage value, which depends on the level in the reservoir. If at any hour, the price is above the usage value, she produces at capacity. Otherwise, she does not produce. This rule does not guarantee the absence of ex post regret. She may produce early in the year and miss out on higher prices later in the year. But the rule is ex ante optimal. The difficulty of course is to compute the usage value. This is where the math is hidden. Fortunately, it is by now a well-studied problem, and computing capabilities are progressing so that operators can compute better and better estimates.

Consider now the problem of controllable technologies. They can compute the usage values, hence the production by the storage technologies. They can thus compute the residual demand.

We are not exactly back to the previous case, however. There is a fixed-point problem: the reservoir operator computes the usage value using its expectation of prices. For any hour (state of the world), controllable production is dispatched against residual demand, which produces a price. This price must be consistent with the expected price used to compute the usage value.

However, the logic is the same, and the main results hold. On the horizontal segments of the supply curve, the price is set by the variable production costs of the existing technologies. On the vertical segments, including on-peak, the price is set by the value of the marginal Megawatt-hour consumed. The time-weighted average price is equal to the long-run marginal cost of the baseload technology.

Additional free-entry conditions corresponding to the storage technologies hold. Consider, for example, a battery that stores and discharges energy. The battery purchases energy off-peak, and resells it on-peak. The variable cost of production is close to zero, although there is energy lost in the process. In the long-term equilibrium, batteries are installed up to the point where the expected spread between on-peak and off-peak prices, corrected for the energy loss, is equal to the fixed cost. The relevant on-peak versus off-peak spread depends on the storage technology. For example, if a battery can store energy for a day, the spread between the on-peak price at the end of the day and the off-peak price in the middle of the night is relevant.

Consider now pumped storage plants, which pump water into a high reservoir off-peak and release it into a low reservoir on-peak. The same economic logic applies: when the on-peak versus off-peak spread is high, investors finance pumped storage plants. However, the deployment of pumped storage plant is limited by the availability of suitable sites. Depending on the availability of suitable sites and the technology mix, we could reach a situation where the spread exceeds the fixed cost of the last feasible site.

### 2.2.4 Intertemporal Linkages and Operating Reserves

The foregoing story does not specify how the system adjusts in real time to supply or demand shocks. We have simply said that in each hour, demand is equal to supply, since limited storage is available today. We have therefore implicitly assumed that production can adjust perfectly to changes in demand.

In practice, it is not that simple. Most production facilities have limited abilities to adjust. For example, the ramp-up rate determines the speed at which a power plant can increase its production, and the ramp-down rate the speed at which it can decrease its production. Ramp-up and ramp-down capabilities vary by technologies, some being more flexible than others. In addition, starting up or shutting down a power plant is costly.

Furthermore, supply and demand are subject to sudden random shocks. For example, a power plant may "trip" and suddenly stop production, the wind may exceed the acceptable speed, hence wind turbines may suddenly shut down, and a large user may unexpectedly stop his production process, hence his electricity consumption. This would not be a problem if production facilities had unlimited ability to adjust, but adjustment is limited.

How do power systems cope with the presence of random shocks and limited ramp-up and ramp-down flexibility? First, system operators and power producers make production decisions based not on a single hour but on a stream of hours: production at hour $(t+1)$ is partially determined by production at hour $t$. Second, system operators create operating reserves.

**Unit commitment** Until now, the analysis has ignored start-up and shut-down costs and ramp-up and ramp-down rates, hence has ignored linkages between hours. When they are introduced, the simple and, one hopes, intuitive arguments and properties presented above no longer hold.

For most power plants, the dispatch decision can be decomposed in two related decisions: decide today whether to turn the plant on tomorrow, a decision known as unit commitment, and then decide how much to produce for every hour tomorrow, taking into account the ramp-up and ramp-down rates, as well as the minimum production level required by the machine. Exceptions to this rule are nuclear plants, whose shut down and restart costs are so high that the decision to turn on a plant is taken for months, not only for a day, and storage hydro plants, which can be turned on and off almost immediately. The necessity to commit units today for tomorrow causes power markets to be organized around two settlements.

This two-stage unit commitment problem is significantly harder to resolve analytically. Fortunately, a branch of applied mathematics, called operations research, is dedicated to solving these kinds of problems. The economic intuition is basically unchanged, although the analysis is much richer.

**System operator and power exchange** Power markets require a system operator to physically control supply and demand in real time. This feature is unique to the power industry. In most other markets, physically delivery of the underlying commodity is decentralized: no single central entity controls the production of all oil fields, the consumption of all oil refineries, and the movement of all oil tankers; rather, every market participant optimizes the physical movements of its own assets.

In the power industry, the system operator has her fingers on the switch(es): in real time, she is allowed to turn power plants on or off, and curtail customers. This function is conceptually different from the power exchange (PX), which organizes wholesale markets, that is, provides a platform for producers and consumers to sell and buy energy.

A simple example illustrates the articulation between PX and SO. Consider a producer who sells 100 MW for the next day from 4:00 p.m. to 5:00 p.m. into the PX. The next day at 4:30 p.m., the producer's plant suddenly trips and cannot produce at all. To balance the system, that is, to ensure that supply meets demand, the SO must turn on another plant. She then requests the defaulting producer to pay for this additional energy.

In most markets in the United States, the SO is also the PX. The underlying argument is that engineering constraints perfectly understood and mastered by the SO also structure the work of the PX. In most European markets, the SO and the PX are different, and in fact multiple PXs exist for a single SO. This structure reflects the implicit choice of power markets' designers to place less weight on engineering and physical constraints.

**Operating reserves and other ancillary services** In the above example, the SO procures 100 MW at very short notice, which suggests she must keep operating reserves to meet this, and other, contingencies.

Historically, vertically integrated power companies were responsible for supply-demand balance on their service territory. In real time, they balanced domestic demand and supply and managed imports and exports. To achieve this objective, they maintained appropriate levels of operating reserves.

In the restructured power industry, this task is performed by system operators, who procure these operating reserves and other ancillary services using a variety of procurement mechanisms, mostly from power producers. For example, the SO in Alberta provides a very clear description of the issues and the different types of reserves, reproduced in box 2.2.3.

Demand response also starts providing these services. Electric car batteries are expected to become an important provider of operating reserves. Different companies are designing software that enables plugged and idle electric vehicles to inject energy into or stop pulling from the grid at very short notice. These microreserves are then aggregated and sold in the operating reserve market by the SO.

Operating reserves are extremely important operationally, as without them the lights would go out. From an economic perspective, they can be treated as additional demand, for example, the system operator demands operating reserves in the same way a user demands electric power. The main difference is that generators providing operating reserves do not receive the market price unless they are actually called to produce. Instead, they receive the market price minus their variable cost, so they are indifferent between producing electricity and providing reserves. Thus operating reserves do not modify the standard peak-load pricing story.

**Box 2.2.3**
Operating Reserves in Alberta

Operating reserves are used by the Alberta Electric System Operator to maintain system reliability and to ensure power is available when we need it. Because electricity can not be stored and saved for when it is required, power supplied must always be equal to power being consumed. To achieve this balance, our system controllers constantly monitor the demand in the province and match it with available supply. Operating reserve is used by the system controller to ensure that this supply-demand balance is achieved seamlessly.

In Alberta, operating reserves are categorized as regulating reserves, spinning reserves, and supplemental reserves, where each type of reserve performs a unique function.

Owing to the size and complexity of the Alberta Interconnected Electric System, the balance between generation (supply) and consumption (demand) is not instantaneous; often there is a lag while generation is catching up to supply or while generation is decreasing in response to lower demand. Regulating reserves instantaneously provide the power difference between supply and demand required during that lag period.

Spinning and supplemental reserves (collectively referred to as contingency reserves) are used to maintain the balance of supply and demand when an unexpected system event occurs. These reserves provide capacity the system controller can call on with short notice to correct any imbalance. These reserves can come from the supply side (generators) or from the demand side (load curtailment by turning off large electrical consumers to reduce demand immediately; note that load can only provide supplemental reserve). Spinning reserves are the fastest- acting contingency reserve. Generators or loads providing spinning reserves are synchronized to the grid (the turbine is spinning but not generating power); this unique feature allows the reserve to be provided very quickly. In addition to the ability to respond very quickly, spinning reserves also provide frequency support to the Alberta Interconnected Electric System. Supplemental reserves, on the other hand, are not required to be synchronized to the grid and are slower to respond when called upon.

The amount of operating reserve the Alberta Electric System Operator procures is determined by reliability standards set by the Western Electricity Coordinating Council, of which the Alberta so is a member.

## 2.2.5 Three Time Horizons of Security of Supply

Policy makers and practitioners often mention security of supply when discussing the electricity industry, usually in sentences such as "the government will guarantee security of supply." This term is confusing, since it mixes three different time horizons, hence three different notions. It is essential to disentangle them.

### 2.2.5.1 Energy security of supply

Energy security of supply is the ability of a region (e.g., a country, a group of countries, or a state in a federal country) to secure its long-term supply of primary energy. It is measured in units of energy: Terawatt-hours (TWh), ton of oil equivalent (TOEs), or Gigajoules (GJ).

The simplest form of security of supply is to own the primary energy required to fuel the economy for the foreseeable future. A more sophisticated approach is to have long-term contracts or agreements with "friendly" foreign governments.

For example, when the U.S. president Franklin Roosevelt met King Abdulaziz of Saudi Arabia onboard the U.S. Navy cruiser *Quincy* on February 14, 1945, he entered into a security of supply agreement: the United States would protect Saudi Arabia, and in return Saudi Arabia would continue to export its oil to the United States. This agreement has remained in force up to 2016. During the 1970s, Saudi Arabia was a force stabilizing the OPEP (Yergin 1990).

In 1973, faced with the oil crisis and the risk of running out of oil, the French government launched the nuclear electricity program to guarantee the security of its power supply. The primary fuel, uranium, was procured from current or former colonies (New Caledonia in the Pacific Ocean and Niger in Africa).

In 2010 the British nuclear renaissance was a consequence of the reduction in British-owned natural gas reserves in the North Sea. Planners in Whitehall realized that their gas-fired power plants would soon rely on Russian gas, and they disliked the prospect. Better, they thought, to have home-grown energy.

Can competitive firms guarantee energy security of supply? Theoretically, yes. In practice, however, probably no. While private firms sign long-term commercial contracts, these contracts are often linked to broader alliances between countries. Energy security of supply falls squarely within the government's perview.

#### 2.2.5.2 Generation adequacy

As we have seen, when demand does not respond to price, which was the case for most of the twentieth century, generation capacity is determined to meet demand for the next year or the next five years at an agreed-on reliability criterion. The capacity that satisfies the generation-adequacy standard is measured in units of electric power (usually GW). It is routinely estimated and reported by TSOs.

Generation adequacy is often confused with security of supply in the public discourse. This is misleading. The former ensures that, on average, rolling blackouts do not exceed an agreed-on level, for example, three hours per year. If the generation-adequacy standard is not met, rolling blackouts may reach ten hours during a very cold winter in Europe (or a very warm summer in the United States), or brownouts may occur. This is, of course, unpleasant. No one likes to be deprived of electric heating precisely when the temperature drops. But it is much less dramatic than having to ration all users for the entire winter because the gas reserves are insufficient.

#### 2.2.5.3 System reliability

System reliability is the ability for the system to react in real time to unforeseen circumstances, for example, a sudden loss of generation, load increase, or any event that materially affects the power flows.

Remember that power injected in the grid must equal power taken out of the grid (either for consumption or dissipated through losses or stored) at all times. Excess demand imbalance causes a frequency drop, and reciprocally excess supply causes a frequency increases. If this deviation exceeds a certain level, parts of the power system automatically shutdown. This is an uncontrolled blackout. This must be distinguished from an organized rolling blackout. In the latter, rationing is organized. The underlying economics is that it is more efficient (ex ante) to curtail customers for a few hours than to build an additional power plant. An uncontrolled blackout, on the other hand, is an economic calamity. Buyers and sellers are prevented from executing economic transactions. One specificity of blackouts is that recovery takes time and effort. A power grid is not simply switched back on.

To guarantee the reliability of the power grid, system operators have put in place protocols and practices. As previously discussed, SOs procure operating reserves to guarantee system reliability. In another example, discussed in chapter 6, maximum power allowed to flow on transmission lines is lower than the physical capacity of the line, so that the system is able to operate even if one large component (e.g., power plant or an other transmission line) suddenly fails. What matters for reliability is the ramp rate (up or down). It is expressed in MW per minute.

The economics of reliability constitutes an area for fertile future joint work by economists and engineers. First, most reliability protocols and practices are very static, having been designed before the condition of system components could be measured and analyzed in real time. As sensors and remote-control devices are installed on the power grid, these reliability protocols and practices will become dynamic, hence more efficient. Second, in the near future, demand response and batteries will become a nonnegligible reality in power markets and will contribute to system reliability. Robust rules to include them will have to be designed.

#### 2.2.5.4 Additional thoughts

**Link between the three time horizons** The three time horizons are not as separate as the previous discussion may suggest. They feed continuously into one another. For example, generation adequacy can be estimated the day before or even two hours before actual dispatch, hence it resembles system reliability. However, they are different notions, expressed in different units. Poor security of supply leads to long episodes of shortages, usually administered through rolling blackouts. Inadequate generation adequacy leads to more rolling blackouts than is efficient. Inadequate reliability leads to unmanaged system blackouts.

**Comparison with other industries** Observers sometimes argue that reliability is unique to the power industry: if there is a shortage of planes in Chicago, it does not affect travelers on the East Coast. The distinction between capacity adequacy and system reliability helps clarify this point.

Adequacy is the quality of having sufficient infrastructure to accommodate the next anticipated spike in demand, for example, having sufficient planes in Chicago next summer during peak flying season.[6] If planes are in short supply next summer in Chicago, prices will go up, some travelers will not be able to fly on their preferred date, and some may find themselves on the wrong end of overbooking. This is, indeed, unlikely to affect passengers on the East Coast.

On the other hand, as all U.S. air travelers know, if thunderstorms hit Chicago in the summer, as they often do, flights in and out of Chicago will be delayed, possibly canceled, which will most definitely affect travelers on the East Coast and on the West Coast, as well. The consequences of a local reliability issue are less severe for the air travel system than for the power grid. In particular, total blackouts are extremely infrequent. The difference, however, is a matter of degree, not principle.

#### 2.2.5.5 Impact of renewable energy sources on the three horizons of security of supply

**Security of supply** Renewable energy sources have a positive impact on energy security of supply, since wind and solar energy are produced "locally," or, more exactly, cannot be expropriated by neighboring countries. This point is well understood, and the argument is used by proponents of RES.

**Generation adequacy** Theoretically, entry of RES has no impact on long-term generation adequacy, since the generation fleet adapts to match the residual demand. In reality, adaptation is painful and costly, because assets fully-depreciated need to be retired. In Europe and some parts of North America, the speed and extent of RES entry has surprised most observers and existing producers. The residual-demand curve is shifting rapidly and requires equally rapid adaptation of the generation mix.

The most vivid example is the so-called duck chart (figure 2.13), which represents the projected evolution of the residual demand curve (also called the net load) for a typical day in California from 2013 to 2020. Taken together, the curves look like a duck, hence the chart's name. Between 2013 and 2020, the California Independent System Operator expects residual demand to be significantly reduced between hours seven and seventeen hours, leading to overcapacity in these hours: less baseload or mid-merit plants are required in the long-term equilibrium.

Net load then increases rapidly between hours 17 and 20, requiring a significant ramp-up capability, which will be provided by a combination of flexible generation assets, storage, and demand response. This last point is very important, and is sometimes overlooked: the

6. A common misconception is to assume that the Wednesday before Thanksgiving is the busiest air travel day in the United States. It turns out to be very low on the list. The Sunday after Thanksgiving, on the other hand, was the second busiest air-travel day in the United States in 2015.

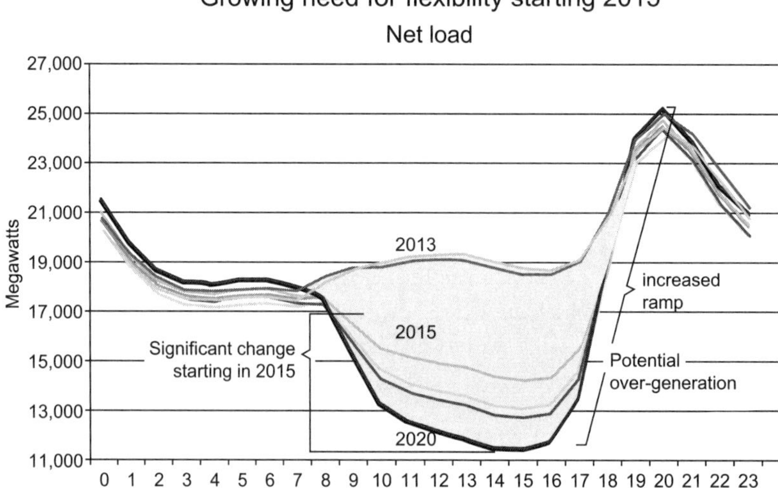


**Figure 2.13**
Projected evolution of the residual demand curve (also called net load) for a typical day in California from 2013 to 2020. Residual demand is significantly reduced between hours 8 and 17 hours. It then increases rapidly between hours 17 and 20, requiring a significant ramp-up capability.

duck chart implicitly assumes demand at every hour is exogenously given and must be met, thus it treats the demand for ramp-up capability as an engineering problem. A more modern view of power markets it that dynamic price will lead to higher prices, hence lower demand at hour 20, and lower prices, hence higher demand between hours 8 and 17, thus mitigating the need for ramp-up capability. It will also reward flexibility of generation assets and remunerate storage assets, whether centralized or decentralized. The policy response to the duck chart challenge should be dynamic pricing, not subsidies to flexible generation assets. This point is revisited in chapter 9.

**System reliability** Most observers expect maintaining system reliability will become more complex in the future, as production from renewable energy sources is more volatile in the short term than production from incumbent generation technologies: neither the wind blows nor the sun shines with perfect regularity.

Recent engineering studies for Europe suggest that the availability of these operating reserves should not be an issue: as discussed in chapter 8, given their nature, European RES will lead to a higher share of mid-merit and peaking plants, which will be able to provide the required flexibility.

However, the costs of providing these reserves is not negligible. It varies significantly by system and by market penetration of RES. Gowrisankaran et al. (2016) shows that the cost of intermittency for solar power in Arizona is around \$12/MWh if the SO adjusts its reserve levels to cope with the fluctuations (particularly short-term changes) in solar output, but is far higher if the SO continues with traditional levels of reserves. A well-connected country like Denmark can increase its operating reserves by trading power with its neighbors, although at a cost of between 4 and 8 percent of the value of the energy produced (Green and Vasilakos 2012).

The engineering challenge is to increase operating reserves, and the policy one is to do so at the least possible cost. Dynamic pricing of operating reserves appears to be the best approach, which constitutes a very fertile field of further research.

## 2.3 The Simple Equations behind the Story

The analytical framework is the same throughout this book. Therefore, we invest significant time in this chapter discussing its different components. Most readers, even those not versed in analytics, would benefit from reading this section. The last sections of this chapter are more technical and would interest mostly analytically inclined readers.

### 2.3.1 States of the World, Demand, and Supply

**States of the world** Electric power demand varies from one hour to the next, as illustrated in figure 2.1. For a given hour, it also varies randomly. For example, I expect demand tomorrow at 5:00 p.m. to be higher than tomorrow at 3:00 a.m., During to a "structural" variation captured in figure 2.1. However, I do not know exactly today what the demand will be at 5:00 p.m. tomorrow, which depends on the temperature, a random variation.

To capture these two sources of variation, we introduce the notion of state of world, which corresponds to a particular realization of demand for a particular hour. The number of possible states of the world is infinite, and these are indexed by $\theta \in (0, +\infty)$. $F(\theta)$ and $f(\theta) = F'(\theta)$ are the ex ante cumulative and probability density functions of state $\theta$. The probability distribution $F(\theta)$ can be understood (and computed) by estimating the distribution of possible demand realization for every hour of the year. This produces a (possibly infinite) set of hourly demand realizations. The probability distribution $F(\theta)$ is the distribution of this set.

This approach has two limitations. First, it blurs the distinction between structural and random variations. This is acceptable for the simple analysis we present but would be inappropriate for more sophisticated analyses, for example, including intertemporal linkages or pricing demand volatility. Second, uncertainty affects both demand and supply conditions. Here, only demand uncertainty is explicitly modeled, since including production uncertainty does not modify the economic insights.

### Demand

**Assumption 1.** All customers have the same underlying demand $D(p, \theta)$ in state $\theta$, where $p$ is the electric power price, up to a scaling factor.

Assumption 1 greatly simplifies the derivations while preserving the main economics insights. Inverse demand is $P(Q, \theta)$ defined by $D(P(Q, \theta), \theta) = Q$, and gross consumers surplus is $S(p, \theta) = \int_0^{D(p,\theta)} P(q, \theta)dq$.

**Assumption 2.** Inverse demand $P(Q, \theta)$ is downward sloping: $\partial_Q P(Q, \theta) < 0$ and increasing with the state of the world $\partial_\theta P(Q, \theta) > 0$.[7]

Downward sloping inverse demand is a standard economic assumption. It "orders" units of consumption: the first units consumed are higher valued. Inverse demand increasing with the state of the world orders states of the world. It leads to simpler notations but is not essential for the argument.

Throughout this text, I use a linear inverse demand with constant slope as an example: $P(Q, \theta) = a(\theta) - bQ$ where $b > 0$ guarantees $\partial_Q P(Q, \theta) < 0$, and $a'(\theta) > 0$ "orders" states of the world by increasing demand.

Linear inverse demand is consistent with empirical estimations (see, e.g., Patrick and Wolak 1997) and yields simple expressions, in particular, when considering imperfect competition (chapters 3 and 4). I use a particular specification of $a(\theta)$ and $f(\theta)$ to derive the estimates presented in this chapter. Other authors, for example Borenstein (2007) and Allcott (2011) use isoelastic demand $P(Q, \theta) = a(\theta)Q^{-\eta}$ where $\eta > 0$ is the constant demand elasticity.

**Supply** A single production technology is available, characterized by variable cost of production per unit $c$ and hourly fixed cost of production per unit $r$, both constant and expressed in €/MWh. Production technology exhibits constant long-run returns to scale: in the long run, producing two Megawatts during one hour costs exactly twice as much as producing one Megawatt during one hour.

Demand and production technologies jointly satisfy the following conditions regarding

**Condition 1.** Value of the first and last unit.

1. Consuming the first unit generates positive net surplus in the short term in every state of the world: $P(0, \theta) > c \ \forall \theta \geq 0$, and in the long-term on average $\mathbb{E}[P(0, \theta)] > (c + r)$.

2. Inverse demand falls below the marginal production cost for very large output: $\lim_{Q \to +\infty} P(Q, \theta) < c$.

7. Partial derivatives are expressed using the notation $\partial$, for example, $\partial_Q P(Q, \theta) = \frac{\partial P}{\partial Q}(Q, \theta)$ and $\partial_{Q\theta} P(Q, \theta) = \frac{\partial^2 P}{\partial Q \partial \theta}(Q, \theta)$.

Condition 1 guarantees existence of positive optimal capacity: it is economically efficient to produce and consume at least one MWh (Condition 1.1), and not efficient to produce and consume an infinite volume (Condition 1.2).[8]

### 2.3.2 Socially Optimal Consumption and Investment

The previous notation enables us to formalize the intuition presented in section 2.1. The objective is to maximize the expected hourly net surplus, subject to the constraint that demand is always lower than capacity $K$. Formally, this is written as

$$max_{p(\theta),K}\mathbb{E}[(S(p(\theta),\theta)-cD(p,\theta))]-rK$$
$$st:\ \ D(p,\theta)\leq K\ \forall\theta\quad(\lambda(\theta)),$$

where $\mathbb{E}[.]$ is the expectation over the states of the world, $p(\theta)$ is the electricity spot price, and $\lambda(\theta)$ is the Lagrange multiplier of the constraint $D(p,\theta)\leq K$ in state $\theta$. The Lagrangian is

$$\mathcal{L}=\mathbb{E}[(S(p(\theta),\theta)-cD(p,\theta))+\lambda(\theta)(K-D(p,\theta))]-rK.$$

**Optimal consumption** The first-order derivative with respect to price is

$$\begin{aligned}\frac{\partial\mathcal{L}}{\partial p(\theta)}&=\frac{\partial S}{\partial p(\theta)}-(c+\lambda(\theta))\frac{\partial D}{\partial p(\theta)}\\&=(p(\theta)-(c+\lambda(\theta)))\frac{\partial D}{\partial p(\theta)},\end{aligned}$$

observing that $\frac{\partial S}{\partial p(\theta)}=p(\theta)\frac{\partial D}{\partial p(\theta)}$. Thus the first-order condition with respect to price is

$$p^*(\theta)=c+\lambda(\theta).$$

We thus distinguish two periods. Off-peak,

$$D(p(\theta),\theta)<K\Leftrightarrow\lambda(\theta)=0\Leftrightarrow p^*(\theta)=c.$$

As illustrated on the horizontal portion of the supply curve in figure 2.4, the off-peak price is the variable production cost of power $c$, and production $Q(\theta)$ is determined by $P(Q(\theta),\theta)=c$.

Since demand is increasing in states of the world, on-peak is characterized by $\theta\geq\hat{\theta}_0(K,c)$, where $\hat{\theta}_0(K,c)$ is the lowest state of the world such that demand at price $c$ is equal to or larger than installed capacity $K$. Formally, $\hat{\theta}_0(K,c)$ is the first state of the

8. Technical conditions such as Condition 1 are named differently from Assumptions (such as Assumptions 1 and 2). The former merely provide boundaries on the values of parameters, or on the shape of functional forms. There is no reason to expect them not to be met. The latter are more structural and may not be met in practice.

world such that

$$D(c, \hat{\theta}_0(K, c)) \geq K \Leftrightarrow P(K, \hat{\theta}_0(K, c)) \geq c,$$

since inverse demand $P(Q, \theta)$ is decreasing in its first argument. Two cases must be distinguished. If $P(K, \theta) > c$ for all $\theta \geq 0$, then $\hat{\theta}_0(K, c) = 0$. Otherwise, $\hat{\theta}_0(K, c)$ is uniquely defined by $P(K, \hat{\theta}_0(K, c)) = c$ and is increasing in both arguments.

Then, on-peak, we have

$$D(p(\theta), \theta) = K \Leftrightarrow p^*(\theta) = P(K, \theta) \Leftrightarrow \lambda(\theta) = P(K, \theta) - c > 0.$$

As illustrated in the vertical portion of the supply curve in figure 2.4, demand is set by installed capacity $K$, and the wholesale price is $P(K, \theta)$. The shadow price of the capacity constraint is the operating margin $(P(K, \theta) - c)$.

Assumption 2 guarantees that the necessary conditions determine the unique maximum.

**Optimal capacity** The first-order condition with respect to $K$ is

$$\frac{\partial \mathcal{L}}{\partial K} = \mathbb{E}[\lambda(\theta)] - r = 0 \Leftrightarrow \mathbb{E}[\lambda(\theta)] = r.$$

Increasing capacity by one unit generates no net surplus if capacity is not constrained and net surplus $(P(K, \theta) - c)$ otherwise. Therefore, the marginal social value of capacity is

$$\Psi_0(K, c) = \int_{\hat{\theta}_0(K,c)}^{+\infty} (P(K, \theta) - c) f(\theta) d\theta = \mathbb{E}[(P(K, \theta) - c)^+], \tag{2.1}$$

where $(x)^+ = max(x, 0)$. The function $\Psi_0(K, c)$ is continuous and decreasing in both arguments by inspection and is represented by the shaded area under the price curve and above the cost $c$ in figure 2.6.

Since $P(0, \theta) > c \; \forall \theta \geq 0$ by condition 1.1, $\hat{\theta}_0(0, c) = 0$, hence

$$\Psi_0(0, c) = \int_0^{+\infty} (P(K, \theta) - c) f(\theta) d\theta = \mathbb{E}[P(K, \theta)] - c.$$

Since $\mathbb{E}[P(0, \theta)] > (c + r)$ by Condition 1.1, $\Psi_0(0, c) > r$. Since $\lim_{Q \to +\infty} P(Q, \theta) < c$ by Condition 1.2, $lim_{K \to +\infty} \hat{\theta}_0(K, c) = +\infty$; hence $lim_{K \to +\infty} \Psi_0(K, c) = 0 < r$.

Since $\Psi_0(K, c)$ is continuous in its first argument, $\Psi_0(0, c) > r$, and $lim_{K \to +\infty} \Psi_0 (K, c) = 0 < r$, the intermediate value theorem guarantees the existence of a $K^* > 0$ such that

$$\Psi_0(K^*, c) = r. \tag{2.2}$$

Since $\Psi_0(K, c)$ is decreasing in its first argument, the optimal capacity $K^*$ defined by equation 2.2 is unique.

Off-peak, as long as capacity is not constrained, price equals marginal cost, hence marginal capacity generates no economic value. On-peak, when capacity is constrained, price exceeds marginal cost. The optimal capacity is such that the marginal social value capacity is exactly equal to the marginal capacity cost $r$.

We verify that equation (2.2) implies that the average hourly price is equal to the hourly long-run marginal cost of power:

$$\begin{aligned}\mathbb{E}[p^*(\theta)] &= c + \mathbb{E}[p^*(\theta) - c] \\ &= c + \mathbb{E}[(p^*(\theta) - c)\mathbb{I}_{\{p^*(\theta)=c\}}] + \mathbb{E}[(p^*(\theta) - c)\mathbb{I}_{\{p^*(\theta)>c\}}] \\ &= c + r\end{aligned}$$

**Equivalence optimum and perfect competition outcome** If competition is perfect, producers bid their variable production cost up to their capacity into the wholesale market. Off-peak, offered energy exceeds demand, hence the wholesale price is set at this variable production cost. On-peak, demand is precisely equal to the offered capacity, hence the wholesale price is set by the value of the marginal Megawatt-hour consumed. Finally, equation (2.2) is also a free-entry condition (Joskow and Tirole 2007).

**A specific case** Suppose inverse demand is linear with constant slope, $P(Q, \theta) = a(\theta) - bQ$, where the function $a(.)$ is increasing and $b > 0$. Integrating the function $\Psi_0(K, c)$ by parts yields

$$\Psi_0(K, c) = \int_{\hat{\theta}_0(K,c)}^{+\infty} \partial_\theta P(K, \theta)(1 - F(\theta))d\theta = \int_{\hat{\theta}_0(K,c)}^{+\infty} a'(\theta)(1 - F(\theta))d\theta, \qquad (2.3)$$

since the boundary terms are equal to zero. Thus equation (2.2) uniquely determines the first on-peak state of the world $\theta^*$ such that

$$\int_{\theta^*}^{+\infty} a'(\theta)(1 - F(\theta))d\theta = r.$$

Then the optimal capacity is uniquely determined by condition

$$P(K^*, \theta^*) = c \Leftrightarrow bK^* = a(\theta^*) - c.$$

## 2.4 Completely Inelastic Demand

We now relax the assumption that demand is elastic. Specifically, we assume that all consumers face a constant retail price $p^R$, hence demand in state $\theta$ is $D(p^R, \theta)$.

**Rationing** As previously discussed, it is not efficient to build capacity such that demand is always met. Since demand cannot be met in all states of the world, the SO must sometimes curtail a fraction of demand. We now describe the formalism of curtailment.

Denote $\gamma \in [0, 1]$ the *serving ratio*: $\gamma = 0$ means full curtailment, while $\gamma = 1$ means no curtailment. For state $\theta$, $\mathcal{D}(p, \gamma, \theta)$ is the demand for price $p$ and serving ratio $\gamma$, and $\mathcal{S}(p, \gamma, \theta)$ is the gross consumer surplus. By construction, $\mathcal{D}(p, 1, \theta) \equiv D(p, \theta)$ and $\mathcal{S}(p, 1, \theta) \equiv S(p, \theta)$.

Since curtailed customers are homogeneous, the SO has no basis to discriminate. Thus curtailment proceeds by geographic zones, and

$$\mathcal{D}(p^R, \gamma, \theta) = \gamma D(p^R, \theta).$$

$\mathcal{S}(p^R, \gamma, \theta)$ depends on the customers' information structure. Suppose first rationing is perfectly anticipated, for example, the SO announces the day before the exact duration and location of the curtailment. As in Joskow and Tirole (2007) it is reasonable to assume that the fraction $\gamma$ of customers who will not be curtailed consumes normally; hence it receives surplus $S(p^R, \theta)$ per customer; and that the fraction $(1 - \gamma)$ of customers who will be curtailed does not attempt to consume; hence it receives no surplus. At the aggregate level, this yields

$$\mathcal{S}(p^R, \gamma, \theta) = \gamma S(p^R, \theta).$$

If rationing is not perfectly anticipated, customers who end up not being curtailed may refrain from consuming and hence receive no surplus. Conversely, consumers who end up being curtailed may engage in electricity consuming activity (e.g., step into an elevator); hence they derive a negative surplus when power is cut. Joskow and Tirole (2007) illustrate in a simple example how the net surplus $(\mathcal{S}(p, \gamma; \theta) - p\mathcal{D}(p, \gamma, \theta))$ depends on the information consumers hold about the outage.

**Value of Lost Load** Absent rationing, the marginal surplus associated with a unit increase in consumption is $\frac{\frac{\partial S}{\partial p}}{\frac{\partial D}{\partial p}}(p, \theta)$: if the price increases by $\delta p$, surplus decreases by $\frac{\partial S}{\partial p}\delta p$, and consumption by $\frac{\partial D}{\partial p}\delta p$, thus the net surplus per unit change is $\frac{\frac{\partial S}{\partial p}}{\frac{\partial D}{\partial p}} = p$.

For rationed customers, Joskow and Tirole (2007) formally define "the marginal surplus associated with a unit increase in supply to [rationed] consumers." Formally, it is expressed by the value of lost load (VoLL) as

$$v(\bar{p}, \gamma, \theta) = \frac{\frac{\partial \tilde{S}(\bar{p}, \gamma, \theta)}{\partial \gamma}}{\frac{\partial \tilde{D}(\bar{p}, \gamma, \theta)}{\partial \gamma}}:$$

if the serving ratio increases by $\delta\gamma$, surplus increases by $\frac{\partial \tilde{S}}{\partial \gamma}\delta\gamma$, and consumption by $\frac{\partial \tilde{D}}{\partial \gamma}\delta\gamma$, thus the net surplus per unit change is $\frac{\frac{\partial \tilde{S}}{\partial \gamma}}{\frac{\partial \tilde{D}}{\partial \gamma}}$.

If rationing is perfectly anticipated, the VoLL is

$$v(p^R, \gamma, \theta) = \frac{\frac{\partial \mathcal{S}}{\partial \gamma}}{\frac{\partial \mathcal{D}}{\partial \gamma}}(p^R, \gamma, \theta) = \frac{S(p^R, \theta)}{D(p^R, \theta)} > p^R.$$

The VoLL depends only on the state of the world. In particular, it does not depend on the serving ratio. It is always higher than the price.

This expression formalizes the intuition discussed previously. Since rationing is proportional, all units are uniformly curtailed, those with high value and those with low value. The VoLL is thus the average surplus per unit curtailed, higher than the marginal value. This relation does not always hold, in particular if rationing is efficient.

**Assumption 3.** Rationing is perfectly anticipated.

Assumption 3 greatly simplifies the analysis since the VoLL no longer depends on the serving ratio; in particular, it produces analytically tractable solutions and delivers most of the economic insights. It is reasonable for OECD countries, where SO*s* know a day ahead if whether rationing is possible and warn consumers.

If demand is linear with constant slope and since rationing is anticipated and proportional, the VoLL takes a simple form:

$$\begin{aligned} v(p^R, \theta) &= \frac{S(p^R, \theta)}{D(p^R, \theta)} = \frac{\int_0^{D(p^R,\theta)} (a(\theta) - bq)dq}{D(p^R, \theta)} = \frac{\left[\left(a(\theta) - \frac{b}{2}q\right)q\right]_0^{D(p^R,\theta)}}{D(p^R, \theta)} \\ &= a(\theta) - \frac{b}{2} D(p^R, \theta) = a(\theta) - \frac{b}{2}\frac{a(\theta) - p^R}{b} = \frac{a(\theta) + p^R}{2}. \end{aligned}$$

Derivations presented below assume the SO knows exactly the VoLL in every state of the world. While this assumption is highly unrealistic, it constitutes a useful analytical benchmark, since effects observed cannot be attributed to a poor estimate of the VoLL. As discussed previously, in reality, the SO uses her best estimate of the average VoLL and prioritizes curtailment by geographic zones (using criteria such as economic activity, political weight, and network conditions).

**Condition 2.** Properties of the VoLL and constant retail price $p^R$.

1. The constant retail price is higher than variable cost and lower than the value of the first unit in all states of the world: $c < p^R < P(0, \theta)\ \forall \theta \geq 0$.
2. Average VoLL exceeds the long-run marginal cost of production

$$\mathbb{E}[v(p^R, \theta)] > (c + r).$$

Condition 2.1 puts bounds on the admissible values of the retail price. Since $p^R < P(0,\theta) \Leftrightarrow D(p^R,\theta) > 0$, it implies that demand at the retail price is always positive. Since the VoLL is endogenous in this text, Condition 2.2 is a condition on the demand function $D(p,\theta)$. Condition 2 mirrors Condition 1 and is sufficient to guarantee existence and unicity of the optimal capacity.

**Optimization program** The optimization program then becomes

$$max_{\gamma(\theta),K}\mathbb{E}[\gamma(\theta)(S(p^R,\theta) - cD(p^R,\theta))] - rK$$
$$st:\ \gamma(\theta)D(p^R,\theta) \leq K\ \forall\theta\ (\lambda(\theta)),$$

hence the associated Lagrangian is

$$\mathcal{L} = \mathbb{E}[\gamma(\theta)(S(p^R,\theta) - cD(p^R,\theta)) + \lambda(\theta)(K - \gamma(\theta)D(p^R,\theta))] - rK.$$

The first-order derivative with respect to the serving ratio $\gamma(\theta)$ is

$$\frac{\partial\mathcal{L}}{\partial\gamma(\theta)} = S(p^R,\theta) - (c+\lambda(\theta))D(p^R,\theta) = (v(p^R,\theta) - (c+\lambda(\theta)))D(p^R,\theta).$$

Off-peak, $\lambda(\theta) = 0$. Since $v(p^R,\theta) \geq p^R > c$ by Condition 2.1 and $D(p^R,\theta) > 0$, $\frac{\partial\mathcal{L}}{\partial\gamma(\theta)} > 0$, it is the case that $\gamma^*(\theta) = 1$. This formalizes the intuition that curtailment is not required off-peak. Production $Q(\theta)$ is determined by $Q(\theta) = D(p^R,\theta) \Leftrightarrow P(Q(\theta),\theta) = p^R$.

On-peak arises when demand at price $p^R$ is equal to or larger than installed capacity: $D(p^R,\theta) \geq K \Leftrightarrow P(K,\theta) \geq p^R \Leftrightarrow \theta \geq \hat{\theta}_0(K,p^R)$. For $\theta \geq \hat{\theta}_0(K,p^R)$, $\gamma^*(\theta)$ is uniquely defined by $\gamma^*(\theta)D(p^R,\theta) = K$. Since $D(p^R,\theta) > 0$, the first-order condition $\frac{\partial\mathcal{L}}{\partial\gamma(\theta)} = 0$ yields

$$\lambda(\theta) = v(p^R,\theta) - c.$$

The shadow value of capacity $\lambda(\theta)$ is discontinuous at $\theta = \hat{\theta}_0(K,p^R)$, as it jumps from $\lambda(\theta) = 0$ off-peak to $\lambda(\theta) > 0$ for the "lowest" on-peak state of the world. This is normal: as soon as demand at the retail price $p^R$ equals installed capacity, rationing occurs, and the value of electricity jumps to the VoLL.

The first-order derivative with respect to $K$ then yields

$$\frac{\partial\mathcal{L}}{\partial K} = \int_{\hat{\theta}_0(K,p^R)}^{+\infty} (v(p^R,\theta) - c)f(\theta)d\theta - r.$$

The same argument as before proves existence and unicity of the optimal capacity $K^*$. The marginal value of generation capacity when all customers face a constant price $p^R$ is

$$\bar{\Psi}_0(K,p^R,c) = \int_{\hat{\theta}_0(K,p^R)}^{+\infty} (v(p^R,\theta) - c)f(\theta)d\theta. \tag{2.4}$$

The marginal value is decreasing in its first argument by inspection. Since $P(0,\theta) > p^R \ \forall\theta \geq 0$ by Condition 2.1, $\hat{\theta}_0(0, p^R) = 0$ hence

$$\bar{\Psi}_0(0, p^R, c) = \int_0^{+\infty} (v(p^R,\theta) - c) f(\theta) d\theta = \mathbb{E}[v(p^R,\theta)] - c > r,$$

by Condition 2.2. Since $\lim_{Q\to+\infty} P(Q,\theta) < c < p^R$ by Conditions 1 and 2, $lim_{K\to+\infty}\hat{\theta}_0(K, p^R) = +\infty$, hence $lim_{K\to+\infty}\bar{\Psi}_0(K, p^R, c) = 0 < r$.

Thus, the optimal capacity $K^*$ is the unique solution to

$$\bar{\Psi}_0(K^*, p^R, c) = r \Leftrightarrow \int_{\hat{\theta}_0(K^*,p^R)}^{+\infty} (v(p^R,\theta) - c) f(\theta) d\theta = r. \tag{2.5}$$

Since the VoLL does not depend on installed capacity (since rationing is perfectly anticipated), capacity enters equation (2.5) only to determine the duration of curtailment.

Equation (2.5) is equivalent to rule 2-3.1 in Stoft (2002, 139). Suppose the *SO* does not know the VoLL, rather uses a constant approximation of the VoLL, denoted $v$. Assuming that $v \gg c$, and denoting $h$ the expected number of hours when curtailment occurs, the above equation becomes:

$$v \times \frac{h}{8{,}760} = r \Leftrightarrow v \times h = 8{,}760 \times r,$$

which is relation 2-3.1 in Stoft (2002). It is illustrated on the left-panel of figure 2.8.

## 2.5 A More Realistic Peak-Load Pricing Story

### 2.5.1 Demand and Supply

**Demand** In reality, two categories of consumers exist. Some large industrial users buy directly form the wholesale spot markets. Retailers purchase energy on the wholesale spot markets and resell it to their customers, either at a retail price indexed on the spot price or at a constant retail price.

This dichotomy is formally represented by introducing the fraction $\alpha > 0$ of consumers who face and react to the wholesale spot price $p(\theta)$ in state $\theta$. Since the spot price is called the real-time Price (RTP), these are called RTP consumers. The remaining fraction $(1-\alpha)$ of consumers face a constant two-part pricing scheme ("constant-price" consumers), with retail price $p^R$ per *MWh*, constant across all states of the world, and connection charge $A$ per year. Since all consumers have the same load profile up to a scaling factor by Assumption 1, the fraction $\alpha$ is constant across states of the world.

We make the following assumption:

**Assumption 4.** The SO has the technical ability to curtail "constant-price" customers while not curtailing $RTP$ customers.

Assumption 4 is unrealistic today, as the SO can only organize curtailment by zone and cannot differentiate by type of customer. However, those conditions it will be met when smart meters are rolled out.

Since price-reactive customers face the wholesale spot price $p(\theta)$ in state $\theta$, absent curtailment, supply demand balance yields

$$Q(\theta) = \alpha D(p(\theta), \theta) + (1-\alpha) D(p^R, \theta)$$

$$\Longleftrightarrow$$

$$p(\theta) = \rho(Q(\theta), \theta) = P\left(\frac{Q(\theta) - (1-\alpha)D(p^R, \theta)}{\alpha}, \theta\right). \tag{2.6}$$

The function $\rho(Q, \theta)$ is derived from $P(Q, \theta)$ by a rotation around the point $(D(p^R, \theta), p^R)$. It is steeper than the inverse demand curve, since a fraction $(1-\alpha)$ of customers do not respond to spot prices; hence they have vertical demand.

For example, suppose inverse demand is linear with constant slope. If no rationing occurs, $\rho(Q, \theta)$ is also linear:

$$\rho(Q, \theta) = a(\theta) - b\left(\frac{Q - (1-\alpha)\frac{a(\theta)-p^R}{b}}{\alpha}\right) = a(\theta)\left(1 + \frac{1-\alpha}{\alpha}\right) - \frac{1-\alpha}{\alpha} p^R - \frac{bQ}{\alpha}$$

$$= \frac{a(\theta) - (1-\alpha)p^R - bQ}{\alpha}.$$

**Supply** Different generation technologies are available, indexed by $n = 1, \ldots, N$. $c_n$ is the variable cost, and $r_n$ is the hourly capacity cost (i.e., annual capacity cost expressed in €/MW/year divided by 8,760 hours per year) of technology $n$, both expressed in €/MWh. Generation technologies are ordered by increasing variable cost: $c_n > c_m \; \forall \, n \geq m$. There is a trade-off between capacity and variable costs: if a technology produces at higher variable cost, then its capacity cost is lower, that is, $r_n < r_m \; \forall \, n \geq m$.

Sufficient conditions on the costs $\{c_n, r_n\}_n$, presented later in the text, ensure that technologies 1 to $N$ are used at the optimum. The installed capacity of technology $n$ is $k_n \geq 0$, and the cumulative installed capacity up to and including technology $n$ is

$$K_n = \sum_{m=1}^{n} k_m.$$

### 2.5.2 Optimal Price, Rationing, and Dispatch

The optimal dispatch and investment program has been derived for example by Borenstein and Holland (2005) and Joskow and Tirole (2007). Joskow and Tirole (2007) shows that it is never optimal to curtail price-reactive customers. The total consumer surplus and demand in state $\theta$ are therefore:

$$\begin{cases} \tilde{S}(p(\theta), p^R, \gamma(\theta), \theta) = \alpha S(p(\theta), \theta) + (1-\alpha)\gamma(\theta) S(p^R, \theta) \\ \tilde{D}(p(\theta), p^R, \gamma(\theta), \alpha, \theta) = \alpha D(p(\theta), \theta) + (1-\alpha)\gamma(\theta) D(p^R, \theta). \end{cases}$$

The optimization program is then

$$\begin{aligned} &\max_{p(.), p^R, \gamma(.), u_n(.), k_n} \mathbb{E}[\tilde{S}(p(\theta), p^R, \gamma(\theta), \theta) - \sum_{n\geq 1} c_n u_n(\theta) k_n] - \sum_{n\geq 1} r_n k_n \\ &st: \forall \theta \geq 0 \ \tilde{D}(p(\theta), p^R, \gamma(\theta), \theta) \leq \sum_{n\geq 1} u_n(\theta) k_n \ (\lambda(\theta)), \end{aligned} \tag{2.7}$$

where $u_n(\theta) \in [0, 1]$ the dispatch ratio of technology $n$ and $\lambda(\theta) \geq 0$ the Lagrange multiplier in state $\theta$. The Lagrangian is

$$\mathcal{L} = \mathbb{E}[\tilde{S} - \sum_{n\geq 1} c_n u_n(\theta) k_n + \lambda(\theta) \left( \sum_{n\geq 1} u_n(\theta) k_n - \tilde{D} \right)] - \sum_{n\geq 1} r_n k_n,$$

and the first-order derivatives are

$$\begin{cases} \frac{\partial \mathcal{L}}{\partial p(\theta)} = \alpha (p(\theta) - \lambda(\theta)) \partial_p D(p(\theta), \theta) \\ \frac{\partial \mathcal{L}}{\partial u_n(\theta)} = (\lambda(\theta) - c_n) k_n \\ \frac{\partial \mathcal{L}}{\partial \gamma(\theta)} = (1-\alpha)(v(p^R, \theta) - \lambda(\theta)) D(p^R, \theta) \\ \frac{\partial \mathcal{L}}{\partial p^R} = (1-\alpha) \mathbb{E}[\gamma(\theta)(\partial_p S(p^R, \theta) - \lambda(\theta) \partial_p D(p^R, \theta))] \\ \frac{\partial \mathcal{L}}{\partial k_n} = \mathbb{E}([\lambda(\theta) - c_n] u_n(\theta)) - r_n. \end{cases} .$$

We examine each condition in turns.

**Wholesale spot price** The condition $\frac{\partial \mathcal{L}}{\partial p(\theta)} = 0$ yields $p^*(\theta) = \lambda(\theta)$: price-reactive customers pay the opportunity cost of electricity in each state.

**Optimal dispatch** The derivative $\frac{\partial \mathcal{L}}{\partial u_n(\theta)}$ yields the dispatch rule

$$u_n^*(\theta) = \begin{cases} 1 & if \ c_n < p^*(\theta) \\ 0 & if \ c_n > p^*(\theta). \\ \frac{\tilde{D} - \sum_{m<n} k_m}{k_n} & if \ c_n = p^*(\theta) \end{cases}$$

Technology $n$ produces at capacity (respectively does not produce) if its opportunity cost is lower than the price (respectively exceeds the price) in state $\theta$. If technology $n$ is marginal, that is, price setting, energy balance sets the dispatch ratio. $p^*(\theta) = c_n > 0$ is therefore the wholesale spot power price.

The optimal dispatch rule yields the staircase supply curve represented for three technologies in figure 2.11. Using the convention $c_{N+1} \to +\infty$, the steps of the staircase are formally defined for $1 \leq n \leq N$ by $\upsilon_n = \{\theta : c_n < p(\theta) < c_{n+1}\}$ for the vertical segments and $h_n = \{\theta : p(\theta) = c_n\}$ for the horizontal ones. For $n = 1, \ldots, N$, for $\theta \in h_n$, technology $n$ is the marginal technology, thus $p^*(\theta) = c_n$. A price-reactive customer consumes $D(c_n, \theta)$; hence total production $Q^*(\theta)$ is such that $\rho(Q^*(\theta), \theta) = c_n$.

For $n = 1, \ldots, (N-1)$, for $\theta \in \upsilon_n$, all technologies up to $n$ produce at capacity, and the price is set by the intersection of the elastic inverse demand curve and the vertical supply curve: $p^*(\theta) = \rho(K_n, \theta)$.

Finally, for $\theta \in \upsilon_N$, total generation produces at capacity. Rationing may occur, and is discussed below.

**Optimal rationing rule** To describe the economics of rationing, a technical condition and a definition are needed:

**Condition 3.** Properties of the VoLL

1. The VoLL is nondecreasing as the state of the world increases but increases less than the spot price absent curtailment: $0 \leq \partial_\theta v(p^R, \theta) < \partial_\theta \rho(K, \theta)$.
2. At $\theta = 0$, the VoLL is lower than the value of the first unit:

$v(p^R, 0) < \rho(0, 0)$.

Condition 3 is met if demand is linear with constant slope ($v(p^R, \theta) = \frac{a(\theta)+p^R}{2}$) or isoelastic ($v(p^R, \theta) = \frac{p^R}{1-\eta}$).

Define $\bar{\theta}(K_N)$ as the lowest state of the world such that the wholesale spot price absent curtailment $\rho(K_N, \theta)$ is larger than or equal to the VoLL $v(p^R, \theta)$. Three situations may occur. First, if $\rho(K_N, \theta) \geq v(p^R, \theta)$ for all $\theta \geq 0$, then $\bar{\theta}(K_N) = 0$. Second, if $\rho(K_N, \theta) < v(p^R, \theta)$ for all $\theta \geq 0$, then $\bar{\theta}(K_N) \to +\infty$. Otherwise, $\bar{\theta}(K_N)$ is uniquely defined by $\rho(K_N, \bar{\theta}(K_N)) = v(p^R, \bar{\theta}(K_N))$. Condition 3.1 guarantees unicity of $\bar{\theta}(K_N)$ if it exists.

Rationing may occur only when generation produces at capacity, that is, in state $\theta \in \upsilon_N$ and $p^*(\theta) = \rho(K_N, \theta)$. As long as the wholesale spot price set by price-responsive customers is lower than the VoLL, that is, as long as $\rho(K_N, \theta) < v(p^R, \theta) \Leftrightarrow \theta < \bar{\theta}(K_N)$, the SO serves all constant price customers, since the value they would place on a curtailed Megawatt-hour exceeds the wholesale spot price.

If $\rho(K_N, \theta) \geq v(p^R, \theta) \Leftrightarrow \theta \geq \bar{\theta}(K_N)$ then the SO starts curtailing constant price customers, since the value they place on a curtailed Megawatt-hour is lower than the wholesale

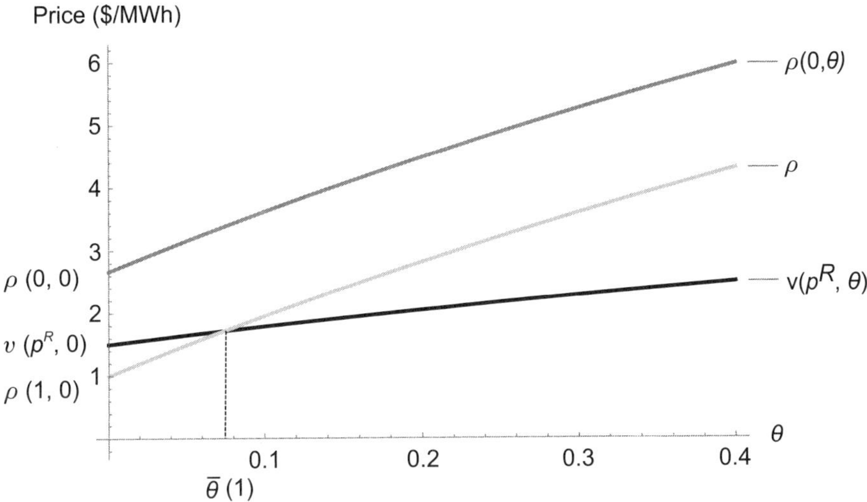


**Figure 2.14**
Functions $v(p^R, \theta)$, $\rho(0, \theta)$, and $\rho(1, \theta)$ if demand and uncertainty are represented by the exponential specification presented in section 2.7.

spot price absent curtailment. Specifically, the SO sets the serving ratio $\gamma^*(\theta)$ to balance supply and demand:

$$K_N = \alpha D(v(p^{R*}, \theta), \theta) + (1-\alpha)\gamma^*(\theta) D(p^{R*}, \theta).$$

and sets the wholesale spot price equal to the opportunity cost of power, which is the VoLL

$$p^*(\theta) = v(p^R, \theta), \tag{2.8}$$

Condition 3.1 eliminates the possibility that the VoLL $v(p^R, \theta)$ and the wholesale spot price absent curtailment cross multiple times. Thus if curtailment starts to occur, it occurs for all $\theta \geq \bar{\theta}(K_N)$. It implies that the VoLL acts as an endogenous cap on prices.

This situation is illustrated in figure 2.14, which represents the functions $v(p^R, \theta)$, $\rho(0, \theta)$, and $\rho(1, \theta)$ for the specification of inverse demand function and uncertainty introduced as an illustrative example in section 2.7.

Since $v(p^R, 0) < \rho(0, 0)$ (Condition 3.2) and $\partial_\theta v(p^R, \theta) < \partial_\theta \rho(K, \theta)$ (Condition 3.1), these curves never cross: $v(p^R, \theta) < \rho(0, \theta)$ $\forall \theta \geq 0$. Thus if cumulative capacity is zero, curtailment always occurs at $\bar{\theta}(0) = 0$. Similarly, $\bar{\theta}(K) = 0$ for all $K$ such that $\rho(K, 0) \geq v(p^R, 0) \Leftrightarrow K \leq \mathcal{D}(v(p^R, 0), 0)$. This is a technical property, required to prove existence of the optimal capacity.

Wholesale price path when curtailment occurs is illustrated in figure 2.14 for $K = 1$. The functions $\rho(1, \theta)$ and VoLL $v(p^R, 0)$ cross for $\bar{\theta}(1) > 0$. The wholesale spot price on peak is thus $\rho(1,\theta)$ for $\theta \leq \bar{\theta}(1)$ and $v(p^R,\theta)$ for $\theta > \bar{\theta}(1)$: since $\rho(1,\theta) > v(p^R,\theta)$ for $\theta > \bar{\theta}(1)$, the VoLL acts as an endogenous cap on wholesale spot prices.

The simultaneous presence of price-reactive and constant-price customers imposes a "no arbitrage" relationship, that is, the price received by producers and paid by price-reactive customers is precisely equal to the VoLL for curtailed constant price customers.

Theoretically, we could curtail all constant price customers ($\gamma = 0$), in which case the price would rise above the VoLL. While this is theoretically possible, it is unlikely to be the equilibrium outcome, hence we do not consider it.

**Optimal retail price** The condition $\frac{\partial \mathcal{L}}{\partial p^R} = 0$ yields

$$\mathbb{E}\left[(p^{R*} - p^*(\theta))\gamma^*(\theta)\frac{\partial D}{\partial p}(p^{R*},\theta)\right] = 0 \Leftrightarrow p^{R*} = \frac{\mathbb{E}\left[p^*(\theta)\gamma^*(\theta)\frac{\partial D}{\partial p}(p^{R*},\theta)\right]}{\mathbb{E}[\gamma^*(\theta)\frac{\partial D}{\partial p}(p^{R*},\theta)]}.$$

As in Joskow and Tirole (2007), the optimal retail price is the weighted average wholesale price, where the weights are the marginal "rationed demand," to take into account the possibility that constant-price consumers may be rationed. The optimal retail price needs not cover the full production cost, since the fixed part of two-part retail price balances the retailers' profits.

If rationing occurs, numerical simulations are in general required to determine $p^{R*}$. A simple solution exists if inverse demand is linear with constant slope $P(q,\theta) = a(\theta) - bq$ and no rationing occurs at the optimum, $\gamma^* = 1$. Then, $\gamma^* \frac{\partial D}{\partial p}(p^{R*},\theta) = -\frac{1}{b}$, and

$$p^{R*} = \mathbb{E}[p^*(\theta)].$$

In a two-part tarif, the optimal retail price is simply the average spot price.

By contrast, Borenstein and Holland (2005) assumes linear retail contracts. The retail price is thus determined by the zero-profit condition:

$$p^R = \frac{\mathbb{E}[p^*(\theta)\gamma^*(\theta)D(p^{R*},\theta)]}{\mathbb{E}[\gamma^*(\theta)D(p^{R*},\theta)]}, \text{ hence is not optimal.}$$

### 2.5.3 Optimal Generation Mix

Finally, conditions $\frac{\partial \mathcal{L}}{\partial k_n} = 0$ yields

$$\mathbb{E}[(p^*(\theta) - c_n)u_n^*] = r_n \Leftrightarrow \mathbb{E}[(p^*(\theta) - c_n)^+] = r_n. \tag{2.9}$$

With a slight abuse of notation, define

$$\begin{aligned}\bar{\Psi}_0(K,p^R,c) &= \int_{\hat{\theta}_0(K,c)}^{\bar{\theta}(K)} (\rho(K,\theta) - c)f(\theta)d\theta + \int_{\bar{\theta}(K)}^{+\infty} (v(p^R,\theta) - c)f(\theta)d\theta \\ &= \mathbb{E}[(min[\rho(K,\theta), v(p^R,\theta)] - c)^+].\end{aligned} \tag{2.10}$$

#### 2.5.3.1 Necessary conditions

We derive later a sufficient condition for existence and unicity of $K_n^*$ solutions to condition (2.9); hence we focus first on the necessary conditions. The optimal investment conditions are summarized by the following:

**Proposition 2.1.** The optimal cumulative capacity $K_N^*$ is characterized by

$$\bar{\Psi}_0(K_N^*, p^R, c_N) = r_N. \tag{2.11}$$

For $n = 1, \ldots, (N-1)$, the optimal cumulative capacity $K_n^*$ is characterized by

$$\bar{\Psi}_0(K_n^*, p^R, c_n) - \bar{\Psi}_0(K_n^*, p^R, c_{n+1}) = r_n - r_{n+1}. \tag{2.12}$$

***Proof.*** For $n = N$, first-order condition (2.9) immediately yields equation (2.11). To derive equation (2.12), we first establish the following relation:

$$\mathbb{E}[(p^*(\theta) - c_n)^+] = \bar{\Psi}_0(K_n, p^R, c_n) - \bar{\Psi}_0(K_n, p^R, c_{n+1}) + r_{n+1}.$$

Starting from

$$\begin{aligned}\mathbb{E}[(p^*(\theta) - c_n)^+] = {} & \mathbb{E}[(p^*(\theta) - c_n)\mathbb{I}_{\{\theta \in v_n\}}] + \mathbb{E}[(c_{n+1} - c_n)\mathbb{I}_{\{\theta \in h_{n+1}\}}] \\ & + \mathbb{E}[(p^*(\theta) - c_n)\mathbb{I}_{\{p^*(\theta) > c_{n+1}\}}],\end{aligned}$$

we first observe that

$$\begin{aligned}\mathbb{E}[(p^*(\theta) - c_n)\mathbb{I}_{\{\theta \in v_n\}}] & = \mathbb{E}[(p^*(\theta) - c_n)\mathbb{I}_{\{p^*(\theta) > c_n\}}] - \mathbb{E}[(p^*(\theta) - c_n)\mathbb{I}_{\{p^*(\theta) \geq c_{n+1}\}}] \\ & = \bar{\Psi}_0(K_n, p^R, c_n) - \mathbb{E}[(p^*(\theta) - c_n)\mathbb{I}_{\{p^*(\theta) \geq c_{n+1}\}}].\end{aligned}$$

Second, adding and subtracting $(c_{n+1} - c_n)$ yields

$$\begin{aligned}\mathbb{E}[(p^*(\theta) - c_n)\mathbb{I}_{\{p^*(\theta) \geq c_{n+1}\}}] & = \mathbb{E}[(p^*(\theta) - c_{n+1})\mathbb{I}_{\{p^*(\theta) \geq c_{n+1}\}}] \\ & \quad + \mathbb{E}[(c_{n+1} - c_n)\mathbb{I}_{\{p^*(\theta) \geq c_{n+1}\}}] \\ & = r_{n+1} + \mathbb{E}[(c_{n+1} - c_n)\mathbb{I}_{\{p^*(\theta) \geq c_{n+1}\}}]\end{aligned}$$

since $\mathbb{E}[(p^*(\theta) - c_{n+1})\mathbb{I}_{\{p^*(\theta) \geq c_{n+1}\}}] = r_{n+1}$.

Finally, combining these expressions yields

$$\begin{aligned}\mathbb{E}[(p^*(\theta) - c_n)^+] & = \bar{\Psi}_0(K_n, p^R, c_n) - \mathbb{E}[(p^*(\theta) - c_n)\mathbb{I}_{\{p^*(\theta) \geq c_{n+1}\}}] \\ & \quad + \mathbb{E}[(c_{n+1} - c_n)\mathbb{I}_{\{p^*(\theta) \geq c_{n+1}\}}] + r_{n+1} \\ & = \bar{\Psi}_0(K_n, p^R, c_n) - \bar{\Psi}_0(K_n, p^R, c_{n+1}) + r_{n+1},\end{aligned}$$

as announced. Inserting this relation into first-order condition (2.9) then yields equation (2.12). □

Equation (2.11) is identical to equation (2.2). It indicates that the total installed capacity is solely determined by the long-run marginal cost of the last technology and the demand function. In particular, the cumulative capacity does not depend on the costs of inframarginal technologies.

As previously mentioned, the VoLL acts as an endogenous cap on prices. When the VoLL is reached, it is preferable to curtail constant-price customers than to serve them. Thus the marginal value of capacity is lower when the VoLL is reached than when it is not. This intuition is formalized by observing that

$$\begin{aligned}\bar{\Psi}_0(K_n, p^R, c_n) &= \mathbb{E}[(\rho(K_N,\theta) - c_N)\mathbb{I}_{\{c_N \leq \rho(K_N,\theta) \leq v(p^R,\theta)\}}] \\ &\quad + \mathbb{E}[(v(p^R,\theta) - c_N)\mathbb{I}_{\{\rho(K_N,\theta) > v(p^R,\theta)\}}] \\ &= \mathbb{E}[(\rho(K_N,\theta) - c_N)\mathbb{I}_{\{c_N \leq \rho(K_N,\theta)\}}] - \mathbb{E}[(\rho(K_N,\theta) \\ &\quad - v(p^R,\theta))\mathbb{I}_{\{\rho(K_N,\theta) > v(p^R,\theta)\}}].\end{aligned}$$

The second expectation is positive by construction, hence $\bar{\Psi}_0(K_n, p^R, c_n) \leq [(\rho(K_N,\theta) - c_N)^+]$.

Intuition for equation (2.12) is that a marginal substitution of technology $(n+1)$ by technology $n$ reduces operating costs; hence it increases net surplus by $(\bar{\Psi}_0(K_n, p^R, c_n) - \bar{\Psi}_0(K_n, p^R, c_{n+1}))$. On the other hand, it increases fixed cost by $(r_n - r_{n+1})$. At the optimum, these two effects exactly balance each other.

Again, the cumulative capacity up to technology $n$ depends only on the costs of technologies $n$ and $(n+1)$ and not on the costs of other technologies. For $n > 1$, the long-term equilibrium capacity of technology $n$ depends only on the costs of technologies $(n-1)$, $n$, and $(n+1)$.

#### 2.5.3.2 Sufficient conditions for existence and unicity

Assuming Conditions 1, 2, and 3 hold, the sufficient condition for existence and unicity of $0 < K_1^* < \cdots < K_N^*$ is fairly intuitive: the first unit is worth producing and consuming, and the long-run marginal costs are increasing with the variable production costs. Formally, we have

**Corollary 2.1.** There exists a unique $K_N^* > 0$ solution to condition (2.11) if and only if

$$\mathbb{E}[v(p^R,\theta)] > c_N + r_N.$$

If the above condition holds, there exists a unique $K_n^*$ solution to condition (2.12) if and only if the long-run marginal costs of technologies are increasing in $n$:

$$\forall n \leq (N-1),\ c_{n+1} + r_{n+1} > c_n + r_n.$$

Furthermore, if

$$\forall n \leq (N-1),\ \xi_n < \xi_{n+1}\ and\ \Psi_0(\xi_n, c_{n+1}) - r_{n+1} > \Psi_0(\xi_n, c_{n+2}) - r_{n+2},$$

where $\xi_n$ is uniquely defined by $\Psi_0(\xi_n, c_n) = r_n$, then $K_n^* \in (0, K_{n+1}^*)$ .

***Proof.*** Consider first the existence and unicity of $K_N^*$.

$$\rho(0,\theta) \geq c_N \Leftrightarrow P\left(\frac{-(1-\alpha)\gamma D(p^R,\theta)}{\alpha},\theta\right) \geq c_N$$
$$\Leftrightarrow \alpha D(c_N,\theta) + (1-\alpha)\gamma D(p^R,\theta) \geq 0,$$

which holds since (a) $P(0,\theta) > c_N \Leftrightarrow D(c_N,\theta) > 0$ (Condition 1), and (b) $P(0,\theta) > p^R \Leftrightarrow D(p^R,\theta) > 0$ (Condition 2). Thus $\rho(0,\theta) > c_N\ \forall\theta \geq 0$, hence $\hat{\theta}_0(0, c_N) = 0$. Since $\rho(0,0) > v(p^R, 0)$ (Condition 3), we have $\bar{\theta}(0) = 0$. Therefore,

$$\bar{\Psi}_0(0, p^R, c_N) = \mathbb{E}[v(p^R,\theta) - c_N] = \mathbb{E}[v(p^R,\theta)] - c_N,$$

and hence

$$\bar{\Psi}_0(0, p^R, c_N) > r_N \Leftrightarrow \mathbb{E}[v(p^R,\theta)] > c_N + r_N,$$

as announced.

For sufficiently large capacity, $\rho(K,\theta)$ is lower than $c_N$, thus $lim_{K\to+\infty} v_N = \{\emptyset\}$, hence $lim_{K\to+\infty}\bar{\Psi}_0(K, p^R, c_N) = 0$, and $\lim_{K\to+\infty} \bar{\Psi}_0(K, p^R, c_N) - r_N = -r_N < 0$.

Since (a) $\bar{\Psi}_0(K, p^R, c)$ is continuous and decreasing in its first argument, (b) $\bar{\Psi}_0(0, p^R, c_N) - r_N > 0$ and (c) $lim_{K\to+\infty}\bar{\Psi}_0(K, p^R, c_N) - r_N < 0$, there exists a unique $K_N^* > 0$ solution to equation (2.11).

Consider now the existence and unicity of $K_n^*$. For $n = 1, \ldots, (N-1)$, $\rho(0,\theta) > c_N > c_n\ \forall\theta \geq 0\ \forall n$, hence $\hat{\theta}_0(0, c_n) = 0$. Therefore,

$$\bar{\Psi}_0(0, p^R, c_n) - \bar{\Psi}_0(0, p^R, c_{n+1}) = \mathbb{E}[v(p^R,\theta) - c_n] - \mathbb{E}[v(p^R,\theta) - c_{n+1}]$$
$$= \mathbb{E}[(c_{n+1} - c_n)] = c_{n+1} - c_n.$$

Thus

$$\bar{\Psi}_0(0, p^R, c_n) - \bar{\Psi}_0(0, p^R, c_{n+1}) > r_n - r_{n+1} \Leftrightarrow c_{n+1} + r_{n+1} > c_n + r_n,$$

which is necessary and sufficient for the existence and unicity of the $K_n^* > 0$ solution of equation (2.12).

Finally, for $n = 1, \ldots, (N-1)$, since $(\bar{\Psi}_0(K, p^R, c_n) - \bar{\Psi}_0(K, p^R, c_{n+1}))$ is decreasing in its first argument, $K_n^* < K_{n+1}^*$is equivalent to

$$\beta_{n+1}(K_{n+1}^*) = \Psi_0(K_{n+1}^*, c_n) - r_n - (\Psi_0(K_{n+1}^*, c_{n+1}) - r_{n+1}) < 0.$$

For $n = N-1$, we have

$$\beta_N(K_N^*) = \Psi_0(K_N^*, p^R, c_{N-1}) - r_{N-1} - (\Psi_0(K_N^*, p^R, c_N) - r_N)$$
$$= \Psi_0(K_N^*, p^R, c_{N-1}) - r_{N-1}$$

by definition of $K_N^*$. Thus

$$\beta_N(K_N^*) < 0 \Leftrightarrow \Psi_0(K_N^*, p^R, c_{N-1}) < r_{N-1} \Leftrightarrow \xi_N = K_N^* > \xi_{N-1}.$$

For $n < N-1$, observe that

$$\beta_{n+2}(\gamma_{n+1}) = \Psi_0(\gamma_{n+1}, c_{n+1}) - r_{n+1} - (\Psi_0(\gamma_{n+1}, p^R, c_{n+2}) - r_{n+2})$$
$$= -(\Psi_0(\gamma_{n+1}, p^R, c_{n+2}) - r_{n+2})$$

by definition of $\gamma_{n+1}$. Then

$$\gamma_{n+1} < \gamma_{n+2} \Leftrightarrow \Psi_0(\gamma_{n+1}, p^R, c_{n+2}) - r_{n+2} > 0 \Leftrightarrow \beta_{n+2}(\gamma_{n+1}) < 0 \Leftrightarrow \gamma_{n+1} > K_{n+1}^*.$$

We also have

$$\beta_{n+2}(\gamma_n) = \Psi_0(\gamma_n, p^R, c_{n+1}) - r_{n+1} - (\Psi_0(\gamma_n, p^R, c_{n+2}) - r_{n+2}) > 0$$

by assumption, thus $\gamma_n < K_{n+1}^*$. Then

$$\beta_{n+1}(K_{n+1}^*) = \Psi_0(K_{n+1}^*, p^R, c_n) - r_n - (\Psi_0(K_{n+1}^*, p^R, c_{n+1}) - r_{n+1})$$
$$< \Psi_0(\gamma_n, p^R, c_n) - r_n - (\Psi_0(\gamma_{n+1}, p^R, c_{n+1}) - r_{n+1}) = 0. \qquad \square$$

#### 2.5.3.3 Link with screening-curve analysis

We now derive formally the screening-curve analysis presented in section 2.2. Suppose demand is completely price-insensitive. The total cost, expressed in €/MW per year, to serve a strip of constant demand lasting $h$ hours using technology $n$ is

$$C_n(h) = 8{,}760 \times r_n + c_n h.$$

Technology $n$ has higher fixed cost and lower variable cost than technology $(n+1)$. As can be seen in figure 2.9, $C_n(h) \leq C_n(h)$ if and only if $h \leq h_n$, where $h_n$ is uniquely defined by

$$C_n(h_n) = C_{n+1}(h_n) \Leftrightarrow 8{,}760 \times r_n + c_n h_n = 8{,}760 \times r_{n+1} + c_{n+1} h_n$$
$$\Leftrightarrow h_n(c_{n+1} - c_n) = 8{,}760(r_{n+1}) \Leftrightarrow \frac{h_n}{8{,}760} = \frac{r_n - r_{n+1}}{C_{n+1} - C_n}.$$

Then

$$h_n \leq 8{,}760 \Leftrightarrow \frac{C_{n+1} - C_n}{r_n - r_{n+1}} \leq 1 \Leftrightarrow C_{n+1} + r_{n+1} \leq C_n + r_n.$$

We now turn to the load-duration curve presented in figure 2.1. For a given number of hours $h$ on the $x$-axis, the $y$-axis of the load duration curve represents the level $Q(h)$ such that demand exceeds $Q(h)$ for precisely $h$ hours.

The function $Q(h)$ is related to the function $G(.)$, the cumulative distribution function of demand, that is, $G(x) = Pr(D \leq x)$. The level $Q(h)$ is defined by $Pr(D \geq Q(h)) = \frac{h}{8{,}760}$, or equivalently $(1 - G(Q(h))) = \frac{h}{8{,}760}$.

We have seen that technology $n$ is preferred over technology $(n+1)$ for $h \geq h_n$. By construction, all demand $D$ lower than $Q(h_n)$ lasts more than $h_n$ hours. Thus technology $n$ is preferred over technology $(n+1)$ for all $D \leq Q(h_n)$, hence cumulative capacity $K_n$ is defined by $K_n = Q(h_n)$. Since $\frac{h_n}{8{,}760} = 1 - G(Q(h_n)) = 1 - G(K_n)$, $K_n$ is defined by

$$(1 - G(K_n))(c_{n+1} - c_n) = r_n - r_{n+1}. \tag{2.13}$$

When demand is inelastic, the price does not matter; hence it can be considered constant. The function $\Psi_0(K, c)$ simplifies to

$$\Psi_0(K, c) = (p - c)Pr(D \geq K) = (p - c)(1 - G(K)).$$

Condition (2.12) thus becomes

$$(p - c_n)(1 - G(K_n)) - (p - c_{n+1})(1 - G(K_n)) = (c_{n+1} - c_n)(1 - G(K_n)) = r_n - r_{n+1},$$

which is exactly condition (2.13).

The screening curve analysis is an application of the general situation described by conditions (2.12) when demand is perfectly inelastic.

### 2.5.4 Properties of the Long-Term Equilibrium

**Net surplus** The net surplus from producing and consuming electricity is

$$W = \mathbb{E}[\alpha S(p, \theta) + (1 - \alpha)\gamma(\theta)S(p^R, \theta) - \sum_{n \geq 1} c_n u_n(\theta)k_n] - \sum_{n \geq 1} r_n k_n.$$

Equation (2.9) yield

$$\mathbb{E}[c_n u_n(\theta)] + r_n = \mathbb{E}[p(\theta)u_n(\theta)]$$

$$\Rightarrow$$

$$\sum_n (\mathbb{E}[c_n u_n(\theta)] + r_n)k_n = \mathbb{E}\left[p(\theta)\sum_n u_n(\theta)k_n\right]$$
$$= \mathbb{E}[p(\theta)(\alpha D(p(\theta), \theta) + (1 - \alpha)\gamma(\theta)D(p^R, \theta))],$$

since aggregate production $(\sum_n u_n(\theta)k_n)$ is equal to demand served $(\alpha D(p(\theta), \theta) + (1 - \alpha)\gamma(\theta)D(p^R, \theta))$ in every state of the world. Thus

$$W = \mathbb{E}[\alpha(S(p(\theta),\theta) - p(\theta)D(p(\theta),\theta)) + (1-\alpha)\gamma(\theta)(S(p^R,\theta) - p(\theta)D(p^R,\theta))].$$

This is a standard result: since the expected wholesale spot price is equal to the long-term marginal cost of production, the net surplus is the gross surplus minus demand, valued at the wholesale spot price. The presence of constant price customers does not affect the result. This expression proves useful in chapters 5 and 8.

When rationing occurs, constant-price consumers receive no surplus, since the wholesale spot price is precisely equal to the VoLL:

$$(S(p^R,\theta) - p(\theta)D(p^R,\theta)) = D(p^R,\theta)(v(p^R,\theta) - p(\theta)) = 0.$$

Therefore, the net hourly surplus simplifies to

$$W = \mathbb{E}[\alpha(S(p(\theta),\theta) - p(\theta)D(p(\theta),\theta)) + (1-\alpha)(S(p^R,\theta) - p(\theta)D(p^R,\theta))\mathbb{I}_{\{\gamma=1\}}].$$

#### 2.5.4.1 Average price

For $n=1$, equation (2.9) yields

$$\mathbb{E}[(p(\theta) - c_1)u_1(\theta)] = r_1.$$

For every state of the world, price $p(\theta)$ satisfies $(p(\theta)-c_1)(1-u_1(\theta))=0$: either $p(\theta)=c_1$ or technology *1* produces at capacity, in which case $u_1(\theta)=1$. Taking the expectation over states of the world,

$$\mathbb{E}[(p(\theta)-c_1)(1-u_1(\theta))] = 0 \Leftrightarrow \mathbb{E}[p(\theta)-c_1] = \mathbb{E}[(p(\theta)-c_1)u_1(\theta)] = r_1$$

$$\Leftrightarrow$$

$$\mathbb{E}[p(\theta)] = r_1 + c_1.$$

Under free-entry, the time-weighted average price is equal to long-run marginal cost of the baseload technology.

#### 2.5.4.2 No curtailment condition

Intuitively, no curtailment occurs when the fraction of price-reactive consumers is higher than a threshold, that is, for $\alpha \geq \alpha_{min}$. Curtailment never occurs if and only if $\rho(K,\theta) < v(p^R,\theta)$ for all $\theta \geq \hat{\theta}_0(K,c_N)$. We have

$$\rho(K,\theta) < v(p^R,\theta) \Leftrightarrow P\left(\frac{K-(1-\alpha)D(p^R,\theta)}{\alpha},\theta\right) < v(p^R,\theta)$$

$$\Leftrightarrow$$

$$\frac{K-(1-\alpha)D(p^R,\theta)}{\alpha} > D(v(p^R,\theta),\theta) \Leftrightarrow \alpha > \frac{D(p^R,\theta)-K}{D(p^R,\theta)-D(v(p^R,\theta),\theta)},$$

since $D(p^R, \theta) - D(v(p^R, \theta), \theta) > 0$. Then, since $D(p^R, .)$ and $v(p^R, .)$ are bounded, $\frac{D(p^R,\theta)-K}{D(p^R,\theta)-D(v(p^R,\theta),\theta)}$ is also bounded. Thus, no curtailment occurs with positive probability for aggregate capacity $K$ if and only if,

$$\alpha \geq \alpha_{min}(K) = max_{\theta \geq \hat{\theta}_0(K, c_N)} \left[ \frac{D(p^R, \theta) - K}{D(p^R, \theta) - D(v(p^R, \theta), \theta)} \right].$$

Hence, curtailment never occurs at the long-term equilibrium if and only if $\alpha \geq \alpha_{min}(K_N^*) \equiv \alpha_{min}$.

## 2.6 Operating Reserves

As previously discussed, SOs must secure operating reserves to balance supply and demand in the event of unforeseen shocks to supply and demand. Borenstein and Holland (2005) proposes a simple yet effective representation of operating reserves. For simplicity, only one type of reserve is considered, the nonspinning one (that is, a plant that is not running but can start up and produce energy within a short agreed-on time frame). Since the plant is not running, the marginal cost of providing reserves is normalized to zero. In reality, SOs run multiple markets for operating reserves, for example, spinning, ten-minute, thirty-minutes. The economic insights are not modified, as long as the no-arbitrage condition presented below holds.

Hogan (2005) proposes that the SO runs a single market for energy and operating reserves, an approach called co-procurement and used in many North American power markets. Generating units that are called to produce receive the wholesale price $p(\theta)$, and generating units that provide operating reserves receive the wholesale price $p(\theta)$ less the marginal cost of generation $c$, assumed to be perfectly known by the SO. Generators are therefore indifferent between producing energy and providing reserves, an essential condition (Borenstein and Holland 2005). When an unscheduled generation outage occurs, operating reserves produce energy and receive the full price $p(\theta)$.

Operating reserves requirements are expressed as a percentage of demand, denoted $h(\theta)$, and taken as given here.[9] Defining the optimal $h(\theta)$ requires advanced network analysis, hence it is beyond the scope of this work. Joskow and Tirole (2007) shows the optimal reserve ratio increases with the state of the world; hence $h(\theta)$ is assumed to be nondecreasing.

As in section 2.5, a fraction $\alpha$ of customers face real-time prices, while a fraction $(1-\alpha)$ faces constant retail price $p^R$. Different from section 2.5, the retail real-time price $w(\theta)$ must be higher than the wholesale price $p(\theta)$ to cover generators' revenues from the operating reserves market. A natural choice is to directly include the cost of reserves in

9. In practice, various metrics for operating reserves are used, including absolute values expressed in MW. Expressing reserves as a percentage of peak demand simplifies the analysis while preserving the main economic intuition.

the retail price faced by price reactive customers:[10]

$$w(\theta) = p(\theta) + h(\theta)(p(\theta) - c)$$

$$\Leftrightarrow$$

$$w(\theta) - c = (1 + h(\theta))(p(\theta) - c). \tag{2.14}$$

Throughout this section, the retail and wholesale prices are assumed to be related by equation (2.14). The notation and model structure are identical to those in the previous sections, except that the superscript $OR$ is added when appropriate.

Only the fraction $\frac{1}{1+h(\theta)}$ of installed capacity is used to meet demand in state $\theta$, hence $\frac{K}{1+h(\theta)}$ and not $K$ is the output appearing in the function $\rho(., \theta)$. The formal proof is presented below.

Since retail customers face the price $w(\theta)$, the social planner's program is

$$\max_{\{w(\theta), \gamma(\theta)\}, K} \mathbb{E}[\tilde{S}(w(\theta), \gamma(\theta), \theta) - c\tilde{D}(w(\theta), \gamma(\theta), \theta)] - rK$$
$$st: \ (1 + h(\theta))\tilde{D}(w(\theta), \gamma(\theta), \theta) \leq K \ (\lambda(\theta)).$$

The associated Lagrangian is

$$\begin{aligned}\mathcal{L} = \mathbb{E}[&\tilde{S}(w(\theta), \gamma(\theta), \theta) - c\tilde{D}(w(\theta), \gamma(\theta), \theta)\\ &+ \lambda(\theta)[K - (1 + h(\theta))\tilde{D}(w(\theta), \gamma(\theta), \theta)]] - rK\end{aligned}$$

and

$$\begin{cases} \frac{\partial \mathcal{L}}{\partial w(\theta)} = \alpha\{w(\theta) - [c + (1 + h(\theta))\lambda(\theta)]\}\frac{\partial D}{\partial w(\theta)} \\ \frac{\partial \mathcal{L}}{\partial \gamma(\theta)} = (1 - \alpha)\{v(p^R, \theta) - [c + (1 + h(\theta))\lambda(\theta)]\}D(p^R, \theta). \\ \frac{\partial \mathcal{L}}{\partial K} = \mathbb{E}[\lambda(\theta)] - r \end{cases}$$

First, off-peak $\lambda(\theta) = 0$ and $\gamma(\theta) = 1$. Then $p(\theta) = c = w(\theta)$. Since there is excess capacity (and by assumption there is no cost in providing reserves), reserves are not paid: the wholesale and retail prices are equal to the variable production cost.

This holds as long as

$$(1 + h(\theta))\tilde{D}(c, 1, \theta) \leq K \Leftrightarrow c \leq \rho\left(\frac{K}{1 + h(\theta)}, \theta\right) \Leftrightarrow \theta \leq \hat{\theta}_0^{OR}(K, c),$$

where $\hat{\theta}_0^{OR}(K, c)$ is uniquely defined[11] by $\rho\left(\frac{K}{1+h(\theta)}, \hat{\theta}_0^{OR}(K, c)\right) = c$.

10. Borenstein and Holland (2005) show it to be the perfect competition outcome.
11. Since $h(\theta)$ is nondecreasing, $m_1(K; \theta) = \rho\left(\frac{K}{1+h(\theta)}; \theta\right)$ is increasing in $\theta$: $\frac{\partial m_1}{\partial \theta} = -\rho_q \frac{Kh'(\theta)}{(1+h(\theta))^2} + \rho_\theta > 0$.

Second, on-peak, if constant-price customers are not curtailed, $(1+h(\theta))\tilde{D}(w(\theta), 1, \theta) = K$; hence $\lambda(\theta) > 0$ and $\gamma(\theta) = 1$. Then supply demand balance yields $w(\theta) = \rho\left(\frac{K}{1+h(\theta)}, \theta\right)$, and the first-order condition yields $\lambda(\theta) = \frac{w(\theta)-c}{1+h(\theta)} = p(\theta) - c > 0$.

Finally, constant-price customers may be curtailed. The retail price is equal to the VoLL $v(p^R, \theta)$, and the serving ratio $\gamma^*(\theta) < 1$ is uniquely defined by $(1+h(\theta))\tilde{D}(v(p^R, \theta), \gamma^*(\theta), \theta) = K$. Then $\lambda(\theta) = \frac{v(p^R,\theta)-c}{1+h(\theta)} = p(\theta) - c$.

Thus the marginal social value of capacity in state $\theta$ is

$$p(\theta) - c = \frac{min\left[\rho\left(\frac{K}{1+h(\theta)}, \theta\right), v(p^R, \theta)\right] - c}{1+h(\theta)},$$

the marginal social value of capacity is

$$\tilde{\Psi}^{OR}(K, p^R, c) = \mathbb{E}\left[\left(\frac{min\left[\rho\left(\frac{K}{1+h(\theta)}, \theta\right), v(p^R, \theta)\right] - c}{1+h(\theta)}\right)^+\right],$$

and the optimal capacity $K^{OR*}$ is then uniquely defined by $\mathbb{E}[\lambda(\theta)] = r$, which yields

$$\tilde{\Psi}^{OR}(K^{OR*} p^R, c) = r. \tag{2.15}$$

Consider now the producers' problem. By construction, producers are indifferent between producing energy and providing reserves. In state $\theta$, they offer $s^n(\theta)$ into the energy cum operating reserves market. $S(\theta) = \sum_{n=1}^{N} s^n(\theta)$ is the total offer. Energy available to meet demand is $Q(\theta) = \frac{S(\theta)}{1+h(\theta)}$. The *SO* then (a) verifies that $s^n(\theta) \leq k^n$, and (b) allocates each $s^n(\theta)$ between energy $q^n(\theta)$ and reserves $b^n(\theta)$. Producer $n$ profit is then

$$\begin{aligned}\pi^n(\theta) &= (q^n(\theta) + b^n(\theta))(p(\theta) - c)\\ &= \frac{s^n(\theta)}{1+h(\theta)}\left(\rho\left(\frac{S(\theta)}{1+h(\theta)}\right) - c\right),\end{aligned}$$

since (a) energy and operating reserves receive same net revenue by construction, and (b) wholesale $(w(\theta))$ and retail $\left(\rho\left(\frac{S(\theta)}{1+h(\theta)}\right)\right)$ prices are linked by equation (2.14). The problem is then isomorphic to standard peak-load pricing, except that $\frac{s^n(\theta)}{1+h(\theta)}$ replaces production $q^n(\theta)$.

## 2.7 An Illustrative Example

Léautier (2014) proposes a simple specification that illustrates the previous analysis. Suppose: (a) inverse demand is linear with constant slope, $P(q, \theta) = a(\theta) - bq$, where $a(\theta) = a_0 - a_1 e^{-\lambda_2 \theta}$ and $b > 0$, and (b) states of the world are distributed according to

$f(\theta)=\lambda_1 e^{-\lambda_1\theta}$, and (c) rationing is anticipated and proportional. As shown below, this specification leads to simple expressions, easy to calibrate.

### 2.7.1 Analytical Expressions

Integrating $\bar{\Psi}_0(K_N, p^R, c_N)$ defined by equation (2.10) by parts yields

$$\bar{\Psi}_0(K_N, p^R, c_N)=\int_{\hat{\theta}_0(K_N,c_N)}^{\bar{\theta}(K_N)} \partial_\theta \rho(K,\theta)(1-F(\theta))d\theta+\int_{\bar{\theta}(K_N)}^{+\infty} \partial_\theta v(p^R,\theta)(1-F(\theta))d\theta$$

since the outer boundary terms are equal to zero and the boundary terms at $\bar{\theta}(K_N)$ cancel one another out because $\rho(K,\bar{\theta}(K_N))=v(\bar{\theta}(K_N))$. Linear demand enables us to further simplify the expression:

$$\begin{aligned}\bar{\Psi}_0(K_N, p^R, c_N)&=\int_{\hat{\theta}_0(K_N,c_N)}^{\bar{\theta}(K_N)} \frac{a'(\theta)}{\alpha}(1-F(\theta))d\theta+\int_{\bar{\theta}(K_N)}^{+\infty} \frac{a'(\theta)}{2}(1-F(\theta))d\theta\\&=\frac{1}{\alpha}\left(\int_{\hat{\theta}_0(K_N,c_N)}^{+\infty} a'(\theta)(1-F(\theta))d\theta-\frac{2-\alpha}{2}\int_{\bar{\theta}(K_N)}^{+\infty} a'(\theta)(1-F(\theta))d\theta\right).\end{aligned}$$

**Equilibrium absent curtailment** If no curtailment occurs at the equilibrium cumulative capacity, the marginal value simplifies to

$$\bar{\Psi}_0(K_N, p^R, c_N)=\frac{1}{\alpha}\left(\int_{\hat{\theta}_0(K_N,c_N)}^{+\infty} a'(\theta)(1-F(\theta))d\theta\right),$$

which generalizes expression (2.3) for $0<\alpha\leq 1$.

The specific functional forms of $a(\theta)$ and $(1-F(\theta))$ enable us to obtain a closed form expression of $\hat{\theta}_0(K_N, c_N)$:

$$\begin{aligned}\bar{\Psi}_0(K_N, p^R, c_N)&=\frac{\lambda_2 a_1}{\alpha}\left(\int_{\hat{\theta}_0(K_N,c_N)}^{+\infty} e^{-(\lambda_1+\lambda_2)\theta}d\theta\right)\\&=\frac{\lambda_2 a_1}{\alpha(\lambda_1+\lambda_2)}e^{-(\lambda_1+\lambda_2)\hat{\theta}_0(K_N,c_N)}.\end{aligned}$$

Introducing $\lambda=\frac{\lambda_1}{\lambda_2}$, equation (2.11) yields

$$e^{-(\lambda_1+\lambda_2)\hat{\theta}_0(K_N^*,c_N)}=\frac{\alpha(1+\lambda)r_N}{a_1}\Leftrightarrow e^{-\lambda_1\hat{\theta}_0(K_N^*,c_N)}=\left(\frac{\alpha(1+\lambda)r_N}{a_1}\right)^{\frac{\lambda}{1+\lambda}}.$$

Absent curtailment,

$$Pr(p(\theta)\geq c_N)=\left(\frac{\alpha(1+\lambda)r_N}{a_1}\right)^{\frac{\lambda}{1+\lambda}}.$$

As expected, the number of on-peak hours increases with the fixed cost $r_N$. It also increases with the fraction of price-responsive customers $\alpha$: when customers are price responsive, the price can rise above marginal cost while remaining at "reasonable" levels. Thus fixed costs can be covered by multiple hours with reasonable prices. By contrast, if no customers are price responsive, the price rises immediately to the VoLL on-peak: prices are higher on-peak, hence the required number of on-peak hours is lower.

The optimal cumulative capacity is then determined by

$$\rho(K_N^*, \hat{\theta}_0(K_N^*, c_N)) = c_N \Leftrightarrow bK_N^* = a_0 - a_1 e^{-\lambda_2 \hat{\theta}_0(K_N^*, c_N)} - ((1-\alpha)p^R + \alpha c_N).$$

Observing that

$$e^{-(\lambda_1+\lambda_2)\hat{\theta}_0(K_N^*, c_N)} = \frac{\alpha(1+\lambda)r_N}{a_1} \Leftrightarrow e^{-\lambda_2 \hat{\theta}_0(K_N^*, c_N)} = \left(\frac{\alpha(1+\lambda)r_N}{a_1}\right)^{\frac{1}{1+\lambda}}$$

yields a closed-form expression of $bK_N^*$:

$$bK_N^* = a_0 - a_1 \left(\frac{\alpha(1+\lambda)r_N}{a_1}\right)^{\frac{1}{1+\lambda}} - ((1-\alpha)p^R + \alpha c_N).$$

Finally, we also obtain a closed-form expression of the on-peak price:

$$\begin{aligned}\rho(K_N^*, \theta) &= \frac{a(\theta) - (1-\alpha)p^R - \left(a_0 - a_1 \left(\frac{\alpha(1+\lambda)r_N}{a_1}\right)^{\frac{1}{1+\lambda}} - ((1-\alpha)p^R + \alpha c_N)\right)}{\alpha} \\ &= c_N + \frac{a_1}{\alpha}\left(\left(\frac{\alpha(1+\lambda)r_N}{a_1}\right)^{\frac{1}{1+\lambda}} - e^{-\lambda_2 \theta}\right) = c_N + \frac{a_1}{\alpha}(e^{-\lambda_2 \hat{\theta}_0(K_N^*, c_N)} - e^{-\lambda_2 \theta}).\end{aligned}$$

**Equilibrium with curtailment** The closed-form expressions provided before enable us to fully characterize $\alpha_{min}$, the threshold fraction of reactive customers above which no curtailment occurs. As shown previously, curtailment never occurs for capacity $K$ if and only if

$$\alpha \geq \frac{D(p^R, \theta) - K}{D(p^R, \theta) - D(v(p^R, \theta), \theta)} = 2\frac{a(\theta) - p^R - bK}{a(\theta) - p^R} = 2\left(1 - \frac{bK}{a(\theta) - p^R}\right).$$

Since $a(\theta)$ is monotically increasing, $\left(1 - \frac{bK}{a(\theta) - p^R}\right)$ is monotically increasing, hence curtailment never occurs if only if $\alpha \geq \alpha_{min}(K) = 2\left(1 - \frac{bK}{a_0 - p^R}\right)$.

Substituting in the expression of $K_N^*$, this is equivalent to

$$a_0 - a_1 \left(\frac{\alpha r_N}{a_1}(1+\lambda)\right)^{\frac{1}{1+\lambda}} - (\alpha c_N + (1-\alpha)p^R) \geq (2-\alpha)\frac{a_0 - p^R}{2}$$

$$\Leftrightarrow$$

$$\alpha \geq \alpha_{\min} = \frac{a_1[(1+\lambda)r_N]^{\frac{1}{\lambda}}}{\left[\frac{a_0+p^R-2c_N}{2}\right]^{1+\frac{1}{\lambda}}}.$$

Thus curtailment occurs if and only if $\alpha < \alpha_{\min}$.

Suppose curtailment occur for capacity $K_N$. The first state of the world in which curtailment occurs, $\bar{\theta}(K_N)$ is defined by

$$c_N + \frac{a_1}{\alpha}(e^{-\lambda_2 \hat{\theta}_0(K_N,c_N)} - e^{-\lambda_2 \bar{\theta}(K_N)}) = \frac{a_0 - a_1 e^{-\lambda_2 \bar{\theta}(K_N)}}{2}$$

$$\iff$$

$$\frac{2-\alpha}{2} e^{-\lambda_2 \bar{\theta}(K_N)} = e^{-\lambda_2 \hat{\theta}_0(K_N,c_N)} - \frac{\alpha}{a_1}\left(\frac{a_0}{2} - c_N\right).$$

Using the specific functional form, the marginal value of capacity is

$$\bar{\Psi}_0(K_N, p^R, c_N) = \frac{a_1}{\alpha(1+\lambda)}(e^{-(\lambda_1+\lambda_2)\hat{\theta}_0(K_N,c_N)} - \frac{2-\alpha}{2} e^{-(\lambda_1+\lambda_2)\bar{\theta}(K_N)}).$$

Since $e^{-(\lambda_1+\lambda_2)\bar{\theta}(K_N)}$ can be expressed as a function of $e^{-\lambda_2 \hat{\theta}_0(K_N,c_N)}$, equation (2.11) yields the value of $e^{-(\lambda_1+\lambda_2)\hat{\theta}_0(K_N,c_N)}$, although no closed-form solution is available.

### 2.7.2 Parameters Calibration

The demand-curve parameters are estimated in two steps: (a) an actual load duration curve, assuming price is constant, is used to estimate $\lambda$ and derive a first set of relationships, and (b) estimates of price elasticity are then used to derive the last relation among parameters. This approach is that of Borenstein (2005), Borenstein and Holland (2005), and Holland and Mansur (2006) and is consistent with the reality of the French power market: in 2009, most customers paid a constant power price, denoted $p_0$. Observed demand fluctuations are the result, therefore, of variations in the states of the world ($a(\theta)$ and $f(\theta)$). As the share of price-reactive demand increases, joint estimation of all parameters will become possible.

**Estimation of $\lambda$ and first set of relationships** As previously, denote $G(.)$ the cumulative distribution of demand, that is, $G(x) = Pr(D \leq x)$. If demand is linear,

$$G(x) = \Pr\left(\frac{a(\theta)-p_0}{b} \leq x\right)$$
$$= \Pr(a(\theta) \leq p_0 + bx) = \Pr(\theta \leq a^{-1}(p_0+bx)) = (F \circ a^{-1})(p_0+bx).$$

Demand measured depends on (a) the state of the world $\theta$ and (b) demand conditional on that state of the world $\theta$. Estimating the distribution $G(.)$ allows us to identify $F \circ a^{-1}$. $F(.)$ and $a(.)$ cannot be identified separately.

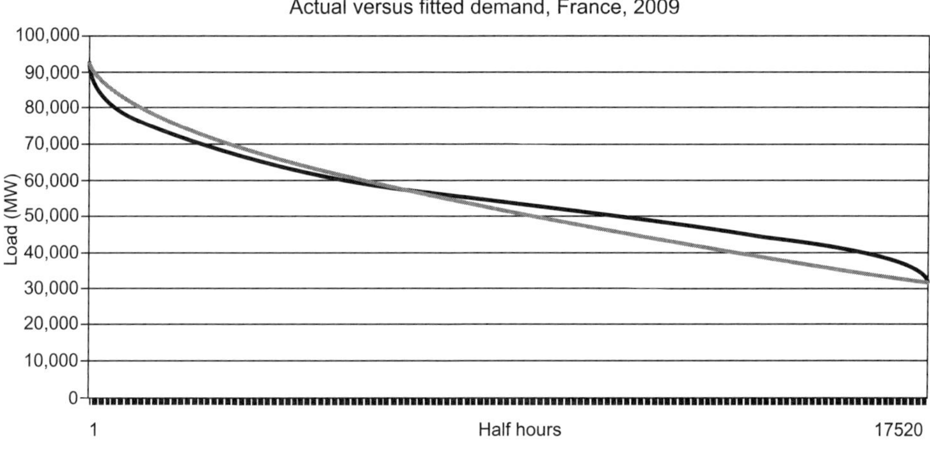


**Figure 2.15**
Actual and fitted demand.

If $a(\theta) = a_0 - a_1 e^{-\lambda_2 \theta}$ and $f(\theta) = \lambda_1 e^{-\lambda_1 \theta}$,

$$G(x) = 1 - \exp\left[\frac{\lambda_1}{\lambda_2} \ln \frac{a_0 - (p_0 + bx)}{a_1}\right] = 1 - \left[\frac{a_0 - (p_0 + bx)}{a_1}\right]^{\lambda}.$$

Then $1 - G(x) = \Pr(load \geq x) = \left[\frac{a_0 - (p_0 + bx)}{a_1}\right]^{\lambda}$ can be estimated from an actual load-duration curve.

Values for $a_0$ and $a_1$ cannot be estimated by maximum likelihood from the data. The minimum and maximum admissible values for load must be set exogenously. We choose these values to be the observed minimum and maximum values for load. Denote $\phi < 1$ the ratio of minimum to maximum demand for price $p_0$ and $Q^{\infty} = \lim_{\theta \to +\infty} Q(p_0, \theta) = \frac{a_0 - p_0}{b}$ the maximum demand. We have

$$\begin{cases} a_0 - bQ^{\infty} = p_0 \\ a_0 - a_1 - b\phi Q^{\infty} = p_0 \end{cases} \Leftrightarrow \begin{cases} a_1 = bQ^{\infty}(1 - \phi) \\ a_0 = p_0 + bQ^{\infty}. \end{cases}.$$

Estimation on 2009 demand in France leads to $Q^{\infty} = 92.4\ GW$ and $\phi = \frac{31.5}{92.4} = 0.34$. The average price paid by customers is $p_0 = 100\ €/MWh$.[12] Then maximum likelihood estimation yields $\lambda = \frac{\lambda_1}{\lambda_2} = 1.78$.

Actual and fitted demand are presented in figure 2.15.

12. French Transmission System Operator (RTE).

**Estimation of $b$ and all other parameters** Since $P(q,\theta)=a(\theta)-bq$, the elasticity of demand $\eta(\delta,\theta)$ for a given price $\delta$ in state $\theta$ is

$$\eta(\delta,\theta)=-\frac{\delta}{a(\theta)-\delta},$$

thus

$$\mathbb{E}[\eta(\delta,\theta)]=-\mathbb{E}\left[\frac{\delta}{a(\theta)-\delta}\right]=-\delta\int_0^{+\infty}\frac{\lambda_1 e^{-\lambda_1\theta}}{a_0-a_1e^{-\lambda_2\theta}-\delta}d\theta.$$

Setting $x=e^{-\lambda_1\theta}$ yields

$$\mathbb{E}[\eta(\delta,\theta)]=-\delta\int_0^1\frac{dx}{a_0-a_1x^{\frac{1}{\lambda}}-\delta}=-\delta\int_0^1\frac{dx}{p_0+bQ^{\infty}-(bQ^{\infty}(1-\phi))x^{\frac{1}{\lambda}}-\delta}.$$

Thus if we know the value of $\mathbb{E}[\eta(\delta,\theta)]$, we can compute $bQ^{\infty}$. Lijesen (2007) provides an up-to-date survey of the empirical literature on price elasticity of electricity, as well as the author is own estimate. From this, I select as a base case $\mathbb{E}[\eta(\delta,\theta)]=-0.01$ at price $\delta=100$ €$/MWh$, which corresponds to the upper estimate from Patrick and Wolak (1997) using UK data, and the lower bound of the Lijesen (2007) estimate on Dutch data. I also run a robustness check with $\eta=-0.1$, which corresponds to Allcott's (2012) estimate for customers who self-selected into an Real-time Price pilot program. The resulting estimates are

$$\begin{cases} \textit{for } \eta=-0.1 \\ bQ^{\infty}=1{,}873\ \text{€}/MWh \\ a_0=1{,}973\ \text{€}/MWh \\ a_1=1{,}236\ \text{€}/MWh \\ \lambda=1.78 \end{cases}, \textit{ and } \begin{cases} \textit{for } \eta=-0.01 \\ bQ^{\infty}=18{,}727\ \text{€}/MWh \\ a_0=18{,}827\ \text{€}/MWh \\ a_1=12{,}360\ \text{€}/MWh \\ \lambda=1.78 \end{cases}.$$

Generation costs are those of a gas turbine, as provided by the International Energy Agency (IEA 2010): $c=$ €$72/MWh$ and $r=$ €$6/MWh$ . They differ slightly from the costs presented in table 2.2.2. The constant retail rate is $p^R=$ €$50/MWh$, taken from Eurostat.[13].

### 2.7.3 Numerical Results

We now present the results of a numerical analysis conducting using this specification.

**High-demand elasticity** Suppose demand is highly elastic, that is, $\eta=0,1$. First, rationing is not required at the optimal capacity when more than 13.9 percentage of demand

13. Table 2 Figure 2 from http://epp.eurostat.ec.europa.eu/statistics_explained/images/a/a1/Energy%_prices_2011s2.xls

**Table 2.2**
On-peak hours and prices, high-demand elasticity

| Share of Real-time Price customers (%) | 14 | 100 |
|---|---|---|
| On-peak hours (number) | 158 | 556 |
| On-peak hours (%) | 1.8 | 6.4 |
| Maximum price (€/MWh) | 997 | 335 |

**Table 2.3**
On-peak hours and prices, low-demand elasticity

| Share of RTP customers (%) | 4 | 100 |
|---|---|---|
| On-peak hours (number) | 16 | 127 |
| On-peak hours (%) | 0.2 | 1.5 |
| Maximum price (€/MWh) | 9,083 | 1,219 |

is price responsive. As indicated in the main text, this is encouraging, since this number appears quite achievable.

Second, if all customers are price reactive (i.e., $\alpha = 100\%$, rightmost column), on-peak represents less than 7 percent of the hours, which is equivalent to a hotel having twenty-three nights per year to recover its fixed cost. The ratio of maximum to off-peak price is 4,6, which is aligned with other industries. Thus if all customers were price reactive and if their demand elasticity was on the high end of the observed values, the outcome of peak-load pricing would be manageable for policy makers.

Unfortunately, only a fraction of customers are price responsive. This then reduces the number of hours during which fixed costs are recovered, hence increase on-peak prices. For example, for $\alpha = 14$ percent, no curtailment occurs, and prices rise to €997/MWh. Historically, that price level has proved high for policy makers.

When rationing is required, price is set at the VoLL. Maximum price achieved is $p_\infty = \frac{a_0 + p^R}{2} = 1,012$ €/MWh.

**Low-demand elasticity** Suppose now that demand elasticity is low (i.e., $\eta = 0,01$). Rationing is not required at the optimal capacity when more that 3.9 percent of demand is price responsive.

Even if all customers are price responsive, peak hours last only 1.5 percent of the hours and lead to maximum price of €1,219/MWh. If only 4 percent of customers are price reactive, rationing is not required, and price rises up to €9,083/MWh, more than 126 times the highest off-peak price. If rationing is required, maximum price achieved is $p_\infty =$ €9,439/MWh.

# 3 From Enron with Love

There was a difference of opinion on the rules of the California System Operator. The rules were not quite clear.
—Jeffrey Skilling, CEO of Enron, testimony to California State Assembly

## 3.1 Prelude: The California Crisis

During the 2000–2001 winter, a major crisis engulfed the California power industry. The market appeared to be gravely disfunctionning: power prices spiked to unprecedented levels, and the California system operator (SO) had to resort to rolling blackouts to balance the system. Since the crisis occurred in an off-peak period during which available generation capacity was expected to vastly exceed demand (peak electricity demand in California occurs in the summer, driven by air-conditioning), misbehavior by power producers and traders was immediately suspected.

Conspiracy theorists were proved right: producers and traders were indeed taking advantage of various flaws and inconsistencies in the market rules to extract abnormal profits. Among various market manipulation strategies, they reduced the production of the power plants they controlled, thus raising prices above the competitive level and generating significant profits for their employers. These ill-gotten gains did not prevent Enron, the leading energy-trading firm active in California at that time, from filing for bankruptcy shortly afterward.

Suspecting exercise of market power is one thing, proving it another. Ironically, Enron's diligence in risk control provided the incriminating evidence. Best practice in risk control requires traders' orders to be recorded. When Enron's tapes were exhumed as part of the various lawsuits surrounding the bankruptcy proceedings, traders' explicit commands to power-plant operators to reduce output were made public. The book *The Smartest Guys in the Room: The Amazing Rise and Scandalous Fail of Enron* (McLean and Elkind 2003) and the movie derived from the book, *Enron: The Smartest Guys in the Room* (Gibney 2005), provide an excellent introduction to the Enron saga, including Enron's contribution to the California power crisis. Rarely do we have the opportunity to observe market manipulation so clearly. I strongly recommend both to all readers.

The California crisis has generated an abundant literature, among both academics and practitioners. Furthermore, it had a significant impact on the evolution of the power industry: by and large, it has halted the restructuring process in the United States. California, which was a leader in deregulation, found itself back to regulation in the aftermath. States that were considering deregulation have concluded that it was prudent to wait—except Texas, which pressed forward.

Different commentators draw different lessons from the crisis. Some believe that the power industry is different: while competition may work reasonably well for potatoes and cars, it is a recipe for disaster for electric power. This view is held by many California officials and consumers' advocates interviewed in the Enron movie, who argue that since "*electric power is supplied by complex and highly developed systems with unusual technical characteristics*" (Joskow and Schmalensee 1983), effectively designing and monitoring power markets is extremely difficult.

While this point is true, and is extensively documented in this chapter and chapter 7, I hold a more nuanced view, which is also consistent with evidence presented in the movie.

First and foremost, the California crisis confirms Plaute's observation that "homo homini lupus est": man is a wolf to his fellow man. When presented with an opportunity to acquire significant riches, traders from Enron and other energy merchants captured it, with little concern for the impact of their behavior on their fellow men and women. This behavior was reproduced a few short years later by employees of the financial industry, Wall Street traders but also retail bankers originating subprime loans, and led to the 2008 financial crisis. Readers interested in this dark side of humanity can watch two highly entertaining movies. *The Big Short*, based on Michael Lewis excellent book of the same title (Lewis 2010), describes how contrarian investors foresaw the mortgage bubble in the United States and took a short position against it. *Margin Call* relates the travails of a (fictitious) Wall Street investment bank upon discovering that it holds a portfolio of toxic mortgage-backed assets.

Taking men's behavior as given, the California crisis can be read in a slightly different light: policy makers designed bad rules, which naturally led to a bad outcome. The rules were extremely complex and sometimes contradictory, as rightly pointed out by Jeff Skilling in this chapter's opening quote. They resulted not from sound analysis but from a bizarre compromise among different political parties and interest groups. When drafting rules, lawmakers were primarily concerned with political deal making and ignored the laws of physics, economics, and human nature. This complexity created numerous arbitrage opportunities, which the traders worked very hard to identify and exploit. One of the most disheartening experiences in my professional life was to observe, time and again, that when given a choice to maximize profits for their company or maximize collective welfare, almost all executives (even Canadians) choose the former. In the words of Milton Friedman (Friedman 1970): "The social responsibility of business is to increase its profits."

The relevant question then becomes, Can we design and monitor power markets to limit firms' ability to exercise market power? This chapter does not offer a simple yes or no answer. Rather, it aims to clarify the underlying economics, so that policy makers are fully aware of the pitfalls and dangers of bad rules.

Section 3.2 reviews the various strategies firms may employ to exercise their market power. As the California saga has made abundantly clear, the power industry is at high risk of falling prey to producers' exercise of market power. Section 3.3 reviews the specificity of the electric power industry that favor the emergence of market power. Section 3.4 presents possible solutions to limit the exercise of market power, drawn from the experience of various market designers around the world.

This chapter concludes by summarizing a few academic articles. Section 3.5 presents different models of imperfect competition in power spot markets. Section 3.6 examines strategic underinvestment. Section 3.7 examines the impact of price caps on investment.

My hope for this discussion is that it will produce three outcomes. First, policy makers will promote demand response, the best long-term check on market power. Second, they and market designers will consider potential market abuse when designing market rules; hence they will make them more robust to manipulations. Finally, policy makers will set up strong and highly quantitative market-monitoring agencies and endow them with wide ranging investigative powers.

Each jurisdiction will then decide whether to create markets for power, depending on the technical characteristics of its power industry (e.g., generation mix), its institutions (e.g., the extent of antitrust practice), and its overall policy objectives.

## 3.2 How Is Market Power Exercised?

The OECD defines market power as

> the ability of a firm (or group of firms) to raise and maintain price above the level that would prevail under competition.... The exercise of market power leads to reduced output and loss of economic welfare.

Market power is defined as a deviation from the perfectly competitive outcome, which is that price should be equal to the cost of the last unit produced and/or the value of the last unit consumed, as presented in chapter 2. Assumptions leading to perfect competition are numerous (and rarely met in practice). Two assumptions matter most. First, producers offer a homogeneous product to consumers, hence output from firm A cannot be differentiated from output from firm B. Second, every producer acts as a price taker, that is, knows that his behavior will not affect the market equilibrium. This assumption is met if every producer is very small and the total production capacity exceeds demand.

### 3.2.1 Short-Term Market Power

The most common form of market power arises from horizontal differentiation among producers, that is, when the first assumption does not hold. This occurs to varying degrees in most industries, from hotels to airplanes to restaurants, in which firms are able to differentiate themselves and thereby to charge their customers more than the marginal cost of the room or the seat or the table they occupy. It does apply to electricity retailing, as discussed in chapter 4. It does not apply to wholesale commodity markets, such as oil, metals, and electric power, in which the product traded is indeed nondifferentiable.

In wholesale power markets, the second assumption does not always hold. Power producers are fully aware that, in some instances, their behavior does affect market equilibrium. They can choose among a variety of strategies to exercise market power. At least three families of strategies are available: (a) withholding output, so that market prices increase, (b) offering prices above their variable production cost, and (c) modifying their entire offer curve. Analytically, we leave the realm of optimization, which produces chapter 3's results, and enter the realm of game theory.

#### 3.2.1.1 Withholding output: Imperfect quantity competition

This strategy was formalized more than a century ago, by the economist-mathematician Antoine-Augustin Cournot (1838). Cournot was the first economist to apply the recently discovered calculus of variations to structure and solve economic problems. He proved that, under certain conditions, competing firms prefer to produce less: their loss of output is more than compensated by higher prices and margins.

Under perfect competition, firms produce an additional unit as long as it generates positive operating profit, that is, up to the point where the value consumers place on the marginal unit, which is the price firms receive for their entire output, is equal to the variable production cost of this marginal unit. This is illustrated on the right in figure 3.1.

Suppose, as Cournot did, that firms behave strategically. Each firm realizes that by producing an additional unit, it incurs the variable production cost and receives the market price for that unit but also reduces the price it receives on its entire output. The marginal revenue is the sum of these two terms. To maximize profit, the firm produces up to the point where marginal revenue is equal to the variable cost of production. The competitive and Cournot equilibria are illustrated in figure 3.1.

This behavior seems natural (if morally objectionable) for a single firm, that is, a monopoly. Cournot's first contribution was to formalize this behavior using calculus. Cournot's second contribution was to prove that it arises as an equilibrium among multiple firms: knowing that my competitor reduces output, I also reduce output. Cournot solved the first Nash equilibrium, more than a century before the concept was formalized by John

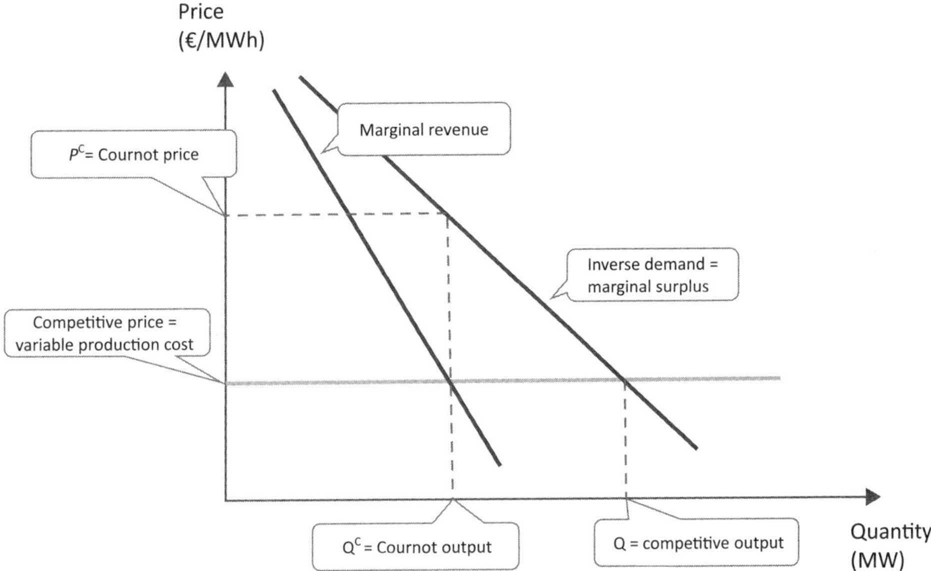


**Figure 3.1**
Perfectly competitive and Cournot outputs and prices. The perfectly competitive output is such that the value of the marginal unit, which is the price, is equal to the variable cost of production of that unit (which is constant in the figure). The Cournot output is such that the revenue generated by the last unit produced, which is the price minus the revenue reduction on all units, is equal to the variable cost of production of that unit. The Cournot price is the value of the last unit produced, which is higher than the competitive price.

Nash. Sadly for Cournot, he was the only economist of his generation with a mastery of calculus and was unappreciated by the profession. He died a bitter man. Cournot's superb contribution to economics was recognized after his death by other mathematically inclined economists, including Alfred Marshall.

Cournot competition assumes that producers compete in quantity: they produce a given output, and an "auctioneer" balances the market, that is, computes the price such that demand at this price precisely equals the offered output. A critique of Cournot competition is its lack of realism: in many markets, "prices are ultimately chosen by firms, not by an auctioneer" (Tirole 1988, 216).

However, this critique does not apply well to the power industry. Since their inception, centralized marketplaces have matched buyers and sellers in electricity markets, thereby acting as auctioneers.

Thus, Cournot competition provides an adequate description of wholesale power markets. Cournot's contribution was well understood by Enron traders, who specifically ordered power plants to reduce output so that prices could rise. More recently, Australian power producers have been found to engage in strategic output withholding in 2015, which proves that Cournot's teachings are practiced today.

Cournot competition yields a simple prediction: a firm's operating margin as a percentage of price (known to economists as the Lerner index), which is the market price minus its marginal cost divided by the market price, is proportional to its market share and to the inverse of the demand price elasticity. The larger the firm, or the less elastic the demand, the higher the margin (as a percentage of price). This prediction has been tested empirically in many industries.

#### 3.2.1.2 Offering prices above variable cost: Imperfect price competition

This strategy was also discovered more than a century ago, in 1897, by the British economist Francis Edgeworth, building on earlier work by the French economist Joseph Bertrand.

Consider two identical firms, competing to serve a constant demand. Suppose both firms have constant (and identical) variable production cost $c$. Bertrand (1883) observed that, if each firm is large enough that it alone can satisfy the entire demand, the only equilibrium is for both firms to sell at their marginal cost: if firm $m$ attempted to sell at a price $p_m > c$, the other could offer a slightly lower price and serve the entire market. Reversing Cournot's analysis, Bertrand proved that competition between two firms is sufficient to drive profits down to zero.

The story, however, does not stop here. Edgeworth (1897) observed that firms have limited production capacity; hence, in many instances, no single firm is able to serve the entire market. Under some circumstances, a producer may find himself pivotal: he is required to balance demand, that is, the other firm's output is lower than demand, and he knows it. Selling at marginal cost is no longer an equilibrium: he could offer a higher price, losing market share, but capturing positive profit.

The specific form of the equilibrium depends on the microstructure of the market. If the market clears using a single price, a single equilibrium price exists, higher than the marginal cost. If every producer sells at his price, the equilibrium is for firms to choose randomly among various prices, which is called a mixed-strategy equilibrium. Edgeworth did not characterize this mixed-strategy equilibrium in 1897. He simply observed that limited capacity defeats the Bertrand result.

Bertrand-Edgeworth competition is also relevant for power markets. Demand is inelastic in the short term, at least until demand response is widespread. On-peak, when demand is near aggregate supply, the highest-cost producers are pivotal; hence they are able to offer and be paid a price significantly higher than their variable cost of production.

#### 3.2.1.3 Combining price and quantity imperfect competition: Supply-function equilibrium

A final strategy combines features of both approaches. In centralized power markets, firms submit supply schedules, that is, a list of combinations of prices and quantities they are willing to produce at these prices. For example, a firm offers to produce 100 MW if the price is \$20/MWh, 150 MW if the price rises to \$30/MWh, and 200 MW if the price is \$40/MWh or higher.

Under perfect competition, firms would simply submit their true production costs and capacities for different units. If competition is imperfect, firms can increase the prices at which they are willing to produce but can also reduce the quantities. For example, the same firm will offer to produce 80 MW if the price is \$25/MWh, 120 MW if the price rises to \$35/MWh, and 160 MW if the price is \$45/MWh or higher.

This form of imperfect competition, known as the supply-function equilibrium (SFE), has been formalized first by Paul Klemperer and Margaret Meyer (1989). It provides an accurate description of imperfect competition in centralized power markets and has spawned a rich academic literature, reviewed later in this chapter. Supply-function equilibrium strategies are solutions of systems of differential equations. Except for few very simple cases, they cannot be characterized analytically and are approximated through numerical simulations. Furthermore, an infinity of equilibrium strategies exists.

#### 3.2.1.4 Which strategy do producers use?

The three strategies described above are analytically very distinct: Cournot producers compete in quantity facing a price-responsive demand; capacity constrained Bertrand-Edgeworth firms compete in price facing constant demand, and supply-function equilibria arise as firms compete by offering entire offer curves facing uncertainty on future demand.

In practice, however, I believe that when firms exercise market power, they use a combination of all strategies: they reduce output to increase the number of on-peak hours, during which they find themselves pivotal. They submit offer curves higher then their marginal cost whenever they have an opportunity to profitably do so.

Can power markets participants actually compute and coordinate on strategies of such sophistication and complexity as supply-function equilibria? This question is, of course, an important field of academic research. I believe the answer will be yes, for a simple reason: firms hire economic graduates, who have been exposed to models of imperfect competition, including supply-function equilibrium, during the course of their studies, and task them with maximizing the value of their portfolio of assets. Since firms submit supply schedules for every hour (or half hour) for every day, they have ample opportunity to coordinate on various strategies. I am not suggesting that all firms participating in power markets are colluding to increase power prices. I am merely observing that coordinating on sophisticated strategies is possible.

### 3.2.2 Long-Term Market Power

Long-term market power is strategically reducing capacity, that is, strategically underinvesting in new assets, or mothballing or shutting down existing plants, to durably increase prices. It can be viewed as an extension of Cournot competition: instead of (or in addition

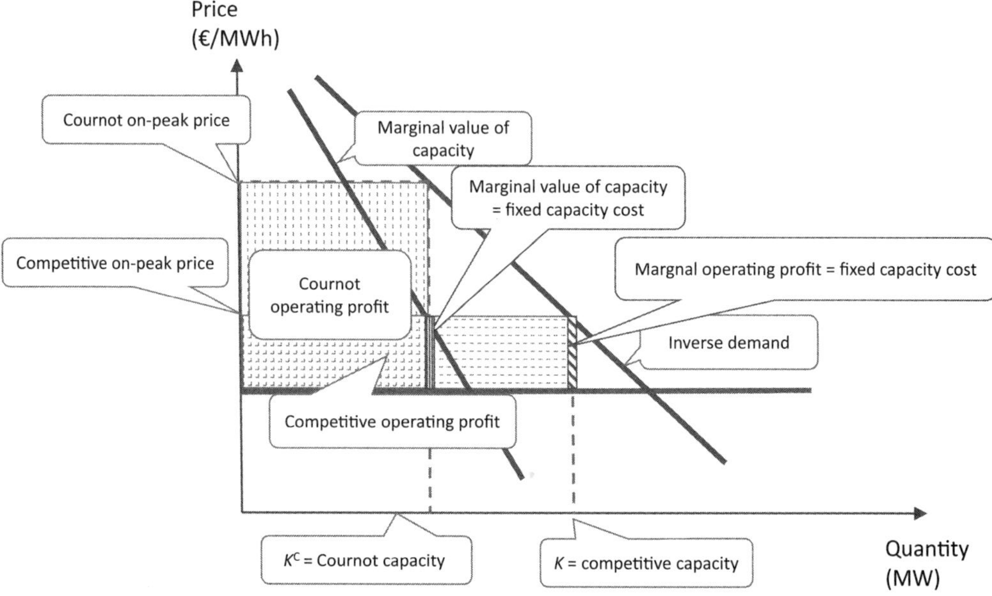


**Figure 3.2**
Strategic capacity reduction.

to) reducing output in the spot market (that is, producing less than the available capacity), firms reduce capacity when they make their investment decision. This then creates a natural and durable scarcity in the market.

This strategy is illustrated in figure 3.2, using the Cournot model and assuming a single on-peak state of the world for simplicity. As discussed in chapter 2, perfectly competitive producers build capacity up to the point where the marginal operating profit is equal to the fixed capacity cost. This is the free-entry condition.

A strategic producer, on the other hand, realizes that by building an additional unit of capacity, he receives the operating margin for that unit but also reduces the price he receives on his entire capacity. The value of a marginal unit of capacity is the sum of these two terms. To maximize his profit, he invests up to the point where the value of the marginal unit of capacity is equal to its fixed cost.

Strategic capacity reduction is well defined analytically. It appears robust to antitrust scrutiny: it enables firms to behave perfectly competitively in the spot markets yet capture oligopoly profits. If the market monitor were to launch an inquiry into the level of capacity, producers could simply respond that actual demand exceeds their expectation. For these reasons, it should be a serious threat.

Yet I have not seen evidence of strategic capacity reduction. Since the 1990s, the power industry, like all commodity industries, has been characterized by boom-bust investment cycles: when demand grows, producers invest gleefully, the boom part of the cycle. If growth falls below expectation, even slightly, installed capacity exceeds demand, depressing prices. Firms have to shut down unprofitable assets—the bust part of the cycle. As the balance between supply and demand is restored, prices increase. If demand growth resumes, producers invest again, and a new cycle starts.

The boom-bust investment cycle is not specific to the power industry. Daniel Yergin (1990) shows it has plagued the oil industry since its inception in the late 1850s. In fact, Standard Oil's monopoly was Rockefeller's attempt to control it. It can also be observed in other commodities, such as steel, aluminum, and chemicals. For this reason, strategic capacity reduction does not appear to be a serious concern in the power industry.

### 3.2.3 Multimarket Strategies

Power producers compete not in a single market but in multiple markets. The possibilities for strategic behavior are thus much richer. This multiplicity of markets makes detection of strategic behavior more difficult.

Three examples illustrate the situation. First, as discussed in chapter 2, most power markets are organized around two settlements to accommodate physical constraints of the generation assets: for every trading period, a day-ahead market determines a day-ahead market clearing price, and a real-time market determines a real-time price. Thus energy for the same hour can be settled at two different prices; hence power traders can choose to buy (or sell) energy in one market or the other and arbitrage the price difference between these markets. Additionally, they can behave strategically in one market to affect price in other market; for example, they can reduce output in the day-ahead market, which will then increase price in the day-ahead market, and possibly also in the real-time market.

Vertical integration is a second example of multimarket strategies: electric power producers are often vertically integrated with a retail business. This gives rise to specific concerns, which are discussed in chapter 4.

Spatial differentiation constitutes a final example of multimarket strategy. As discussed in chapter 6, power markets are in fact multiple "nodes" linked by transmission lines. A power trader may generate at different nodes on the grid. Modifying his behavior at one node affects market price at this node but also at other nodes. These strategies are discussed in chapter 7.

## 3.3 Why Is Exercise of Market Power a Serious Issue in Power Markets?

Three specific factors facilitate short-term market power in the power industry: (a) inelastic demand, (b) transmission constraints, and (c) the complexity of the rules. Other factors are the small number of active market participants and the frequency of repeated interactions. While they undoubtedly matter, they appear to be less specific to the power industry; hence they are not discussed here.

### 3.3.1 Non-Price-Responsive Demand

Most customers are equipped with standard meters, that measure their consumption in kilowatt-hours, but provide no information on where the consumption actually occurred, for example, I know I consumed 5 MWh last year, but not how much was on-peak or off-peak. Customers equipped with standard meters pay a flat price for power, for example, a price constant across days of the year and hours of the day and more generally independent of supply and demand conditions. Their consumption depends on their needs at a given hour, not on the opportunity cost of power at that hour.

A few utilities have introduced advanced tariffs, but these are the exception rather than the norm. For example, many U.S. utilities have experimented with time-of-use (ToU) rates, that is, different rates for predefined off- and on-peak periods. The French utility, EDF, implemented in the 1980s a dynamic rate structure, very close to the peak-load pricing structure discussed in chapter 2: customers would pay a low rate for most of the year, and for a fixed number of days per year, the utility would charge significantly higher rates. Customers would be informed ahead of time that the next day would be a "red" day.[1] In both cases, customers were equipped with different meters.

In the regulated world, offering fixed rates to customers was socially inefficient: too much consumption occurred on-peak, when the constant rate was lower than the opportunity cost of power, and too little off-peak. As a result, too much capacity had to be installed to meet demand. However, it was the norm.

**Non-price-responsive demand facilitates strategic behavior** When markets opened in the early 1990s, non-price-responsive demand created an additional problem. A producer (or a group of producers) could withdraw output or offer high prices, since customers (or their suppliers) had no choice but to accept this high price to satisfy their load. One of the risks (from a producer's point of view) when increasing price is that customers will buy less of your product. In the late 1990s, electricity demand responded very weakly to a price increase. Therefore, power producers were not limited by demand responsiveness; since the volume sold remained constant, a one dollar increase in price translated exactly into a one dollar increase in profit. For prices, the sky was the limit! Since demand did not limit high prices, regulators imposed price caps. One personal anecdote presented in box 3.1 illustrates this situation perfectly.

Technology has since progressed, and real-time metering— and control—is cost effective in many instances. However, as discussed later in this chapter, policy makers are reluctant to promote price-responsive demand, and most electricity demand still does not respond to prices.

1. It is not surprising that EDF adopted a rate structure close to the optimum: EDF's CEO in the 1980s was Marcel Boiteux, the economist who characterized the optimum in the early 1950s.

**Box 3.1**
A very expensive Santa

> I happened to visit San Francisco shortly before Christmas 2000, at the height of the California power crisis. On the way from the airport to the hotel, we drove through a residential neighborhood. Families were proudly displaying Christmas decorations. It was mid-afternoon, no one was at home, and yet, Santa, Rudolph, and all the elves were fully lit, as were the Christmas trees. Residential customers were using power as usual, blissfully unaware that PG&E, their supplier, was paying $1,000 for every MWh. Shortly afterward, PG&E filed for bankruptcy protection. It is thus no surprise that economists from Californian universities have pursued an ambitious research agenda extolling the virtues of demand response (reviewed in chapter 5).

**Non-price-responsive demand makes detection of strategic behavior more difficult**
As discussed in chapter 2, a direct consequence of inelastic demand is that prices can legitimately rise to a few thousand euros per Megawatt-hour for a few hours per year, more than 100 times the average price. This property makes it particularly difficult for market monitors to identify the exercise of market power. Very high price spikes are required and hence should be allowed. The challenge for regulators is to ascertain which spikes are consistent with fixed cost coverage and which spikes reflect exercise of market power.

As shown in chapter 2, the perfectly competitive price should on average cover the long-term marginal cost; hence the operating margin (earnings before interest, taxes, depreciation, and amortization, or EBITDA) should be €60,000/MW per year for a peaking turbine.

This provides us with a benchmark against which to compare the average EBITDA. However, this average has to be taken over very long periods. The EBITDA may be close to zero for a low-demand year (e.g., a mild winter in Europe or a cool summer in North America) but very high during a high-demand year. It is perfectly legitimate for peaking turbines to achieve EBITDA of €120,000/MW in a high-demand year, if these occur every other year. This makes it very hard to detect exercise of market power.

As demand becomes more price-responsive, spikes will become lower and detection will become easier.

### 3.3.2 Transmission Constraints Isolate Portions of the Market into Submarkets

One additional challenge specific to the power industry is the presence of transmission constraints, which split the market into submarkets. A producer located on Long Island, in the state of New York, may be small in its domestic market (the state of New York) but may hold significant market power when Long Island is isolated from the rest of the system. Contrary to other industries, building a large portfolio is not required to be able to impact electricity prices.

The impact of transmission constraints on strategic behavior is discussed in details in chapter 7. A short summary of the main issues is presented here.

The first difficulty is that the constraints, hence the submarkets, are transient. For example, at some hours (usually off-peak) the grid is uncongested, and all of Europe can be considered a single market. At other hours, the grid is heavily congested, and Europe is de facto broken into a dozen submarkets. In some extreme cases, one particular station must run for the system operator to operate the system reliability; hence this plant is a monopoly in its sub-market.

The second difficulty is that congestion on the transmission grid may be the result of producers' strategic behavior, that is, producers may find it optimal to increase (or decrease) their output to artificially constrain specific interconnections.

System operators use financial forward contracts, called financial transmission rights (FTRs), to manage congestion. The third difficulty is that producers may strategically use these FTRs to enhance their market power.

### 3.3.3 Complex Power Market Rules Are Sometimes Conducive to Market Power

To accommodate the physical reality of producing power, rules in power markets are often extremely detailed and complex. This complexity has the undesirable side effect of facilitating the exercise of market power. I present below two examples, then attempt to explain why "bad" rules are sometimes designed.

#### 3.3.3.1 California's 1996 market design

One striking example of a design that badly misfired is the obligation that the California legislative body imposed on retailers (called load-serving entities, LSEs in the United States) to source their power from the wholesale spot market (that is, the day-ahead market). These entities were forced to divest some of their generation assets and were prevented from entering into long-term supply contracts with the new owners of these assets. Market designers' objective was to increase liquidity of the wholesale spot market, as they were rightly concerned that if LSEs sourced most of their load from their own generation or long-term contracts, very little volume would be traded in the wholesale spot market, and the price signal would be less reliable.

On the other hand, LSEs were selling power to their customers at a fixed price. Surprising as this may seem with hindsight, the utilities had fought for that provision. The prevailing view in the mid-1990s was that restructuring would lead to wholesale prices lower than the energy component of the regulated rate. Generation assets would no longer be able to fully cover their capital costs through power sales in the wholesale market and would thereby be stranded. Incumbent utilities therefore lobbied hard to have the right to continue to sell power at the regulated rate until the remainder of the capital cost was recovered.

This might seem an odd policy decision, since it delayed for a few years customers' enjoyement of lower power prices, which was the main expected benefit of liberalization.

Furthermore, one may argue that investors in utilities knew they were exposed to regulatory risk, such as liberalization. This risk explained why utilities stock yielded higher returns than Treasury bills. In other words, one could argue that, since investors were happy to receive for decades the risk premium, they should have no problem when the risk materializes. Still, at the time, lawmakers thought stranded cost recovery was an acceptable price to pay for incumbent utilities to accept liberalization.

The challenge was to determine at the time of market opening the stranded cost owed to incumbent utilities, which was the difference between the regulated revenues they would have received and the future power prices over the remaining life of the assets. This required agreeing on a forecast of future power prices a few decades ahead. Unsurprisingly, it occupied lawmakers, lobbyists, and economic consultants for a long (and happy) period in the 1990s.

Power prices were very low during the first two years of market opening in California. There were instances of spot price hitting zero for a few consecutive hours, when inflexible generation exceeded demand. San Diego Gas and Electric (which later became Sempra) recovered all of its stranded costs and was allowed to charge market-based prices to its customers. The other two LSEs, Southern California Edison (SCE) and Pacific Gas and Electric (PG&E), had not completed their stranded cost recovery by the summer of 2000 and were still charging their customer a fixed rate for energy.

Therefore, the market design had created a naked short position in the spot market for the LSEs: every day, they came into the wholesale spot market to buy their load (which, as box 3.1 illustrated, did not respond to price) at the wholesale spot price, to resell at a fixed price to their customers. As long as supply outstripped demand, California was a buyers' market, and the utilities were fine. By the spring of 2000, however, demand growth, coupled with lack of investment for a decade, had made California heavily dependent on imports. A low rainfall winter reduced imports, hence tightening the supply-demand equilibrium in California. The market became a sellers' one. Power marketers knew that LSEs had no choice but to buy their load from them in the spot market. It was, as one trader put it, "almost too easy to make money." Forced to buy power at $1,000/MWh and sell it at around $50/MWh, LSEs quickly faced severe financial hardship: PG&E filed for bankruptcy protection, a fate SCE only narrowly avoided.

Taken separately each part of the design made some sense: increasing liquidity in the spot market is not unreasonable, and a case can be made that recovering stranded costs through a transitionary fixed rate is acceptable. Taken together, these two parts have contributed to an unmitigated disaster.

#### 3.3.3.2 Two-settlement and predispatch

We previously discussed the two-settlement system, present in many markets in North America, which gives producers additional opportunity for strategic behavior.

Another example (in the same vein) comes from New Zealand. Producers, retailers, and consumers on the wholesale market submit indicative bids and demand curves to the system operator, who then uses these data to compute the market equilibrium price and quantities at every node on the grid. The exercise is repeated every two hours until "gate closure," which occurs two hours before the actual dispatch. On the one hand, these pre-dispatches are extremely useful, as they enable producers and customers to adjust their positions ahead of time. For example, a consumer who realizes tomorrow's price will be high may decide to turn off one assembly line or to turn on an auxiliary generator. These pre-dispatches therefore contribute to system security. On the other hand, they provide producers with an opportunity to test different strategies and to coordinate to maximize profits. I do not have any evidence that such behavior occurs. I am merely observing that this particular feature plays a useful role … but also potentially facilitates collusion. Invited by the New Zealand Electricity Authority, a team of researchers led by Estelle Cantilla of ULB is currently investigating this question.

#### 3.3.3.3 Why were "bad" rules designed in the late 1990s?

Designers for any market face a trade-off between creating simple and transparent rules, and tailoring the rules to reflect the physics of the product being traded. Electricity markets designers have often given preference to the latter for various reasons, some better than others.

When power markets were designed, in the late 1980s in Great Britain, in the mid-1990s in the United States and in the late 1990s early 2000s in Europe, they were replacing the vertically integrated structure that had been the industry's paradigm during the professional life of all people involved. At the risk of oversimplification, policy makers were primarily discussing market designs with power engineers, who were not too keen on restructuring, being very comfortable with the system the way it was. A primary—and legitimate—concern was a system's reliability, that is, ensuring that the lights remain on. Thus many engineering practices were simply turned into market rules, for example, the day-ahead unit commitment process became a day-ahead market. Not enough thought was given to the incentives generated by these rules, in particular, whether they could facilitate the exercise of market power.

In addition, market design occurred in the heated world of give-and-take, coalition-building politics, not in the pure air of academia. Market participants expanded significant lobbying effort to make sure the rules favored their interests, with little regard for the collective welfare.

Thus market designers in the late 1990s faced a daunting task: policy makers, who had entrusted regulators or system operators with the task of designing markets, were issuing sometimes contradictory directions; there was no academic consensus on specific market design issues; engineers, who were the only ones who actually knew the system, were not enthusiastic about reform; and market participants were vigorously lobbying for rules to their advantage. It is not surprising that some rules ended up creating arbitrage opportunities for some participants.

As discussed in the next section, the challenges facing market designers today remain significant.

## 3.4 How Can Exercise of Market Power Be Mitigated?

As the previous discussion illustrates, mitigating exercise of market power is a vital challenge in the restructured power industry. What should policy makers do? First, it is essential they understand that price caps do not constitute a long-term solution. Second, three recommendations emerge: encourage price-responsive demand, invest in aggressive quantitative market monitoring, and design robust rules.

### 3.4.1 Price Caps Are Not a Long-Term Solution

The first tool policy makers deploy to protect consumers from market power is to cap wholesale spot prices. Formal caps are in place in most U.S. markets (Texas being the exception). In Europe the cap in the day-ahead market is set at €3,000/MWh. In addition, as observed by Paul Joskow (2007), SOs' practices when demand is close to supply de facto prevent prices from spiking. For example, SOs sign contracts with reliability-must-run plants and pay them the contract price when they are called. This limits the price increase, precisely when the system is under stress, hence the information contained in prices is the most valuable.

A price cap protects customers from extremely high prices, hence offering insurance. Imposing a price cap is therefore an expedient and effective solution. However, it creates two problems.

#### 3.4.1.1 No demand response, precisely when it is the most needed

**Price caps lead to rationing** When the cap is reached, customers face a constant price by construction, hence they no longer adjust their demand. Administrative rationing is required to adjust demand to available supply. This is well known in many other industries: a price cap leads to rationing. For example, rent control, which was in effect when I was studying in Cambridge, Massachusetts, led to rationing of housing.

**Endogenous and exogenous price caps** It is important to distinguish a price cap from the value of lost load (VoLL). As discussed in chapter 2, if demand is not sufficiently elastic, there may be instances when the SO must curtail a fraction of customers. When this occurs, the wholesale power price is administratively set at the VoLL, which represents the SO's best estimate of the value a curtailed customer would place on a marginal Megawatt-hour. Theoretically, the VoLL should vary with the state of the world, usage, and the conditions and duration of rationing. However, in practice, SOs use a single number for all states of the world. Thus the VoLL acts as an endogenous price cap.

Imposing an exogenous price cap below the VoLL (e.g., at $1,000MWh, while the VoLL is estimated at $20,000/MWh) leads to the same outcome: the wholesale spot price is capped, and customers are curtailed. However, the causality is reversed: the exogenous price cap causes rationing, even if all customers are price reactive. On the other hand, price reaching the VoLL is a consequence of rationing, itself caused by insufficient price reactivity of demand.

#### 3.4.1.2 Missing money

As discussed in chapter 2, prices need to rise to very high levels for a few hours to cover capital costs. By construction, a price cap prevents prices from reaching these levels, thus creating a "missing money problem," which leads to underinvestment. Initially the term *missing money* was applied to generation investment. It also applies to demand response: since prices are capped, market participants have lower incentives to invest in demand-response technologies. Price caps thus also create a dynamic inefficiency on the demand side. Policy makers need another instrument, a capacity mechanism (reviewed inchapter 9), to guarantee that adequate generation and demand response capacity is installed. In addition, policy makers may design support mechanisms specifically aimed at demand response, reviewed in chapter 5.

#### 3.4.1.3 Optimal price cap

The impact of price cap as net surplus is more subtle than it appears. First, imposing a price cap leads to rationing, which decreases net surplus. Second, a price cap impacts equilibrium capacity. The standard argument is that imposing a cap reduces profit, hence investment. This is true if the industry is perfectly competitive, in which case imposition of a cap is hard to justify.

If competition is imperfect, under reasonable conditions, there exists at least one cap that leads to higher installed capacity than does no cap. The intuition for this surprising result is that, if competition is imperfect, increasing a price cap has two opposite effects on investment incentives. First, increasing the price cap increases operating profits, hence the marginal value of generation capacity. This is the standard missing-money argument. Second, increasing capacity generates a higher positive operating margin when the price is equal to the cap than when the price is close but not quite equal to the cap. Increasing the cap reduces the probability that the cap is binding, hence the marginal value of capacity. The first effect dominates for low price caps, the second for high caps. Therefore, there exists an optimal price cap, which leads to higher capacity than does no cap.

#### 3.4.1.4 Price cap versus offer cap

Some markets (e.g., ISO New England) impose a cap on offers, not on wholesale prices: producers cannot bid more than the cap. The spot price may exceed the cap when it is set by bids from customers, as in figures 2.4 and 2.6, or by the SO at the VoLL if curtailment is required, as in figure 2.8. If implemented properly, an offer cap does not distort demand nor cause missing money. However, it does not counter effectively strategic output withholding.

### 3.4.2 Three Recommendations

To address problems of exercise of market power, I make three recommendations. First, policy makers should strive to increase the share of customers facing and responding to real-time prices. Second, market monitors should build highly quantitative and economically sophisticated competencies, to remain on par with power traders. Market monitors should also be endowed with wide-ranging investigative authority. Third, policy makers and market designers should create simple and robust rules. The good news is that progress is being achieved on these two recommendations. The third recommendation, however, is more difficult to implement.[2]

#### 3.4.2.1 Price-responsive demand

Increasing the share of customers who respond to wholesale spot prices is probably the most effective approach to permanently and durably reducing the exercise of market power. When demand decreases as price increases, producers face a trade-off between higher margin and lower output. As we have seen, the more elastic demand, the steeper this trade-off, and the less damaging imperfect competition is.

The good news is that demand response is now feasible in most restructured markets: wholesale spot prices signal the opportunity cost of electricity for every minute, advanced metering systems enable customers or their suppliers to record their actual consumption for every minute, and information and remote control technologies enable customers or their suppliers to remotely and automatically increase or decrease their electricity usage. Despite its technical feasibility, demand response is still well below its potential. Chapter 5 discusses this issue in detail.

Even if all customers faced and responded to the wholesale spot price, market power would remain an issue. Additional remedies are required.

#### 3.4.2.2 Highly quantitative, sophisticated, and aggressive market monitoring

Exercise of market power is a concern in many industries. Starting with the breakup of John Rockefeller's Standard Oil in 1911, governments have launched numerous lawsuits and intervened in numerous industries over the years. Meanwhile, economists have developed theories to characterize and statistical approaches to quantify the exercise of market power.

As the previous discussion has shown, power markets need constant and highly sophisticated monitoring. At the onset of restructuring, a large skills gap existed between most market participants and market-monitoring agencies. The former employed hundreds of highly

2. This section and the next owe much to discussions with the senior team of ISO New England in the United States and the Electricity Authority in New Zealand. However, the views expressed here are my own.

trained, highly quantitative, and highly incentivized analysts to design and implement highly sophisticated trading strategies. Again, the movie *Enron: The Smartest Guys in the Room* provides a vivid illustration of that point. Market-monitoring agencies were slower to catch up. One senior partner at McKinsey and Company summarized the skills imbalance accurately, although coarsely: "Regulators are paid $60,000 per year. Traders who are paid more than $200,000 per year are always going to run circles around them."

This imbalance has been partly reduced. Market-monitoring agencies have "quanted up" and now also employ quantitative analysts, who comb through hundreds of thousands of data points looking for suspicious patterns. Furthermore, regulators have the authority to have firms and individuals who are engaged in unlawful market manipulations prosecuted. This is proving effective: in the summer of 2013, following an investigation by the Federal Energy Regulatory Commission in the United States, JPMorgan Chase agreed to pay $410 million in penalties and disgorgements to rate payers in an out-of-court settlement (Federal Energy Regulation Commission 2013).

Still, market monitoring is a never-ending battle. Trading firms will always hire more quantitative talent and pay them better than market-monitoring agencies. Traders will always have stronger incentives than market monitors.

This issue is of course not unique to power markets and is present, for example, in financial markets. I believe the problem is slightly easier for power markets, since robust counterfactuals can be built, that is, market-monitoring agencies can compute with reasonable accuracy the perfectly competitive prices and quantities. On the other hand, the rules governing the operations of power markets are extremely complex, an issue we turn to next.

#### 3.4.2.3 Simple market rules

Policy makers and market designers should strive to keep market rules simple and robust to market manipulation. This recommendation sounds obvious, but putting it into practice is hard. At the onset of restructuring, rules were poorly designed, in particular in California. How has the environment surrounding market design evolved since then? Some factors have remained constant, some evolutions have improved market design, and some have impacted it adversely.

**Constant factors** Reliability concerns, hence engineers' voices, are still important in designing power markets. The lights have not gone out yet, but they could. Massive RES entry is profoundly transforming the generation mix and increases the volatility of demand met by "incumbent" technologies. This raises new concerns about generation and operating-reserves adequacy: Who produces when the wind does blow? How do we cope with a sudden drop in photovoltaic generation, caused for example by a large cloud?

Another constant factor is that market participants still have their own best interests at heart when commenting on proposed design changes.

**Evolutions improving market design** Market designers today have learned from their elders' mistakes. Most observers agree that preventing market power, or at least not creating opportunity for the exercise of market power, is an essential concern when designing power markets. Systems operators and regulatory agencies now employ staff experienced in analyzing power markets, some trained in economics, and better able to design robust markets. This is a significant and positive development.

**Evolutions adversely impacting market design** Policy makers are probably more involved in electricity market design today than they were in the late 1990s, for two reasons: First, the California disaster has taught policy makers (especially members of the executive branch) that rolling blackouts, or more generally a well-publicized failure of power markets, were extremely damaging for reelection. All remember that California governor Gray Davis was forced out of office in the middle of this term by a popular recall, triggered in large part by the power crisis, even though he was not governor when the law was passed and was not active during the restructuring process. Second, as mentioned in chapter 1, electricity is essential to our efforts to reduce our carbon footprint. Any policy maker who aspires to reduce carbon emissions has to produce policies that impact electricity markets.

Why do I consider such policies to adversely impact market design? After all, policy makers are elected officials; hence their involvement is a sign of democracy. While this is true, experience suggests that the give-and-take and compromise process consubstantial with policy making may not be appropriate when designing markets.

A final evolution adversely impacting market design is that any market design initiative will change existing rules, thus creating new winners and losers. Even though the total effect may be beneficial for market participants collectively, the losers will lobby hard against any change to the status quo. This resistance to change explains partly the difficulty of moving to real-time pricing (RTP): users who consume significantly on-peak are today subsidized by users who consume mostly off-peak. Moving to RTP removes this subsidy. Not surprisingly, the first group objects vehemently to the move (Borenstein 2005).

Another example is the rebalancing of network charges. Users are becoming progressively more heterogeneous in their use of the transmission and distribution networks: some consumers invest in decentralized generation, which reduces their purchase from the grid and sometimes leads them to resell to the grid. This change in usage pattern should lead to a change in rate structure. However, users likely to lose from that change vehemently oppose it.

### 3.4.3 Concluding Thoughts

The previous discussion illustrates a tension at the heart of public policy: perfect markets are formidably efficient, that is, they deliver desired policy outcomes at the lowest cost for societies. Yet, perfect markets do not exist. In some countries, for example, Great

Britain, policy makers and their agencies attempt to continuously improve market design by modifying (and it is hoped, improving) the rules, making them more and more detailed and prescriptive at every iteration. Other countries, for example, New Zealand, opt for a light and stable market design.

Markets are designed by policy makers, or agencies acting on their behalf, under the influence of interested parties. Market participants often have much better information than agencies on the impact of market rules and much stronger incentives to see a favorable outcome. They will therefore expand considerable resources to make markets imperfect.

This argument is reminiscent of Stigler's (1971) seminal article on the theory of economic regulation, which provides convincing evidence (at least for most economists) that most entry regulations are designed, to benefit the regulated firms themselves, not the public at large. Similarly, many detailed rules are designed to favor one group of stakeholders, not to improve the overall efficiency of markets.

For this reason, I urge readers to keep in mind that, with market design as with many other human pursuits "The perfect is the enemy of the good." A simpler design may perform better than a sophisticated one.

## 3.5 Academic Models of Imperfect Competition in Spot Markets

Economists use three families of models to represent imperfect competition. Cournot competition produces the simplest derivations. If inverse demand is linear (or close to linear), the imperfect competition game admits a unique pure strategy equilibrium. Bertrand-Edgeworth competition assumes inelastic demand and explicitly recognizes the constraints imposed by fixed existing production capacity. This richness comes at the cost of complexity: in most cases only mixed-strategy equilibria exist, although they can be characterized analytically. Finally, competition as supply-function equilibrium (SFE) accurately describes the actual bidding process in power markets, at the cost of even greater complexity: in most cases, there exists an infinity of SFEs, characterized by a system of related differential equations. Except in a few specific cases, computational methods are required to describe the equilibrium strategies. These three families of models are presented below.

### 3.5.1 Cournot Competition

We use the same notation as in chapter 2. $N \geq 1$ producers compete to produce electricity and sell it into the wholesale spot market. Customers have identical demand profiles, up to possibly a scaling factor (Assumption 1), and demand varies across states of the world. The wholesale spot price is determined by the equilibrium between imperfectly competitive supply and demand and varies across states of the world.

Retailers purchase the good on the wholesale market and resell it to customers. In the short term, the state of the world $\theta$ is fixed. Retailers pass on the wholesale price to their customers.

In state $\theta$, producer $n$ produces $q^n(\theta)$ while the aggregate output is $Q(\theta) = \sum_{n=1}^{N} q^n(\theta)$. Her operating profit is

$$\pi^n(q^n, Q) = q^n(P(Q,\theta) - c).$$

The assumptions and conditions of chapter 2 continue to hold. An additional condition is required to ensure existence and unicity of a Cournot equilibrium:

**Condition 4.** Conditions on inverse demand. For all values of the aggregate output $Q \geq 0$, for all values of the individual output $q \geq 0$, and for all states of the world $\theta \geq 0$:

1. Marginal revenue is decreasing:

$$2\partial_Q P(Q,\theta) + q\partial_{QQ} P(Q,\theta) < 0.$$

2. Marginal revenue is increasing as states of the world increase:

$$\partial_\theta P(Q,\theta) + q\partial_{Q\theta} P(Q,\theta) > 0.$$

3. Inverse demand and all its derivatives admit a finite limit as $\theta \to +\infty$.

A Nash equilibrium is such that each firm chooses its output to maximize its profit, given the others firms' output. Assuming the equilibrium is interior, a necessary condition is

$$\frac{\partial \pi^n}{\partial q^n} = 0 \Leftrightarrow P(Q,\theta) + q^n \partial_q P(Q,\theta) = c. \tag{3.1}$$

As seen in chapter 2, if competition is perfect, output is such that the value of the marginal unit $P(Q,\theta)$ is precisely equal to its marginal cost $c$. A Cournot oligopolist reduces output compared with perfect competition: if she produces an additional $\delta q^n$, she incurs the additional variable production cost $c\delta q^n$ and receives additional revenues $P(Q,\theta)\delta q^n$, but she also reduces her revenue on her entire output by $q^n \partial_q P(Q,\theta)\delta q^n$. Her marginal revenue is the sum of these latter two terms. To maximize her profit, she produces up to the point where her marginal revenue is equal to her variable cost of production. This is equation (3.1). The competitive and Cournot equilibrium are illustrated in figure 3.1.

Condition 4 guarantees the existence of at least one solution for every state of the world. Since the profit function is strictly concave, this solution is unique. Thus there exists a unique equilibrium, which is symmetric. Aggregate output is defined by:

$$P(Q,\theta)+\frac{Q}{N}\partial_Q P(Q,\theta)=c. \tag{3.2}$$

Denote $Q^C(\theta)$ the unique solution of equation (3.2). Condition 4 guarantees that $Q^C(\theta)$ is increasing in the state of the world and that the Cournot equilibrium price $p^C(\theta)=P(Q^C(\theta),\theta)$ is nondecreasing in the state of the world.

Suppose now producers have different marginal costs $c_n$ for $n=1, \ldots, N$. Producer $n$'s first-order condition is

$$P(Q,\theta)+q^n\partial_Q P(Q,\theta)=c_n \Leftrightarrow \frac{p(\theta)-c_n}{p(\theta)}=\frac{q^n(\theta)}{Q}\left(-\frac{Q}{P(Q,\theta)}\frac{\partial P(Q,\theta)}{\partial Q}\right)=\frac{s^n(\theta)}{\eta(Q,\theta)},$$

where $s^n=\frac{q^n}{Q}$ is firm $n$'s market share and $\eta(Q,\theta)=-\frac{\partial D(p,\theta)}{\partial p}\frac{P(Q,\theta)}{Q}$ is the elasticity of demand. This is inverse elasticity rule: in state $\theta$ firm $n$'s Lerner index $\left(\frac{p-c_n}{p}(\theta)\right)$ is proportional to her market share and inversely proportional to demand elasticity.

Cournot competition is extremely simple analytically. Furthermore, in the early 1980s, two game theorists proved that, if the demand function is concave and rationing of unserved customers is efficient, quantity competition can be reinterpreted as a two-stage game: in the first stage, firms choose their production capacity; then, in the second stage, they compete in price, given these capacity constraints (Kreps and Scheinkman 1983). For these reasons, Cournot competition is often used by economists. As mentioned earlier in this chapter, it is a legitimate representation of wholesale spot power markets since a centralized market sets the price. For example, Enron traders strategically reduced output à la Cournot during the California crisis.

## 3.5.2 Bertrand-Edgeworth Competition

### 3.5.2.1 Set-up

Another approach to capture imperfect competition is to represent price competition in the presence of capacity constraints. Consider, for example, a symmetric duopoly: for $i=1,2$, each firm has constant variable production cost $c_i$ and capacity $k_i$. Price-inelastic load $l$ is distributed on $[\underline{l},\bar{l}]$ following cumulative distribution function $G(.)$. Suppose that $\bar{l}>max(k_1,k_2)$: in some states of the world, both producers are required to serve demand, and that $\bar{l}\leq k_1+k_2$: demand is never rationed. Since demand is price inelastic, the SO imposes a price cap $\bar{p}>max(c_1,c_2)$.

Capacity-constrained price competition captures a realistic feature of power markets: in general, no single supplier is able to serve all demand alone. The discussion below draws heavily from Fabra et al. (2006), with one major simplification: I limit the analysis to symmetric producers, which enables me to derive the main results simply. I also omit the discussion of technical points (e.g., the continuity of the cumulative distribution function of bids in mixed-strategy equilibrium). The interested reader is referred to Fabra et al. (2006) for a richer and more comprehensive analysis.

Denote $c = c_1 = c_2$ the producers' marginal cost, and $k = k_1 = k_2$ their production capacity. If $l \leq k$, any single producer can serve the entire demand. Bertrand competition drives price to marginal cost and profits to zero.

In the rest of this analysis, suppose $k < l \leq 2k$. The equilibrium depends on the microstructure of the market. Fabra et al. (2006) consider a centralized market, where the SO runs an auction, and examine both an uniform price and a pay-as-bid auction.

#### 3.5.2.2 Dispatched quantities

Observe that $l \leq 2k \Leftrightarrow l - k \leq k$. In both auctions, the lowest bidder produces his entire capacity $k$, while his rival produces the remaining and smaller quantity $(l-k)$. In case of a tie, they share demand equally. Mathematically, the dispatched quantity for producer $i$ who bids $b_i$ while his rival bids $b_j$ is

$$q_i(b_i, b_j, l) = \begin{cases} k \; if \; b_i < b_j \\ \frac{l}{2} \; if \; b_i = b_j \\ l - k \; if \; b_i > b_j \end{cases} .$$

#### 3.5.2.3 Uniform price auction

In a uniform price auction, both players are paid the highest accepted price. The equilibria are characterized by the following:

**Theorem 3.1.** For the uniform price auction, there exists an infinity of asymmetric pure-strategy equilibria characterized by $b_i = \bar{p}$ and $b_j = b$, where $b \leq \hat{b} = c + (\bar{p} - c)\frac{l-k}{k}$.

There also exists a unique symmetric mixed-strategy equilibrium where bids are distributed on $[c, \bar{p}]$ according to the cumulative distribution function

$$\xi(b) = \left(\frac{b-c}{\bar{p}-c}\right)^{\frac{l-k}{2k-l}}.$$

***Proof*.** Consider first the pure-strategy equilibrium candidate. We have $\pi_i = (\bar{p} - c)(l - k) \leq (\bar{p} - c)k = \pi_j$. Suppose first $l < 2k$, hence $\frac{l-k}{k} < 1$ and $\hat{b} < \bar{p}$. Firm $i$ cannot offer a bid higher than the price cap $\bar{p}$. If it offers a lower bid, it has to offer $b_i = b - \varepsilon$ to produce the large quantity $k$. Its profit is then $(b - \varepsilon - c)k$, which is lower than $\pi_i$ since $(b - c)k \leqslant (\hat{b} - c)k = (\bar{p} - c)(l - k) = \pi_i$. The high bidder firm $i$ has no incentive to deviate. Consider the low bidder firm $j$. It has no incentive to reduce its bid, since it would reduce profits. If firm $j$ offers $b_j > \hat{b}$, firm $i$ profitably undercuts firm $j$ and receives $(b_j - c)k > (\hat{b} - c)k = \pi_i$. Firm $j$ profit is then $(b_j - c)(l - k) \leqslant (\bar{p} - c)(l - k) \leq (\bar{p} - c)k = \pi_j$, since $\frac{l-k}{k} < 1$. The low bidder firm $j$ has no incentive to deviate either. If $l = 2k$, one firm bids $b_i = \bar{p}$, and the other one bids any bid $b$.

Consider now a symmetric mixed-strategy equilibrium. Denote $\xi(.)$ the distribution of bids. Suppose producer $i$ bids $b_i = b$. If $b < b_j$, she produces the large quantity $k$ and receives price $b_j$; if $b > b_j$, which occurs with probability $\xi(b)$, she produces $(l-k)$ at price $b$. The case $b = b_j$ occurs with probability zero, hence is discarded. Thus

$$\pi^i(b) = \int_b^{\bar{p}} k(b_j - c)\xi'(b_j)db_j + (l-k)(b-c)\xi(b).$$

Since the firm plays a mixed-strategy equilibrium, it is indifferent between any bid on $(c, \bar{p})$, hence $\frac{d\pi^i}{db} = 0$, which yields

$$-k(b-c)\xi'(b) + (l-k)[(b-c)\xi'(b) + \xi(b)] = (l-k)\xi(b) - (2k-l)(b-c)\xi'(b) = 0.$$

Solving this first-order linear differential equation yields

$$\xi(b) = A(b-c)^{\frac{l-k}{2k-l}}.$$

Since $lim_{b \to \bar{p}}\xi(b) = 1$, we have $A = \frac{1}{(\bar{p}-c)^{\frac{l-k}{2k-l}}}$, hence the result. □

In any of the pure-strategy equilibria, both firms are paid the maximum price $\bar{p}$. One firm produces at full capacity, while the other, which bids higher, produces only the remaining output.

For the mixed-strategy equilibrium, the expected profit is $\pi(b) = \pi(\bar{p}) = (l-k)(\bar{p}-c)$: on average, firms are able to capture the maximum margin on the small quantity, a feature that occurs often when a residual quantity must be produced by the "losing" firm.

#### 3.5.2.4 Pay-as-bid auction

In this auction, each player is paid her bid. The equilibrium is characterized by the following:

**Theorem 3.2.** 3.2 No pure-strategy equilibrium exists. There exists a unique mixed-strategy equilibrium where bids are distributed on $[\underline{b}, \bar{p}]$ according to the cumulative distribution function

$$\xi(b) = \frac{k}{2k-l}\frac{b-\underline{b}}{b-c},$$

where

$$\bar{p} - \underline{b} = \frac{2k-l}{k}(\bar{p}-c).$$

***Proof.*** First, no asymmetric pure-strategy equilibrium exists: the lowest bidder could always profitably increase her bid. Second, consider a symmetric pure-strategy equilibrium $b$. If $b > c$, one producer can increase her profit by bidding $(b-\varepsilon) > c$ for $\varepsilon > 0$ arbitrarily small: the gain in market share exceeds the margin loss. This is the standard Bertrand argument. However, $b = c$ is not an equilibrium either: a firm can slightly increase its bid, to capture positive margin on the small quantity, hence positive profit.

A symmetric mixed-strategy equilibrium is characterized by its cumulative probability distribution $\xi(.)$ and its support $[\underline{b}, \bar{b}]$. Standard arguments show that $\underline{b} > c$ and $\bar{b} = \bar{p}$. Suppose producer $i$ bids $b_i = b$. His bid is the lowest, with probability $Pr[b_j \geq b] = (1 - \xi(b))$, in which case he produces his entire capacity $k$. Otherwise, he produces the residual quantity $(l - k)$. His expected profit is therefore

$$\pi(b) = [k(1 - \xi(b)) + (l - k)\xi(b)](b - c) = [k - (2k - l)\xi(b)](b - c).$$

By definition of a mixed-strategy equilibrium, a producer is indifferent between all bids. In particular,

$$\pi(b) = \pi(\underline{b}) \Leftrightarrow [k - (2k - l)\xi(b)](b - c) = k(\underline{b} - c) \Leftrightarrow \xi(b) = \frac{k}{2k - l}\frac{b - \underline{b}}{b - c}.$$

Then,

$$\xi(\bar{p}) = 1 \Leftrightarrow \bar{p} - \underline{b} = \frac{2k - l}{k}(\bar{p} - c).$$

Observe that

$$\pi(b) = \pi(\bar{p}) = (l - k)(\bar{p} - c).$$

The expected profit is identical in the uniform price and pay-as-bid auctions.

### 3.5.3 Supply-Function Equilibria

In centralized power markets, for example in the United States, producers bid a supply schedule, that is, a nondecreasing function $q(p)$ specifying the quantity $q$ the firm is willing to produce at price $p$. The appropriate strategic "variables" in an equilibrium are therefore neither quantities (Cournot), nor prices (Bertrand-Edgeworth); rather, they are supply functions.

The theory of supply-function equilibrium (SFE) was developed by Klemperer and Meyer (1989). It was applied to the power industry by Richard Green and David Newbery (1992), just after the opening of electricity markets in England and Wales. Green and Newbery (1992) is probably one of the most important academic articles concerning the power industry, clearly illustrating the potential for exercise of market power.

Supply-function equilibria are well suited to represent imperfect competition in power markets. However, as discussed below, no closed-form solution exists, except in a few simple cases. Furthermore, in general, multiple equilibria exist. Since these features limit their applicability, a large academic literature has derived conditions under which closed-form solutions exist or the set of equilibria can be narrowed down.

I present below the basic argument, following Green (1996), which provides a gentle introduction to SFEs. A rich and growing literature on SFEs exists. For the interested reader, Newbery (2008) offers a good summary of results, including stability and uniqueness of equilibria; Holmberg and Newbery (2010) presents policy implications with a particular focus on computations of losses in net surplus. More recently, Holmberg et al. (2013) provide an essential theoretical advance, showing how step-functions SFEs, which accurately represent bids in wholesale power markets, converge toward continuous SFEs, which are the solution commonly used.

Demand in state $\theta$ is $D(p,\theta)$. $N$ producers compete in a centralized market by submitting a nondecreasing continuous function $q^n(p)$, which specifies the quantity $q^n$ they are willing to supply at price $p$. The functions do not depend on $\theta$: producers submit bids in the day-ahead market, before all the information on real-time supply and demand conditions has been revealed. The distribution of states $\theta$ in this context is thus slightly different from the rest of this text, as it includes only the day-ahead uncertainty. We continue to assume that $\partial_p D(p,\theta) < 0$ and $\partial_\theta D(p,\theta) > 0$. We also assume that the slope of the demand curve is constant across states of the world—$\partial_{p\theta} D(p,\theta) = 0$—but may vary with output, that is, $\partial_{pp} D(p,\theta)$ may be different from zero.

Producers have increasing and (weakly) convex total cost $C^n(q^n)$, which allows for different specifications.

For price $p$ in state $\theta$, producer $n$ faces residual demand

$$R^n(p,\theta) = D(p,\theta) - \sum_{m \neq n} q^m(p).$$

Assuming the firm produces exactly this residual demand, its profit is

$$\pi^n(p,\theta) = p[D(p,\theta) - \sum_{m \neq n} q^m(p)] - C^n(D(p,\theta) - \sum_{m \neq n} q^m(p)).$$

An equilibrium is a set of functions $\{q^n(p)\}_n$ such that the function $q^n(p)$ is the best response to the other functions $\{q^m(p)\}_{m \neq n}$. The direct approach would be to apply variational calculus and characterize $q^n(p)$ as the solution of Euler equation.

An indirect approach is to find the profit-maximizing price in state $\theta$. We have

$$\frac{\partial \pi^n}{\partial p} = [D(p,\theta) - \sum_{m \neq n} q^m(p)] + [p - C^{n'}(D(p,\theta) - \sum_{m \neq n} q^m(p))] \times \left[ \partial_p D(p,\theta) - \sum_{m \neq n} \frac{dq^m}{dp} \right].$$

Assuming $D(p,\theta)$ is well behaved, $\frac{\partial \pi^n}{\partial p} = 0$ defines a unique profit-maximizing price, hence output for firm $n$. Rearranging the first-order condition yields

$$q^n(p) = [p - C^{n'}(q^n(p))] \left[ \sum_{m \neq n} \frac{dq^m}{dp} - \partial_p D(p,\theta) \right]. \tag{3.3}$$

Since $\partial_{p\theta} D(p,\theta)=0$, the output $q^n(p)$ defined by equation (3.3), does not depend on the state of the world $\theta$. The pair $(p, q^n(p))$ forms a point on the profit-maximizing supply function of firm $n$.

Therefore, the SFEs are characterized by a set of $N$-related differential equations, that is, every supply function $q^n(p)$ is a function all the other supply functions $\{q^{n'}(p)\}_{n' \neq n}$. As shown below, specific results can be obtained for specific set of assumptions.

**The simplest case** The simplest case is a symmetric $N$-firm oligopoly, with linear demand with constant slope $D(p,\theta)=\alpha(\theta)-\beta p$ and constant marginal cost $c \geq 0$. Assume the equilibrium is symmetric, that is, all producers bid the same supply function.

Following Newbery (2008), equation (3.3) becomes

$$q=[p-c]\left[(N-1)\frac{dq}{dp}+\beta\right] \Leftrightarrow (N-1)\frac{dq}{dp}-\frac{q}{p-c}=-\beta.$$

For $N=2$, the general solution of the associated homogeneous equation is

$$q(p)=Ae^{\int \frac{dx}{x-c}}=A(p-c),$$

where $A$ is an integration constant. Applying the variation-of-parameters method, we search for a particular solution of the nonhomogeneous equation $y(p)=\kappa(p)(p-c)$. We must have

$$\kappa'(p)=-\frac{\beta}{p-c} \Leftrightarrow \kappa(p)=-\beta ln(p-c) \Leftrightarrow y(p)=-\beta(p-c)ln(p-c).$$

Therefore, there exist an infinity of equilibria supply functions for $N=2$, given by

$$q(p)=(A-\beta ln(p-c))(p-c).$$

The constant $A$ depends on how competitive the industry is.

For $N>2$, the general solution of the associated homogeneous equation is

$$q(p)=Ae^{\int \frac{dx}{(N-1)(x-o)}}=A(p-c)^{\frac{1}{N-1}},$$

where $A$ is an integration constant. A particular solution of the nonhomogeneous equation is

$$y(p)=-\beta\frac{p-c}{N-2},$$

hence there exist an infinity of equilibria supply functions for $N>2$, given by

$$q(p)=A(p-c)^{\frac{1}{N-1}}-\beta\frac{p-c}{N-2}.$$

This simplest example illustrates the challenge in deriving SFEs.

## 3.6 Imperfect Competition at the Investment Stage

Economic analysis of imperfectly competitive investment in the power industry is more recent and starts in the mid-2000s. This reflects changing concerns by policy makers and academics. When the power industry was restructured in the 1990s, there was excess capacity in most markets. In fact, the presence of excess capacity was used by reformers as evidence of the failure of regulation, hence a justification for reform. Thus the performance of the spot market, and the ability of agents to manipulate these markets, were the primary focus of policy makers' and academics' attention.

By the late 1990s, excess capacity had disappeared in most markets, and capacity adequacy became a concern for policy makers. Economists then developed models of investment. Most are structured as two-stage games under uncertainty: firms choose their capacity in the first stage, at which point demand that will prevail is still undetermined. Then firms compete in the spot market, constrained by their capacity.

With the benefit of hindsight, it seems almost strange not to include this investment stage: as discussed in chapter 2, producers' operating margin under perfect competition is generated exclusively on-peak when they produce at capacity. This gives them strong incentive to reduce investment to increase the number of on-peak hours. Furthermore, since investment is observable by all competing firms, collusion is easier to sustain. Proving collusion is more difficult: firms can claim that demand growth turned out higher than expected. Despite this, models of imperfectly competitive investment in the power industry did not appear in the academic literature until 2005 and later, more than fifteen years after the onset of industry restructuring.

This section presents the simple and elegant model of imperfectly competitive investment followed by spot-market competition developed by Gregor Zöttl (2011). An attractive feature of this model is its intuitive convergence toward the perfectly competitive benchmark when the number of firms tends toward infinity. We then discuss briefly other models.

### 3.6.1 Cournot Competition with Uncertain Demand

As before, $N$ producers compete. They now play a two-stage game: in stage 1, firm $n$ installs capacity $k^n$; in stage 2, it produces $q^n(\theta) \leq k^n$ in state $\theta$ and sells it entirely on the spot market. Firms compete à la Cournot in the spot markets, facing inverse demand $P(Q, \theta)$ and constrained by their capacity $k^n$. Stage 2 can be interpreted as a repetition of multiple states of the world over a given period (e.g., one year), drawn from the distribution $F(.)$. All customers respond to wholesale spot prices, that is, $\alpha = 1$, using chapter 3's notation.

The game is solved by backward induction: firms first compute spot-market profits from a Nash equilibrium in each state of the world $\theta$, given installed capacities $(k^1, \ldots, k^N)$; then they make their investment choice in stage 1 based on the expectation of these spot-market profits.

$Q(\theta) = \sum_{n=1}^{N} q^n(\theta)$ and $K = \sum_{n=1}^{N} k^n$ are the aggregate production in state $\theta$ and aggregate installed capacity.

#### 3.6.1.1 Off- and on-peak spot-market equilibrium

Two mutually exclusive situations must be considered in describing the structure of the spot-market equilibrium:

• Off-peak, equilibrium output is lower than installed capacity and firms play a standard Cournot game with marginal cost $c$. Aggregate production in state $\theta$ is the (unconstrained) Cournot output, denoted $Q^C(\theta)$, as previously defined by equation (3.2).

• On-peak, equilibrium output is equal to installed capacity; No firm can produce beyond capacity; hence no upward deviation is possible. A downward deviation is not profitable, either: since capacity is lower than the (unconstrained) Cournot output, impact of the volume reduction exceeds the impact of price increase. Wholesale price is $P(K,\theta)$, and individual profit is $\frac{K}{N}(P(K,\theta)-c)$.

The Cournot critical state of the world, denoted $\hat{\theta}_N^C(K,c)$, is the lowest state of the world such that the (unconstrained) Cournot output is equal to or larger than the aggregate capacity $K$. Since the marginal revenue is decreasing,

$$K \leq Q^C(\theta) \Leftrightarrow P(K,\theta) + \frac{K}{N}\partial_Q P(K,\theta) \geq c.$$

If $P(K,\theta) + \frac{K}{N}\partial_q P(K,\theta) \geq c \forall \theta \geq 0$, then $\hat{\theta}_N^C(K,c) = 0$. Otherwise, $\hat{\theta}_N^C(K,c)$ is uniquely defined by

$$P(K,\hat{\theta}_N^C(K,c)) + \frac{K}{N}\partial_Q P(K,\hat{\theta}_N^C(K,c)) = c.$$

Condition 4 guarantees that $\hat{\theta}_N^C(K,c)$ is increasing in both arguments, and all states $\theta \geq \hat{\theta}_N^C(K,c)$ are on-peak.

Finally, observe that $lim_{N\to+\infty}\hat{\theta}_N^C(K,c) = \hat{\theta}_0(K,c)$: as $N$ tends to infinity, competition becomes perfect. The lowest on-peak state of the world is $\hat{\theta}_0(K,c)$ as was the case in chapter 2.

#### 3.6.1.2 Equilibrium investment

We then have:

**Proposition 3.1.** There exists a unique symmetric equilibrium capacity $K^C$ characterized by

$$\Psi_N^C(K^C,c) = \int_{\hat{\theta}_N^C(K^C,c)}^{+\infty} (P(K^C,\theta) + \frac{K^C}{N}\partial_Q P(K^C,\theta) - c) f(\theta) d\theta = r. \quad (3.4)$$

***Proof.*** The proof can be found in Zöttl (2011) and proceeds in three steps. First, Zöttl (2011) proves that no asymmetric equilibrium exists and that, if a symmetric equilibrium exists, it is characterized by equation (3.4). Second, he proves existence and unicity of a solution of equation (3.4). Finally, he proves that the solution is indeed an equilibrium, that is, that no unilateral deviation yields higher profit. □

A marginal increase in capacity yields marginal value $\Psi_N^C(K^C, c)$. To see that, observe first that increasing capacity has no impact off-peak; hence the marginal value of capacity includes only on-peak states of the world. Second, a capacity increase $\delta K$ increases (expected) operating margin by $\left(\int_{\hat{\theta}_N^C(K^C,c)}^{+\infty}(P(K,\theta)-c)f(\theta)d\theta\right)\delta K$. However, it also reduces the equilibrium price, hence the (expected) operating margin on the entire capacity by $\left(\int_{\hat{\theta}_N^C(K^C,c)}^{+\infty}\frac{K}{N}\partial_Q P(K,\theta)f(\theta)d\theta\right)\delta K$. Summing these two effects yields the marginal value. At the equilibrium, this marginal value is precisely equal to the capacity cost. This is illustrated in figure 3.2.

Equation (3.4) mirrors equation (3.2): firms internalize the impact of their decision—capacity in equation (3.4), output in equation (3.2)—on the equilibrium price.

In this double Cournot model, firms exercise their market power in the off-peak spot markets by reducing output. Since output reduction is not profitable for them on-peak, they do not exercise their market power in the on-peak spot markets. Rather, they exercise their market power at the investment stage: they underinvest, thus increasing the probability of on-peak states of the world.

**Corollary 3.1.** If inverse demand is linear with constant slope, equilibrium investment is

$$K^C(N) = \frac{N}{N+1}K^*,$$

where $K^*$ is the optimal capacity defined by equation (2.2).

***Proof.*** Remember from section 2.7 that, when demand is linear with constant slope,

$$\Psi_0(K,c) = \int_{\hat{\theta}_0(K,c)}^{+\infty} a'(\theta)(1-F(\theta))d\theta,$$

where $\hat{\theta}_0(K,c)$ is uniquely defined by $a(\hat{\theta}_0(K,c)) - c = bK$.

Since demand is linear with constant slope, integrating by parts yields

$$\Psi_N^C(K^C,c) = \int_{\hat{\theta}_N^C(K^C,c)}^{+\infty} a'(\theta)(1-F(\theta))d\theta$$

and algebraic manipulation proves that $\hat{\theta}_N(K,c)$ is uniquely defined by

$$a(\hat{\theta}_N^C(K,c)) - c = \frac{N+1}{N}bK \Leftrightarrow \hat{\theta}_N^C(K,c) = \hat{\theta}_0\left(\frac{N+1}{N}K, c\right).$$

Therefore

$$\Psi_N^C(K,c) = \Psi_0\left(\frac{N+1}{N}K, c\right).$$

Thus equilibrium capacity $K^C$ is uniquely defined by

$$\Psi_N^C(K^C,c) = \Psi_0\left(\frac{N+1}{N}K^C, c\right) = r \Leftrightarrow K^C = \frac{N}{N+1}K^*.$$

□

If demand is linear with constant slope, equilibrium capacity can be simply and intuitively derived from the optimal capacity. We verify immediately that $lim_{N\to+\infty}$ $K^C(N) = K^*$.

### 3.6.2 Alternative Models of Investment

The double Cournot model developed by Zöttl (2011) produces a very simple and elegant characterization of equilibrium aggregate capacity when strategic underinvestment is possible. This section discusses extension of this model and other models of strategic investment.

**Extension to the double Cournot model** A precursor article, Murphy and Smeers (2005) examines a double Cournot game, similar to Zöttl (2011).

Zöttl (2011) extends his analysis to multiple production technologies. When a finite number of production technologies is available, Zöttl (2011) proves that, if the costs satisfy a sufficient condition, the equilibrium capacity for technology $n < N$ is the unique solution of

$$\Psi_n^C(K_n^C, c_n) - \Psi_n^C(K_n^C, c_{n+1}) = r_n - r_{n+1}, \tag{3.5}$$

where the function $\Psi_n^C(K,c)$ is defined by

$$\Psi_n^C(K,c) = \int_{\hat{\theta}_N^C(K,c)}^{+\infty} (P(K,\theta) + \frac{K}{N}\partial_Q P(K,\theta) - c) f(\theta)d\theta.$$

Equation (3.5) is the equivalent of equation (2.12) when $N$ producers play a double Cournot game. It converges toward equation (2.12) when $N \to +\infty$. Zöttl (2011) also extends the analysis to a continuum of technologies.

**Other static models** Fabra et al. (2011) expands the analysis, developed in Fabra et al. (2006) and presented above, to include an investment stage in the game.

**Dynamic models** All models presented above consider a two-period game: firms first decide on their installed capacity, then participate in the spot markets, given their installed capacity. In reality of course, generation assets last for more than one period; hence the proper analysis requires a dynamic model. Garcia and Shen (2010) extend the Fabra et al. (2011) model to multiple periods.

## 3.7 Impact of Price Cap on Investment and Net Surplus

The double Cournot model developed by Zöttl (2011) lends itself to the analysis of the impact of price cap. The analysis presented here supposes that all customers face and react to spot prices, as in Zöttl (2011). It is derived from Léautier (2018).

### 3.7.1 Price Cap

To limit the exercise of market power, policy makers impose a cap $\bar{p}$ on the wholesale spot prices. In power markets, for example, the price cap may be a formal cap or the result of operational practices that depress prices (see Joskow 2007).

Most results presented below are driven by the fact that admissible prices caps lie in an interval bounded below and above. The price cap must be higher than $(c+r)$, the long-term marginal cost of production; otherwise it would block any investment. At the other extreme, an effective price cap must be binding with positive probability. The highest admissible price cap, denoted $p_{\infty}$, is therefore the highest price achieved. Thus, the set of admissible price caps is the segment $[(c+r), p_{\infty}]$.

The highest price achieved depends on the installed capacity, which, in turn, depends on the price cap. The highest admissible cap is thus a fixed point. From Condition 4 inverse demand is bounded, that is, $P_{\infty}(Q) = \lim_{\theta \to +\infty} P(Q, \theta)$ exists and is finite for all $Q \geq 0$. From the properties of $P(Q, \theta)$, $P_{\infty}(Q)$ is continuous and decreasing. As shown below, $K^C(\bar{p})$, the aggregate Cournot capacity when the price cap is $\bar{p}$, is a continuous function of $\bar{p}$. Thus for all $\bar{p} \geq (c+r)$, $P_{\infty}(K^C(\bar{p})) \in [0, P_{\infty}(0)]$. The Brouwer theorem then guarantees existence of a fixed point $p_{\infty}$ such that $P_{\infty}(K^C(p_{\infty})) = p_{\infty}$.

The lowest state of the world for which the inverse demand for output $Q$ is equal to or larger than price cap $\bar{p}$ is denoted $\hat{\theta}_0(Q, \bar{p})$. If the cap $\bar{p}$ and output $Q$ are both small, we may have $P(Q, \theta) > \bar{p}\ \forall \theta \geq 0$; hence $\hat{\theta}_0(Q, \bar{p}) = 0$. Otherwise, $\hat{\theta}_0(Q, \bar{p})$ is uniquely defined by

$$P(Q, \hat{\theta}_0(Q, \bar{p})) = \bar{p}.$$

Since inverse demand is increasing in the state of the world, once the cap is reached for $\hat{\theta}_0(Q, \bar{p})$, it is binding for all states $\theta \geq \hat{\theta}_0(Q, \bar{p})$.

### 3.7.2 Prelude: Price Cap under Perfect Competition

Suppose the SO imposes a cap $\bar{p} \in [(c+r), p_{\infty}]$, although competition is perfect.

#### 3.7.2.1 Perfectly competitive investment

The net surplus for capacity $K$ and price cap $\bar{p}$ in state $\theta$ is $w(K, \bar{p}, \theta) = \gamma(K, \bar{p}, \theta)$ $(S(\bar{p}, \theta) - cD(\bar{p}, \theta))$, and the expected net surplus is

$$W_0(K, \bar{p}) = \mathbb{E}[w(K, \bar{p}, \theta)] - rK.$$

Off-peak, price is equal to the variable cost of production $c$, hence $w(K, \bar{p}, \theta) = S(c, \theta)$ $-cD(c, \theta)$. On-peak starts when demand for price $c$ reaches capacity, that is, for $\theta = \hat{\theta}_0(K, c)$. As long as the cap is not binding, $w(K, \bar{p}, \theta) = S(P(K, \theta), \theta) - cK$. The price cap becomes binding for $\theta \geq \hat{\theta}_0(K, \bar{p})$. Rationing occurs and $w(K, \bar{p}, \theta) = \gamma(S(\bar{p}, \theta) - cD(\bar{p}, \theta))$. Since by construction $\gamma D(\bar{p}, \theta) = K \Leftrightarrow \gamma = \frac{K}{D(\bar{p}, \theta)}$, $w(K, \bar{p}, \theta) = K\left(\frac{S(\bar{p}, \theta)}{D(\bar{p}, \theta)} - c\right) = K(v(\bar{p}, \theta) - c)$. Pulling the pieces together,

$$\begin{aligned} W_0(K, \bar{p}) = &\int_0^{\hat{\theta}_0(K,c)} (S(c, \theta) - cD(c, \theta)) f(\theta) d\theta \\ &+ \int_{\hat{\theta}_0(K,c)}^{\hat{\theta}_0(K,\bar{p})} (S(P(K, \theta), \theta) - cK) f(\theta) d\theta \\ &+ \int_{\hat{\theta}_0(K,\bar{p})}^{+\infty} K(v(\bar{p}, \theta) - c) f(\theta) d\theta - rK(\bar{p}). \end{aligned}$$

Since the surplus $w(K, \bar{p}, \theta)$ is continuous in $\theta$, the marginal value of capacity is

$$\bar{\Psi}_0(K, \bar{p}, c) = \int_{\hat{\theta}_0(K,c)}^{\hat{\theta}_0(K,\bar{p})} (P(K, \theta) - c) f(\theta) d\theta + \int_{\hat{\theta}_0(K,\bar{p})}^{+\infty} (v(\bar{p}, \theta) - c) f(\theta) d\theta. \quad (3.6)$$

Off-peak, firms produce below capacity; hence increasing capacity has no value. On-peak, before the cap is binding, a marginal Megawatt of installed capacity generates net surplus $(P(K, \theta) - c)$ in state $\theta$. When the cap is binding, the value of a marginal Megawatt-hour is the VoLL; hence a marginal Megawatt of installed capacity generates net surplus $(v(\bar{p}, \theta) - c)$ in state $\theta$. Expression (3.6) is similar to expression (2.10): the marginal value of capacity in state $\theta$ is $(P(K, \theta) - c)$ before rationing, and (VoLL$-c$) when rationing occurs. In expression (3.6), the imposition of an exogenous price cap $\bar{p}$ causes rationing, hence VoLL $= v(\bar{p}, \theta)$; while in expression (2.10) the presence of customers facing the constant price $p^R$ causes rationing, hence VoLL $= v(p^R, \theta)$. If the price cap is never binding, $\hat{\theta}_0(K, \bar{p}) \to +\infty$, and expression (3.6) converges toward expression (2.1).

Derivations similar to those presented in section 2.3 prove that $\bar{\Psi}_0(K, \bar{p}, c)$ is decreasing in capacity $K$; hence net surplus is concave as a function of capacity $K$. The equilibrium-installed capacity $K^*(\bar{p})$ is uniquely defined by $\bar{\Psi}_0(K^*(\bar{p}), \bar{p}, c) = r$.

#### 3.7.2.2 Impact of price cap on perfectly competitive investment

The expected net surplus if competition is perfect is $\mathcal{W}_0(\bar{p}) = W_0(K^*(\bar{p}), \bar{p})$. We have:

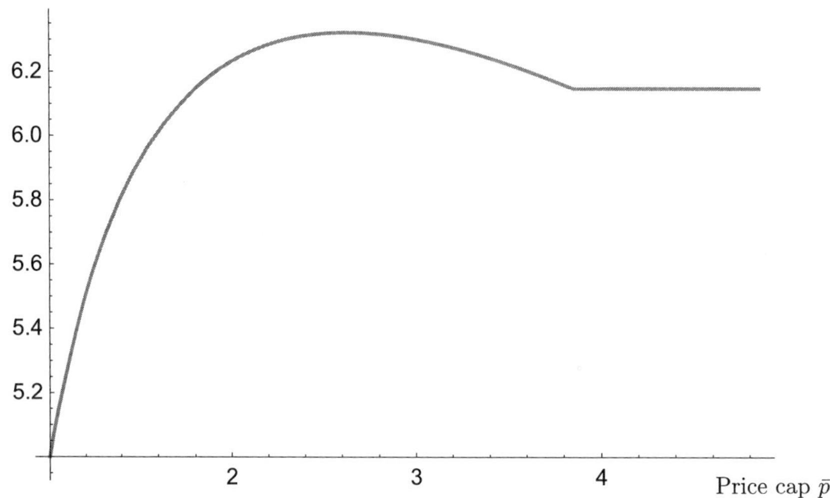


**Figure 3.3**
Equilibrium capacity $K^C(\bar{p})$ if the price cap is always reached on-peak, for $\bar{p} \in [(c+r), (p_\infty + 1)]$. The capacity-maximizing cap is $\hat{p} = 2.6$, and $p_\infty = 3.9$. Equilibrium capacities for different values of the cap are $bK^C(c+r) = 5$, $bK^C(\hat{p}) = 6.3$, and $bK^C(p_\infty) = 6.2$.
*Source:* Léautier (2018).

**Proposition 3.2.** Suppose the SO imposes a price cap $\bar{p} \in [(c+r), p_\infty]$ and competition is perfect. Increasing the price cap increases equilibrium capacity $K^*(\bar{p})$ and expected net surplus $\mathcal{W}_0(\bar{p})$.

***Proof.*** Applying the implicit function theorem to the condition $\Psi_0(K^*(\bar{p}), \bar{p}, c) = r$ yields

$$\frac{dK^*}{d\bar{p}} = \frac{1 - F(\hat{\theta}_0(K^*(\bar{p}), \bar{p}))}{-\int_{\hat{\theta}_0(K^*(\bar{p}),c)}^{\hat{\theta}_0(K^*(\bar{p}),\bar{p})} \partial_Q P(K^*(\bar{p}), \theta) f(\theta) d\theta} > 0,$$

which proves the first point.

Then $\frac{d\mathcal{W}_0}{d\bar{p}} = \frac{\partial W}{\partial K} \times \frac{dK^*}{d\bar{p}} + \frac{\partial W}{\partial \bar{p}} = \frac{\partial W}{\partial \bar{p}}$ since $\frac{\partial W}{\partial K}(K^*(\bar{p}), \bar{p}) = 0$ by definition of $K^*(\bar{p})$. Since the surplus is continuous as a function of the state of the world $\theta$, only the terms in the integrand appear in $\frac{\partial W}{\partial \bar{p}}$; hence $\frac{d\mathcal{W}_0}{d\bar{p}} = \int_{\hat{\theta}_0(K,\bar{p})}^{+\infty} \frac{\partial w}{\partial \bar{p}}(K, \bar{p}, \theta) f(\theta) d\theta$. Then,

$$\frac{\partial w}{\partial \bar{p}} = K \frac{\partial v(\bar{p}, \theta)}{\partial \bar{p}} = K \frac{D \frac{\partial S}{\partial \bar{p}} - S \frac{\partial D}{\partial \bar{p}}}{D^2} = \frac{K}{D} \left( \frac{\partial S}{\partial \bar{p}} - \frac{S}{D} \frac{\partial D}{\partial \bar{p}} \right) = \gamma (\bar{p} - v(\bar{p}, \theta)) \frac{\partial D}{\partial \bar{p}} > 0,$$

hence $\frac{d\mathcal{W}_0}{d\bar{p}} > 0$, which proves the second point.

Increasing the price cap has an unambiguous impact on perfectly competitive equilibrium capacity and net surplus.

Both results are intuitive: a price cap reduces profit, hence investment incentives. Raising the cap thus raises investment incentives. Relatedly, a price cap leads to a loss of net surplus arising from rationing. Raising the cap reduces this loss. This confirms that, regulators should consider setting a price cap if and only if competition is imperfect. Imposing no cap leads to higher equilibrium capacity and net surplus. Under reasonable conditions, these results do not hold under imperfect competition.

### 3.7.3 Imperfectly Competitive Equilibrium Investment

Under perfect competition, the off-peak price is equal to the variable cost of production $c$; hence generation produces at capacity before any cap $\bar{p} \geq c + r$ is reached. If competition is imperfect, off-peak prices exceed the variable cost of production; hence prices may reach the cap before generation produces at capacity. If demand is very inelastic and competition highly imperfect ($N$ small), a price cap close to $(c + r)$ is binding off-peak. The cap is reached on-peak if and only if

$$\hat{\theta}_N^C(K, c) \leq \hat{\theta}_0(K, \bar{p}).$$

Léautier (2018) derives the marginal value of capacity $Q^c(K, \bar{p}, c)$ which is defined piecewise in the $(\bar{p}, K)$ plane, and proves that for all $\bar{p} \in [(c + r), p_\infty]$, there exists a unique symmetric-equilibrium capacity $K^C(\bar{p})$ characterized by $\bar{\Psi}_N^C(K^C(\bar{p}), \bar{p}, c) = r$. If the price cap is always reached on-peak, Zöttl (2011) proves that

$$\begin{aligned}\bar{\Psi}_N^C(K, \bar{p}, c) &= \bar{\Psi}_2(K, \bar{p}, c)\\ &= \int_{\hat{\theta}_N^C(K,c)}^{\hat{\theta}_0(K,\bar{p})} \left( P(K, \theta) + \frac{K}{N} \partial_Q P(K, \theta) - c \right) f(\theta) d\theta\\ &\quad + \int_{\hat{\theta}_0(K,\bar{p})}^{+\infty} (\bar{p} - c) f(\theta) d\theta.\end{aligned}$$

Intuition for $\bar{\Psi}_2(K, \bar{p}, c)$ is the following:

A marginal capacity increase has no impact off-peak. On-peak, it generates additional profit $(P(K, \theta) - c)$ on the marginal unit. If the cap is not binding, it also yields a price reduction on all inframarginal units, hence the net effect is $\left(P(K, \theta) + \frac{K}{N} QP_q(K, \theta) - c\right)$.

When the cap is binding, there is no price reduction, and the net effect is $(\bar{p} - c)$. This yields $\bar{\Psi}_2(K, \bar{p}, c)$. Consider the following simple specification: inverse demand (Léautier 2018) proves that, for this specification, the price cap is always reached on-peak. Equilibrium capacity $K^c(\bar{p})$ is presented in figure 3.4.

### 3.7.4 Capacity-Maximizing Cap

Zöttl (2011) derives a sufficient condition for the existence of at least one interior capacity-maximizing cap: Intuition for result can be obtained by applying the implicit function theorem to condition $\Psi_N^C(K^C(\bar{p}), \bar{p}) = 0$:

$$\frac{dK^C}{d\bar{p}} = \frac{1 - F(\hat{\theta}_0) + \frac{K^C}{N}\partial_q P(K^C, \hat{\theta}_0) f(\hat{\theta}_0) \frac{\partial \hat{\theta}_0}{\partial \bar{p}}}{-\frac{\partial \Psi_2}{\partial K}} \; if \; \Psi = \Psi_2, \tag{3.7}$$

The denominator on the right-hand side is positive, the first term of the numerator is positive, and the second negative. These two terms illustrate the two impacts of a price cap on installed capacity: on the one hand, increasing the cap raises the per unit profit when the cap is binding, which occurs with probability $\left(1 - F\left(\hat{\theta}_0\right)\right)$, hence increases investment incentives. On the other hand, the marginal value of capacity is discontinuously higher after the cap is reached than when the price is close but not quite equal to the cap. Increasing the cap reduces the probability the cap is binding (by a factor $f\left(\hat{\theta}_0\right) \frac{\partial \hat{\theta}_0}{\partial \bar{p}}$), hence reduces the strength of this second effect and the marginal value of capacity.

The discontinuity in the marginal value of capacity can be explained as follows. Suppose first the cap is not binding. If a marginal unit of capacity is added, it receives positive operating margin. On the other hand, since installed capacity increases, market price and operating margin decrease for all capacity units. The impact of this marginal unit of capacity on the value of installed capacity is the sum of both effects. This argument is similar to monopoly pricing or standard Cournot competition: increasing output generates positive margin on the marginal unit, but reduces the margin on all inframarginal units.

Suppose now the cap is binding. If a marginal unit of capacity is added, it receives positive operating margin. The operating margin on all capacity units is unchanged, since adding capacity has no impact on price, which is set at the cap. Thus, the value of a marginal unit of capacity is discontinuously higher when the cap is binding than when it is not.

The first term dominates for price caps close to (c+r), when $\hat{\theta}_0(K^C(\bar{p}), \bar{p})$ is low; hence $(1 - F(\hat{\theta}_0))$ is close to 1. The second dominates for high-price caps, when $\hat{\theta}_0(K^C(\bar{p}), \bar{p})$ is high; hence $(1 - F(\hat{\theta}_0))$ is close to 0. Therefore, $lim_{\bar{p} \to p_\infty} \frac{dK^C}{d\bar{p}} < 0$; hence a cap set at $p_\infty$ does not maximize equilibrium capacity.

### 3.7.5 Net Surplus-Maximizing Cap

The expected net surplus under imperfect competition is $W^C(K, \bar{p})$. It differs from $W_0(K, C)$ since, under imperfect competition, the first on-peak state of the world is no longer $\hat{\theta}(K, C)$. The net surplus under imperfect competition in equilibrium for price cap $\bar{p}$ is $\mathcal{W}^C(\bar{p}) = W^C(K^C(\bar{p}), \bar{p})$. Since $\mathcal{W}^C(\bar{p})$ is continuous on the expected compact set $[(c+r), p_\infty]$, at least one surplus-maximizing cap $p^*$ exists.

We have

$$\frac{d\mathcal{W}^C}{d\bar{p}} = \frac{\partial W^C}{\partial \bar{p}} + \frac{\partial W^C}{\partial K} \times \frac{dK^C}{d\bar{p}}.$$

An increase in the price cap has two impacts on net surplus: a direct impact, through the change in rationing, and an indirect impact, through the change in equilibrium capacity. Under reasonable assumptions, the direct impact is positive. Then, the surplus-maximizing cap p* is higher than the highest capacity-maximizing cap $\hat{p}$, since $\frac{d\mathcal{W}^C}{d\bar{p}}(\hat{p}) = \frac{\partial W^C}{\partial \bar{p}}(\hat{p}) > 0$. Thus to maximize net surplus, market designers have to accept a reduction in equilibrium capacity.

Léautier (2018) derives sufficient conditions for the existence of a surplus-maximizing cap strictly lower than $p_\infty$:

**Proposition 3.3.** If inverse demand is either linear with constant slope or isoelastic, there exists a surplus-maximizing cap strictly lower than $p_\infty$. Proposition 3.3 appears to justify the use of price caps: a regulator can increase net surplus by selecting optional price cap.

However, a simple numerical example constructed in Léautier (2018) and presented in figure 3.4 cautions against underoptimism.

Figure 3.4 shows that the expected net surplus continues to increase after equilibrium capacity has peaked. Surplus-maximizing cap is $p^* = 3.6$, 36 percent higher than $\hat{p}$.

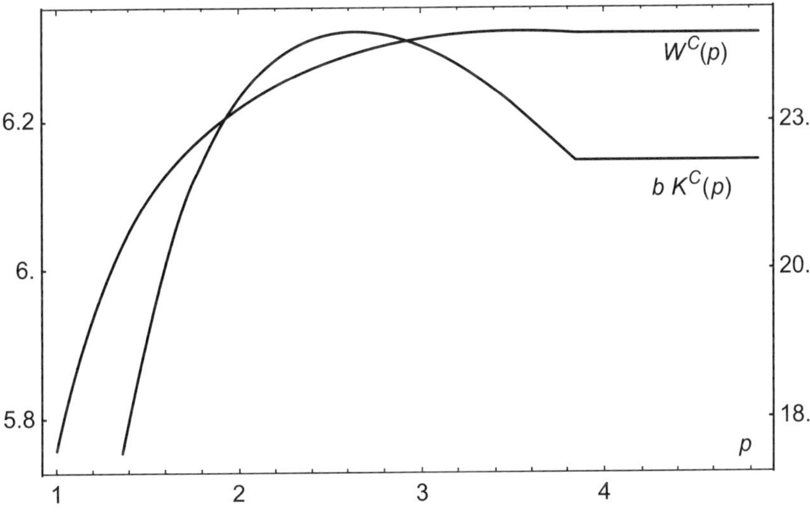


**Figure 3.4**
Equilibrium capacity $bK^C(\bar{p})$ (left scale) and expected net surplus $\mathcal{W}^C(\bar{p})$ (right scale) if the price cap is always reached on-peak, for $\bar{p} \in [(c+r), (p_\infty + 1)]$. The surplus-maximizing cap is $p^* = 3.6$, higher than the capacity-maximizing cap $\hat{p} = 2.6$, and close to the highest admissible cap $p_\infty = 3.9$.
*Source:* Léautier (2018).

Maximum surplus is $\mathcal{W}^C(p^*) = 24.08$, 1.5 percent higher than $\mathcal{W}^C(\hat{p})$. Optimal equilibrium capacity is 1.8 percent lower than maximum capacity.

Figure 3.4 also shows that the expected net surplus is almost "flat" for high values of the price cap. The surplus-maximizing cap $p^*$ is 92.7 percent of the maximum admissible cap $p_\infty$, and the maximum surplus $\mathcal{W}^C(p^*)$ is only 0.1 percent higher than $\mathcal{W}^C(p_\infty) = 24.05$.

This finding weakens the appeal of price cap as policy instruments: the surplus-maximizing cap is very close to the highest admissible cap, and leads to a very small surplus increase compared to no price cap. Policy makers are not interested in setting a seldom-binding cap, hence a price cap low enough to be politically acceptable reduces net surplus. Testing whether a similar result holds for other demand functions and other methods of rationing is an important avenue for further research.

# II RETAIL MARKET

# 4 Selling Power to the People: Retail Competition and Vertical Structure of the Power Industry

## 4.1 Introduction

Electricity retailing is an important activity. This may come as a surprise to most observers, since it (mostly) involves purchasing a commodity on wholesale markets and selling it to retail customers. The retail price is largely determined by the wholesale price, regulated network charges, and taxes. Retailing is considered to be highly competitive, hence (until today) represents only a small fraction of the value of the electric power industry.[1]

Yet for most residential customers, their electric power bill is their only contact with the electric power industry. Their experience with their electricity retailers defines their view on the electric power industry. They are neither aware nor remotely interested in the intricacies of wholesale market design. If they feel their retailer is competent, they support industry restructuring. Otherwise, if their quality of service decreases or if retail price increases, they do not. Since retail customers are also voters, the performance of retail markets matters tremendously to policy makers, despite its small share of the industry's value.

This chapter examines two issues related to retail competition. First, a sizable share of customers are not engaged and do not attempt to switch suppliers, or even to seek for a better rate from the same supplier. An activity that ought to be intensely competitive is therefore only mildly so. How should policy makers respond? Section 4.2 presents empirical evidence of customers' lack of activity to date and discusses possible policy interventions.

Second, significant policy attention has been devoted to the vertical structure of the industry: Should retailers be allowed to own generation assets, hence be vertically integrated? Should producers be allowed to sell a significant share of their output in forward markets and retailers be allowed to purchase forward a significant share of their load? Section 4.3 examines these issues. It concludes that forward commitments such as vertical

1. This statement applies to electricity retailing. Leveraging the customer relationship to sell services may be profitable, but lies outside this book's scope.

integration and forward sales have an ambiguous impact on net surplus: they have benefits, but they also can be detrimental. Policy makers should not ban them outright; rather, they should carefully monitor their impact on industry performance.

Section 4.4 completes section 4.2 by introducing economic models of imperfect competition caused by customers' differentiating between suppliers. Section 4.5 introduces economic models of forward commitment.

## 4.2 Retail Electricity Markets

This section reviews the history of retail markets' opening and empirical evidence of customers' lack of activity to date. It concludes by offering policy recommendations.

### 4.2.1 History of Retail Electricity Liberalization

Retailing was considered a secondary activity by the vertically integrated utilities, whose objective was to deliver sufficient electricity to their users to serve their load. The challenges to meet that objective were mostly engineering related: producing, transporting, and distributing the required Megawatt-hours. Although circumstances vary by country and by utility, most activities considered today "customer care" were performed by the distribution division of the utility: connection and disconnection, repairs, call centers, and so on. Retailing per se was essentially limited to billing and collection, which were also often embedded within the distribution division, and their cost was considered very small. For example, in 1992, "client service" accounted for less than 3 percent of total electricity costs in the United States.

At the onset of restructuring, academics and practitioners were divided on the value of introducing retail competition. Some considered that, since no economies of scale existed in retailing (called supplying in Great Britain), no economic argument justified limiting entry. Their hope was that telecommunication providers, banks, and other retailers would leverage their customer relationships and billing and collection infrastructure to compete vigorously with incumbents, improve quality of service, and reduce prices.

Others considered that, since retailing was a small share of the total cost of serving load, retail competition, while possible, would bring at best little value to consumers. Furthermore, retail competition would increase cost-to-serve as retail companies would engage in costly marketing and advertising expenses and would lead to inefficient duplication of costly billing and collection infrastructure. These observers advocated a single-buyer model, where load-serving entities (LSEs) would purchase electricity on wholesale markets on behalf of their captive customers. For its proponents, this approach would deliver the bulk of gains from restructuring, which were to be found upstream in the generation and wires businesses, while not increasing the cost-to-serve of retailers.

Economic theory cannot offer a clear recommendation. If costs are assumed constant, then duplication of retailing infrastructure is costly, and the single-buyer model leads to higher net surplus. If, on the other hand, one expects that competition will lead to a dynamic reduction in retail costs and creative offering of new services, policy makers should allow retail competition.

Most countries and states ended up allowing full retail competition, often by phases. Large industrial customers were first allowed either to choose their retailer or to purchase directly from the wholesale markets. Then came the turn of medium-size customers, typically small industrial firms or commercial facilities, and finally of small domestic and professional customers. As a result, full retail competition has been in effect for more than twenty years in many U.S. states and European countries.

### 4.2.2 The Duality of Retail Electricity Markets

The single most striking feature of electricity retail markets is their duality: one fraction of customers switch often, looking for the best deal, another remains loyal to its historical supplier. That stickiness by a sizable share of customers makes retail competition imperfect.

Most observers believed that free entry would drive retailing profits to zero: all competing retailers purchase electricity under the same conditions on the wholesale market and have access to the exact same transport and distribution infrastructure under the exact same regulated conditions. Their only differentiating factor is their brand, quality of service, and retail margin. Since electricity is a commodity, it was expected these would not matter much.

While this is true for a fraction of the population, a sizable share of residential customers are sticky: they do not readily change their power supplier. This observation holds—although with regional differences—in the United States, Europe, Australia, and New Zealand. The British electricity market clearly illustrates this duality, as reported in box 4.1 below.

Stickiness arises for two reasons. First, lack of information: a fraction of customers are unaware that switching is possible (apparently, not everyone is interested in the electricity industry), or they perceive that the gains from switching are small, since their power bill is a small fraction of their monthly expenses, or they believe that switching is risky, that is, it may lead to supply disruption.

Second, limited rationality: even when provided with full and certified information, as in the Big Switch study presented in box 4.1, a fraction of customers do not act. "Limited" rationality has been documented in other contexts by a large number of academic studies, summarized, for example, in Ariely (2008).

Since retail electricity is not quite a commodity, retailers can and do exert market power: sticky customers are charged more than it costs to serve them.

### 4.2.3 Possible Policy Responses

Most economists agree that government intervention should at the minimum include working hard to make sure switching is quick, easy, and painless; informing consumers about the value of switching; and monitoring retail markets to mitigate exercise of market power.

**Box 4.1**
Duality in the British electricity retail market

The British energy regulator Ofgem reports that, as of the end of 2017, more than sixty suppliers were active in the domestic retail markets: six large vertically integrated former gas or electricity incumbents and a number of smaller "independent" suppliers, which entered after liberalization, and in most cases are not vertically integrated.

**A small fraction of very active customers** Ofgem reports high switching rates: every month between 2005 and 2012, on average 400,000 domestic customers have switched supplier. This number has come down to slightly more than 200,000 since then. This represents an annual switching rate of 11.1 percent in 2014.

Electricity suppliers compete for the switchers; hence they have invested in expensive—and often creative—marketing campaigns. Furthermore, they have invested in expensive information systems to handle these switchers: open and close accounts, send last bills, and so on. Like all large information technology projects, these investments have had their fair share of problems, which have resulted in billing errors. This has then led to significant customers' discontent.

**A large fraction of inactive customers** On the other hand, Ofgem also reports in the same study that 30 percent of domestic meter points remained with the same supplier from 2002, when the retail market was fully opened, to 2014.

One example is particularly telling: in the spring of 2012, *Which?*, the UK Consumers' Association, and *38 Degrees*, a nonprofit activist organization, launched the "Big Switch" the largest collective energy-switching exercise undertaken in Great Britain: around 150,000 customers were offered an alternative electricity supply offer. The results were studied by economists from the Center for Competition Policy at the University of East Anglia (Deller et al. 2014).

The study is fascinating for three reasons. First, it encompasses a large number of participants: once outliers were eliminated, more than 110,000 decisions to switch or not were analyzed. This large number of observations raises the statistical value of the study's results. Second, participants did not have to search for an offer: the organizers collected the load curves (or their estimates) and ran an auction with participating energy suppliers. Participants were then presented with the best offer resulting from the auction. Thus participants faced no cost of searching for the best offer. The researchers were then able to estimate the impact of the pure switching cost, that is, the inertia associated with the process of switching. They report that out of a total of £16.9 million potential savings, only £5.5 million were effectively captured. Inertia cost these customers more than £10 million.

Finally, in some instances, the auction did not produce the best available offer. In that case, participants were presented with two offers: the auction result and the best offer from the *Switch?* comparison website. This occurred for roughly 50 percent of participants. The researchers were then able to estimate the impact of multiple choices on the switching decisions. After controlling for observable factors (e.g., income, education level, current energy bill), they found that participants presented with two offers instead of one had a lower probability of switching: increased choice seems to have reduced switching.

A more complex question is whether limited rationality justifies additional government intervention. Policy makers can choose the extent of that intervention. I provide below two extremes: the limited approach pursued in New Zealand and the more intrusive approach pursued in Great Britain.

#### 4.2.3.1 Light-handed intervention

The New Zealand electricity authority has decided to limit its intervention to facilitating the switching process and providing information to customers on the value of switching. First, it monitors closely the performance of the switching process. Second, following a review of the electricity industry in the late 2000, the New Zealand government created a consumer switching fund, which ran from November 2010 to April 2014, and endowed it with a small budget of NZ$5 million per year. The information campaign "What's My Number?" was the central program financed by this fund and has since been extended by New Zealand's electricity authority. It hosts a website where consumers can compute their number, that is, the gains likely from switching suppliers. Customers can then contact their new supplier directly or through the website. In addition, the program included a marketing campaign aimed at informing customers that their number was possibly higher than they thought.

#### 4.2.3.2 Heavy-handed intervention

The British regulatory authorities have pursued a more intrusive approach. Following investigations of the retail electricity market, Ofgem, the sectoral regulator, prevented geographic differentiation in 2009, that is, stating that a contract cannot be made available only in a region, and in 2013, Ofgem[2] limited to four the number of contracts that each supplier can offer. The rationale for the ban on geographic differentiation was preventing discrimination. However, as will be discussed in section 4.4, discrimination is sometimes beneficial to customers. Hence, in 2012, Ofgem decided not to extend the ban on geographic differentiation.

Simplification was the rationale for the limit on the number of contracts. However, simplification reduces creativity; hence it may ultimately harm consumers. In June 2016, the Competition and Markets Authority, following its own investigation, suggested that this limit be reversed. Finally, in 2018, the British government introduced a cap on retail price for a large number of customers, its most significant intervention in the power industry since privatization and restructuring in 1990.

#### 4.2.3.3 Policy recommendations

These examples illustrate the tension between economics and policy making. The imposition of a retail price cap is difficult to justify economically when more than sixty suppliers compete. Another argument economists use to advocate light-handed intervention is that

2. Ofgem, August 2013. The retail market review—Implementation of simpler tariff choices and clearer information.

any intervention is bound to have unintended side effects, which may ultimately be harmful to consumers. For example, a price cap may facilitate coordination among suppliers.

On the other hand, policy makers have broader concerns, in particular, the political acceptability of high and increasing power prices. Policy makers and consumers have largely lost their faith in markets since the 2008 financial crisis. This leads the former to intervene on behalf of the latter in markets they perceive are not functioning properly. Furthermore, the propensity to switch is not evenly distributed among the population. The most vulnerable in our society, the least well-off and the old people, are the least likely to take advantage of competition and the most likely to pay more for power than they otherwise could. This led to a backlash against industry restructuring. Intervention in retail power markets is politically legitimate.

The first recommendation is that regulators should provide clarity on the components of the retail price: wholesale electricity price, regulated network charges, taxes and subsidies, and the implicit retail margin, and on their evolutions. In Germany, for example, retail consumers pay around €30/MWh for the Megawatt-hours they consume and around €60/MWh for the RES subsidies. This should be made transparent to customers.

Most interventions have ambiguous effects on net surplus. The second recommendation is that policy makers should choose carefully among the instruments they use to achieve their policy objectives. For example, the growing number of energy poor, who spend a sizable share of their income on energy purchases, is a serious concern in many countries. One approach, adopted in the United States, is to vary the price per unit for different blocks: basic consumption is relatively cheap, but higher consumption, which is deemed less necessary, is more expensive. Another approach, used in France, is to grant energy vouchers to less affluent citizens, which can be redeemed for any energy purchase. The two approaches have the same policy objective in mind. The second one is much less disruptive, hence preferred by economists.

### 4.2.4 The Path Forward

The situation also raises another issue. Consumers are expected to be central to the energy transition. If they have proved reluctant to change supplier to capture large savings, an apparently simple choice, how likely are they to engage into sophisticated real-time optimization of their appliances?

While this concern is legitimate, a case for reasonable optimism can be made. Electricity retailers have so far developed "classical" offers. They have attempted to differentiate themselves on brand; for example, some retailers propose "green" energy, guaranteeing their customers that, for every Megawatt-hour they consume, the retailer purchases one Megawatt-hour of green electricity. Most retailers propose constant-price contracts of varying maturity, although two firms in New Zealand, including the one mentioned at the onset of the next chapter, offer real-time pricing, that is, they guarantee their customers that they pay at every instant the actual price of electricity, not an average price.

Going forward, measurement, communication, and remote control technologies should enable retailers to develop more sophisticated offers, which will not require active customer participation to deliver benefits. Customers will, therefore, face a choice between truly different offers, which may increase their engagement level. Furthermore, as time passes customers will become more used to electricity being provided competitively and will have greater experience choosing among different suppliers for a variety of services. The future may, indeed, be different from the past.

## 4.3 Forward Commitments

In chapters 2 and 3, we have assumed producers sell their energy into wholesale spot markets. This is not always the case. First, most producers are vertically integrated with retailers, that is, they sell a share of their energy to retail customers. Most retail contracts offer a price constant across states of the world for a period of time, for example, a quarter or a year. Therefore, by selling to a retail customer, a producer commits to a sale price for the future.

Second, a producer may sell forward a fraction of its production or enter into a financial derivative that guarantees him a fixed price for its energy. In both cases, the producer enters into a forward commitment, *forward* meaning here "forward in time."

Allowing or discouraging forward commitments is an essential policy decision. Should retailers be allowed to own generation assets, hence to create vertically integrated energy producers and retailers? Should producers be allowed to sell their production in forward markets?

Policies on forward commitments differ across jurisdictions. When power markets were restructured in the United States in the mid 1990s, utilities in New England, PJM, and California were asked to divest an important share of their generation fleet. Furthermore, in California, retailers were prevented from purchasing energy through long-term contracts, a restriction not applied in New England and PJM. In Europe, vertical integration has not been precluded, but it is viewed with suspicion. The European Commission takes a dim view of long-term contracts between electricity producers and retailers. In Great Britain, the Competition and Markets Authority, in the review it conducted of electricity markets in 2015, specifically examined the impact of vertical integration. As it turned out, it concluded it did not harm the market.

Economic analysis confirms that policy makers are correct in forming different opinions on forward commitments: they have benefits but also can be detrimental.

This section reviews the main arguments from the academic literature on forward commitments. Section 4.5 presents a few of the models supporting these arguments.

### 4.3.1 The Pros of Forward Commitments

Forward commitments reduce risks for producers, which leads to higher equilibrium-installed capacity, and for retailers (if they are risk averse and the industry perfectly competitive), which leads to lower retail prices. Both effects increase net surplus. Forward commitments also reduce producers' incentives to exercise market power.

#### 4.3.1.1 Risk reduction

Forward commitments enable producers and retailers to reduce their exposure to electricity price risk, hence increase net surplus.

**Reduction in producers' risk** Consider the model presented in chapters 2 and 3: over a forward period, for example, the coming year, demand is uncertain and varies with weather conditions. Spot prices, hence generators' profits, are therefore also uncertain. Selling energy forward, either to retail customers or through a wholesale forward contract, reduces revenues' volatility.

This does not always translate into an equivalent reduction in operating profits' volatility. Consider, for example, the owner of a combined-cycle gas turbine (CCGT), a widely used technology. If more expensive gas turbines are most of the time the marginal production units of this market, the electricity price is highly correlated to the gas price. Our CCGT is thus naturally hedged: if the gas price increases, so does his supply costs but so do his revenues; hence his operating margins increase only slightly. In this instance, committing to a fixed price to retail customers would increase the producer risk.

Furthermore, contracts with retail customers usually offer much weaker risk reduction than wholesale forward contracts. The latter can extend up to a few years, thus, in effect, insuring a producer against soft wholesale prices for the next quarter (caused, for example, by mild weather) but also against a durable reduction in wholesale electricity prices (caused, for example, by a structural decline in fuel prices). The former are usually revised a few times every year, precisely when wholesale prices fluctuate. They may protect producers against soft wholesale prices for the next quarter. But if wholesale electricity prices durably fall, so do retail prices (as long as retailing is competitive).

The previous discussion suggests that the risk reduction value of a forward commitment varies with the producer's circumstances, market conditions, and the primary source of uncertainty. In the following discussion, we consider that forward commitments do indeed reduce profits' volatility, and examine the impact of reduced volatility on net surplus.

**Reduction in producers' cost of capital** Reduction in the long-term volatility of producers' profits reduces their cost of capital through two channels. First, the covariance impact: following the seminal analysis from Markowitz (1952) and Modigliani and Miller (1958, 1963), classical corporate finance theory suggests that the cost of capital of an asset is determined by the covariance of the returns from this asset with the returns from the market, that is, the investing universe. Léautier and Peluchon (2016) tests this result empirically and finds that for the observed correlation between demand and market returns, forward commitments significantly reduce the cost of capital, in particular, for peaking units.

Second, modern corporate finance theory, exposed, for example, by Tirole (2006), suggests that reducing volatility increases the share of debt available for financing the asset, hence reduces the cost of capital. For example, generation assets backed by twenty-year power purchase agreements are much easier to finance than purely merchant assets.

If the cost of capital is reduced by either of these channels, ceteris paribus equilibrium-installed capacity is increased, hence so is net surplus.

**Reduction in retail prices** Symmetrically, reduced short-term profit volatility of retailers lead to reduced retail prices, which increases consumers' surplus. The argument is formalized by Aid et al. (2011) and is presented in section 4.5. Consider first pure retailers (that is, those with no generation asset) who have a one-year fixed-price contract with their retail customers and will source their energy on the wholesale spot market. If competition is perfect and the retailers are risk averse, the retail price is the expected sourcing cost plus a risk premium.

Suppose now vertical integration is possible. If a technical condition (presented in section 4.5) is met, when one retailer is vertically integrated the risk premium, hence the retail price, decreases. The surprising result is that vertical integration by one supplier is sufficient to drive the retail price down. Forward contracts have a similar (but not quite identical) effect on retail prices.

#### 4.3.1.2 Reduction in producers' market power

**The argument** This other benefit of forward commitment is probably more surprising. Suppose producers compete à la Cournot, as exposed in section 3.6.1. In a seminal article, Allaz and Vila (1993) prove that opening a forward market reduces their exercise of market power, i.e., leads them to produce higher output than absent the forward market.

Allaz and Vila (1993) model a two-stage game: in the first stage, producers sell in the forward market. These forward sales constitute a publicly observable commitment: other producers observe my forward sales, and once I have sold forward, I cannot buy back. In the second stage, producers compete in the spot market, taking into account their and the others' forward sales.

The analysis proceeds in two steps. Consider first the spot market. A producer who has sold a fraction of his output forward, for example 40 percent, captures the increase in the wholesale spot price only on the remaining 60 percent of his output. Thus he has lower incentives to reduce output as a way to increase the spot price than a producer who has not sold any output forward, hence captures the price increase on his entire output.

Consider now the forward market. Allaz and Vila (1993) assume there is no uncertainty; therefore forward sales bring no risk reduction benefit. Intuition suggests that producers do

**Box 4.2**
Technical digression

Forward trading does not universally have a positive impact on spot markets. Mahenc and Salanie (2004) prove that if firms compete in prices (as in section 4.4), the reverse result holds: a forward market leads to higher prices.

Whether firms compete in price or in quantities, ceteris paribus, increasing forward sale reduces equilibrium price in the spot market. Since prices are strategic complements (i.e., a price increase by one firm increases the other firms' price), firms competing in price are able to use forward markets to coordinate on higher prices. In fact, Mahenc and Salanie (2004) prove that, at the equilibrium, producers purchase in the forward market to increase the spot price.

By contrast, since quantities are strategic substitutes (i.e., a quantity increase by one firm decreases the other firms' quantities), firms are hurt by the presence of forward markets.

However, as discussed in chapters 2 and 3, price competition as described by Mahenc and Salanie (2004) does not constitute an adequate representation of competition in wholesale electricity markets. Hence forward contracts, and more generally forward commitments, have a procompetition impact on power producers' behavior.

not sell any volume forward, since they anticipate that their forward sale will reduce their profit in the spot market, and they face no risk, hence they capture no risk-management gain from forward sale.

Yet Allaz and Vila (1993) prove this intuition is incorrect. Producers face a prisoners' dilemma in the forward market: if one producer does not sell in the forward market, the other prefers to sell forward a small quantity. Knowing this, the first producer also sells forward. At the equilibrium, both producers sell forward, and the spot market is more competitive than absent the forward market. Power producers cannot refrain from selling forward, which reduces their profit. As indicated in box 4.2, this result does not hold if producers compete in prices.

**Public commitment to a hedging decision** Allaz and Vila's (1993) result holds if and only if firms publicly commit to their hedging decisions before they select their output (or prices). I believe this assumption holds in reality.

First, financial regulations require firms to publish, in their quarterly statements, a description of their portfolio of forward purchases and sales. While some discretion still exists in disclosure, an outside party can get a close picture of a firm's hedging portfolio. For example, academics have been able to compute the delta equivalent of the forward portfolio for U.S. oil and gas companies (Jin and Jorion 2006). Similarly, electricity suppliers in Britain infer each one another's hedging portfolio from financial statements and other public information.

Second, industrial firms can—and in practice do—commit to a hedging strategy through their risk-management policy. Forward sales and purchases, which require the use of derivatives, are usually handled with extreme caution by boards of directors, concerned about potential speculative behavior by traders. Boards then require management to define and follow a clear hedging strategy. This position is communicated to investors and regulators. Management has then limited discretion to deviate from it.

**Empirical evidence** This positive value of forward commitment is illustrated empirically by Bushnell, Mansur, and Saravia (2008), who test the impact of vertical relationships on equilibrium electricity prices. Specifically, Bushnell, Mansur, and Saravia (2008) model electricity prices in New England, PJM, and California under three scenarios: perfect competition, Cournot competition, including retail commitments, and Cournot competition assuming no retail commitments. Then compare these simulated prices with actual prices.

First, the authors find that the theory is well supported by facts: simulated Cournot prices including retail commitments are extremely close to observed prices. This is somehow surprising as Cournot competition is a very simple representation of imperfect competition, while in reality, firms bid supply functions, hence they play a more sophisticated game. On the other hand, it is encouraging to find that a simple—and highly tractable—model is not too far from reality.

Second, they document the importance of vertical structure—along with horizontal structure—to explain the performance of an industry: they estimate that the production costs with vertical contracts are 59 percent lower than costs without vertical contracts in PJM, and 32 percent in New England: Under Cournot competition, domestic production is reduced and replaced by high-cost import; vertical contracts increase domestic competition, hence reduce the import cost.

### 4.3.2 The Cons of Forward Commitments

Policy makers are sometimes concerned that forward commitments will reduce the liquidity in the spot market. In addition, imperfectly competitive retailers may use forward commitments to increase retail prices. Finally, vertical arrangements may be used to foreclose entry.

#### 4.3.2.1 Reduced liquidity in the wholesale spot market

Consider a vertically integrated producer and retailer. Since the firm uses the energy it produces to serve its customers' load, it does not sell it in the wholesale spot market. Policy makers are concerned that a new entrant retailer will therefore find it difficult to source his customers' needs.

This argument is not as robust as it may seem, as it ignores dynamics. The entrant retailer takes clients away from the vertically integrated utility. Thus the utility has extra energy to sell and will use the wholesale market to do so. Therefore, we should not expect vertical integration to prevent new entrants from sourcing themselves in the spot markets.

A similar point can be made with forward contracts. Consider a producer who has sold forward to a retailer. If a new retailer enters, he will take clients away from the incumber retailer, hence the latter will have spare energy to sell.

A more refined argument is that, if forward commitments are prevalent, liquidity in the wholesale spot market at the time when a retailer contemplates entry is very limited. While retailers may believe that liquidity will be created by their entry, investors may be hesitant. In addition, even if some energy is transacted in the spot market, the volume may be small, which makes the price more volatile and less reliable.

On balance, the liquidity argument is not completely convincing. Liquidity concerns may be better addressed by horizontal separation, that is, by creating multiple smaller firms.

#### 4.3.2.2 Retailers may use forward commitment to increase prices

Léautier and Rochet (2014) examine the impact of forward purchases of input on retailers' behavior. They find that, when risk-averse retailers compete imperfectly in prices, as is the case in the electric power industry, they strategically use their hedging position to soften competition, that is, to increase prices.

Consider two electricity retailers that purchase power on the spot market and sell it at a fixed price to their customers. The retailers have the possibility to purchase a fraction of their expected demand forward, that is, they can fix the cost of a fraction of their purchase.

Consider first the equilibrium in the retail market, taking into account the forward purchase. As in Aid et al. (2011), if a retailer increases his forward purchase, he reduces his expected risk-adjusted cost, hence reduces his price. Since prices are strategic complements, this also reduces his rival's equilibrium price.

Consider now the equilibrium in the forward market. As Mahenc and Salanie (2004) suggest, retailers purchase less than their natural hedging need, thus committing to a price higher than if their realized cost was constant.

This result may seem too theoretical. Forward purchase decisions combine multiple factors. Still, it underlines the strategic value of these decisions and suggests that market monitors should not underestimate them.

#### 4.3.2.3 Forward contracts and foreclosure

Whether forward contracts may be used to foreclose entry to the market has long been debated by economists. Consider a simple industry comprising producers (sellers) and users (buyers). The initial argument was that buyers signing forward contracts with sellers today reduces the size of the market tomorrow, hence it limits entry from new sellers. Thus sellers strategically offer forward contracts today to limit their future competition tomorrow. Market monitors and regulatory authorities should therefore view forward contracts with great suspicion.

In the 1960s, a group of academic economists argued—quite correctly at the time—that most government interventions in markets and industries were not supported by robust microeconomic theory analysis. Since most of these critics of government intervention hailed from the department of economics of the University of Chicago, this school of thought has been called "Chicago critique." In this case, the Chicago critique argued that if a rational buyer willingly enters into a forward contract with a seller, she realizes that this contract limits her future choices; hence she incorporates that effect into the contract price. Thus forward contracts create no damage, and market monitors have no basis for intervention.

In the late 1980s, Aghion and Bolton (1987) developed a simple and elegant microeconomic model proving that, despite buyers' rationality, producers can use forward contract strategically. This article matters greatly for practitioners, as it justifies intervention in vertical arrangements. It is also important for the history of economic thought, as it reverses the Chicago critique.

Aghion and Bolton's (1987) intuition is as follows: consider a single incumbent producer (he) selling one good to a single buyer (she) for delivery tomorrow. The seller and buyer may enter into a forward contract today for delivery tomorrow, that is, a contract price and a breakup fee. A new entrant (he), whose production cost is unknown today, may enter between today and tomorrow, and compete on the spot market tomorrow. This set-up provides a simple description of many economic situations.

All parties behave rationally: the customer realizes that if she enters into a contract today, she may have to pay a breakup fee tomorrow. So does the new entrant as he weighs entry. The incumbent producer is able to use his first-mover advantage to choose the entry rate that maximizes his profits. Ex post, too little entry occurs, that is, more efficient new entrants are blocked out of the market.

Aghion and Bolton (1987) prove this point using simple and elegant algebra, reproduced in section 4.5 below. Nocke and Rey (2016) recently proved that the results hold when multiple producers compete and sell to multiple buyers, a configuration they label "interlocking relationships."

#### 4.3.2.4 Vertical integration may be used to restore market power

Consider the following situation: an upstream monopolist, U, produces a key input for downstream use. There is potential competition in the downstream segment, but it can emerge only if competitors have proper access to U's essential input. The bottleneck owner can therefore alter and even eliminate downstream competition by favoring one downstream firm—for example, a downstream affiliate—and excluding others.

Before examining this situation, it is worth discussing how it applies to the power industry. The transmission and distribution networks are essential inputs. For this reason, they are often vertically separated from the rest of the industry and subject to heavy regulation to ensure all market participants are given fair access. Electric power generation is not an essential input. In fact, the electricity industry was liberalized precisely because generation is believed to be competitive, that is, open to entry. While this is true in the long run, in the short run, depreciated generation assets cannot easily be displaced, because a new entrant finds it more expensive to build a new generation asset than to buy from incumbent producers. Thus electricity produced from existing generation assets can be considered, at least in the short run, as an essential input.

The classical foreclosure doctrine suggests that U has, indeed, an incentive to eliminate downstream competition, that is, to foreclose rival firms, to extend its monopoly power to the downstream segment. If the doctrine holds, policy makers in the United States were correct in requesting that retailers divest their existing generation assets.

In this case again, the Chicago critique challenged the established doctrine by arguing that there is a single final market and therefore only one profit to be reaped, which U can get by exerting its market power in the upstream segment. Therefore, U has no incentive to distort downstream competition.

Hart and Tirole (1990) prove the Chicago critique wrong and restore the microeconomic foundations of the classical doctrine. According to Rey and Tirole (2007), from which this section draws heavily,

> The reconciliation of the foreclosure doctrine and the Chicago School critique is based on the observation that an upstream monopolist in general cannot fully exert its monopoly power without engaging in exclusionary practices.... A bottleneck owner faces a commitment problem similar to that of a durable-good monopolist: once it has contracted with a downstream firm for access to its essential facility, it has an incentive to provide access to other firms as well, even though those firms will compete with the first one and reduce its profits. This opportunistic behavior *ex ante* reduces the bottleneck owner's profit (in the example just given, the first firm is willing to pay and buy less).

Therefore, vertical integration does not aim to increase U's monopoly power; rather, it means to restore it. The Hart and Tirole (1990) analysis provides a sound justification for policy makers and antitrust authorities' concerns about vertical integration.

One policy implication from Rey and Tirole (2007) is worth emphasizing. Nondiscriminatory rules, which aim to restore a playing field between, for example, U's affiliate and other downstream firms, may prove counterproductive, since they contribute to solving U's commitment problem. This intuition is related to the fact that forcing Hotelling duopolists to offer the same price irrespective of customers' location facilitates collusion.

## 4.4 Horizontal Differentiation and Imperfect Competition

Imperfect competition when consumers differentiate among suppliers, which is called horizontal differentiation, has long been studied by economists. The first model was developed in 1929 by Harold Hotelling, who was then associate professor of mathematics at Stanford University. Even in the simplest possible model, simple changes such as preventing price differentiation can have a counterintuitive or ambiguous impact for some customers. This suggests that policy makers need to exercise caution when setting rules for retail markets and attempt to carefully estimate ex post the impact of the rules.

### 4.4.1 Imperfect Price Competition under Horizontal Differentiation

A highly stylized model captures the essence of retail consumers' stickiness. It is developed in Hotelling (1929). Consider two cities, for example Paris (city 1) and Brussels (city 2), linked by a single road. Consumers are spread uniformly on the road, of length normalized to 1. Each customer can consume up to one unit of an homogeneous good and derives surplus $u > 0$ if she consumes, 0 otherwise. One supplier is located in each city, and they have identical variable production cost $c$. Supplier $i = 1, 2$ charges price $p_i$ per unit of the good. Suppliers cannot discriminate spatially, that is, they must offer the same price to all customers. The transportation cost is $t$ per unit of distance, and it is paid by the customers, that is, customers have to go and pick up the good from their supplier. To simplify the analysis, we assume, as in Benabou and Tirole (2016), the following timing: customers first choose among the two suppliers; they then go to the supplier's city (hence they pay the transportation cost); and finally, they decide whether to purchase and consume.

Transportation cost creates an advantage for the "local" supplier: customers close to supplier 1 find it cheaper to purchase from him. Transportation cost thus differentiates between suppliers, despite their producing an homogenous good. The model can be interpreted as a model of brand, and the parameter $t$ measures the cost for a customer to change its supplier.

Consider a customer located at distance $x$ from supplier 1, hence $(1 - x)$ from supplier 2. The total surplus from purchasing one unit from supplier 1 is $u - (p_1 + x \cdot t)$; from supplier 2 it is $u - (p_2 + (1 - x) \cdot t)$. She chooses supplier 1 if and only if

$$p_1 + x \cdot t < p_2 + (1 - x) \cdot t \Leftrightarrow x < \frac{1}{2} + \frac{p_2 - p_1}{2t}.$$

Then, once she has chosen supplier 1, she consumes if and only if

$$u - (p_1 + x \cdot t) > 0 - x \cdot t \Leftrightarrow u > p_1.$$

We first assume this condition is met, and we will verify it later. Firm 1 market share is

$$D(p_1, p_2) = \frac{1}{2} + \frac{p_2 - p_1}{2t}.$$

If both firms charge the same price, that is, $p_1 = p_2$, they each capture half of the market. If firm 1 increases its price, it reduces its market share. On the other hand, if firm 2 increases its price, firm 1's market share increases.

Differentiation then enables firms to capture positive profits. Firm 1's profit is

$$\pi(p_1, p_2) = \left(\frac{1}{2} + \frac{p_2 - p_1}{2t}\right)(p_1 - c).$$

The first-order condition for profit maximization is

$$\frac{\partial \pi}{\partial p_1} = \left(\frac{1}{2} + \frac{p_2 - p_1}{2t}\right) - \frac{1}{2t}(p_1 - c) = 0 \Leftrightarrow 2p_1 - p_2 = c + t.$$

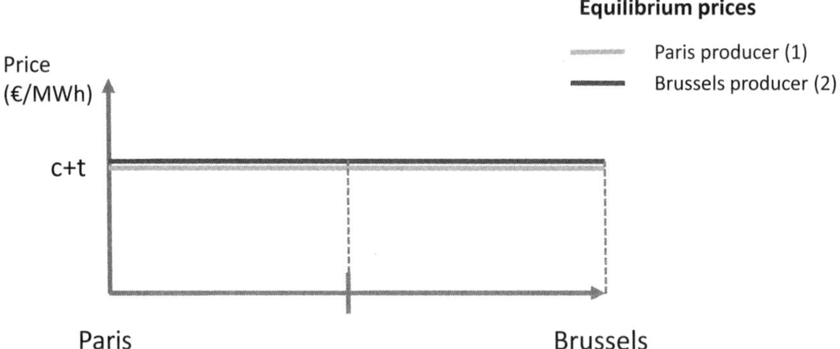


**Figure 4.1**
Hotelling equilibrium retail price when suppliers cannot discriminate among customers.

A similar analysis shows that firm 2 chooses its price $p_2$ to satisfy

$$2p_2 - p_1 = c + t.$$

The unique solution of that system of two equations is

$$p_1 = p_2 = c + t.$$

Each firm captures half of the market and realizes profits $\pi = \frac{t}{2}$. Equilibrium profits increase with the transportation cost, that is, with the preference customers have for their "local" supplier.

We have assumed $u > p_1$. Using the equilibrium value of $p_1$, this is equivalent to $u > c + t$. We assume this condition is met. If it were not, a fraction of demand would not be covered, and the equilibrium would be more complex. The equilibrium price is illustrated in figure 4.1.

### 4.4.2 Price Differentiation within a Market

Consider now the same situation, except that producers pay for transportation and are allowed to charge spatially differentiated prices $p_i(x)$ per unit of the good. The analysis presented below follows Thisse and Vives (1988).

Denote $m_i(x)$ supplier $i$'s marginal cost of producing and delivering the good to customer $x$. Since $m_1(x) = c + x \cdot t$ and $m_2(x) = c + (1 - x) \cdot t$, supplier 1 has a marginal cost advantage if and only if $x \leq \frac{1}{2}$. As before, transportation costs differentiate between suppliers of the homogenous good, this time on marginal costs. The costs are the dashed line in figure 4.2.

Suppliers choose their price to capture positive profits, hence $p_i(x) \geq m_i(x)$. The unique equilibrium is characterized as follows. For $x \leq \frac{1}{2}$, supplier 1 chooses $p_1(x)$ lower than and arbitrarily close to $m_2(x)$, thus at the limit $p_1(x) = m_2(x)$ and serves all customers (since

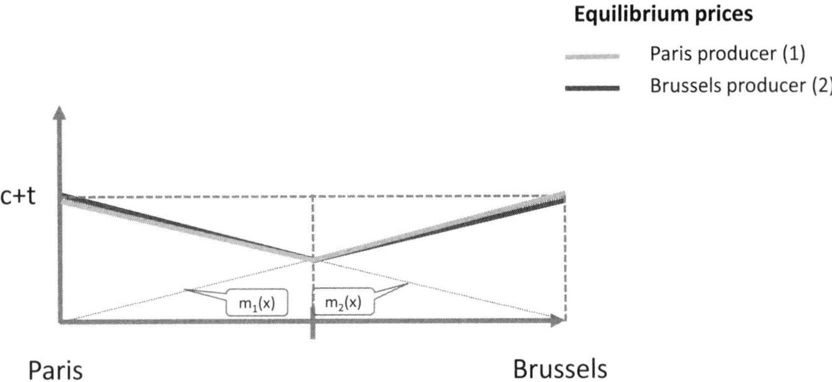


**Figure 4.2**
Costs and equilibrium prices when suppliers are allowed to discriminate among customers, based on location.

$u > c + t$). For $x > \frac{1}{2}$, knowing supplier 2 is cost advantaged and serves the entire market, supplier 1 chooses $p_1(x) = m_1(x)$, and serves no supplier. Supplier 2 follows the same strategy. Equilibrium prices are presented in figure 4.2.

Allowing spatial differentiation forces suppliers to offer lower prices. The intuition for that surprising result is that a constant price acts as a coordination device for suppliers, hence it facilitates collusion. Thus by forcing suppliers to commit to a constant price, policy makers inadvertently facilitate collusion and harm consumers.

### 4.4.3 Price Differentiation across Multiple Markets

The story does not end there. Detailed analysis of the British electricity retail market reveals a slightly different structure. Each regional market is divided into a sticky fraction, which is loyal to the historical supplier, and a competitive fraction, which is open to newcomers. To represent this situation, Crampes and Laffont (2016) consider two suppliers, competing in three different markets:

1. Customers located on the left of city 1. Supplier 1 is a monopoly on this market, which has price-reactive demand $Q_1(p) = a_1 - \frac{p}{2}$.
2. Customers located on the right of city 2. Supplier 2 is a monopoly on this market, which has price-reactive demand $Q_2(p) = a_2 - \frac{p}{2}$, with $a_1 \geq a_2$.
3. Customers located between cities 1 and 2. Suppliers 1 and 2 compete on this market, characterized by transportation cost $t$, which can be positive, in which case competition is imperfectly competitive, or equal to zero.

For simplicity, production cost $c$ is normalized to zero. Suppose first suppliers are allowed to differentiate prices across markets but not within market 3. In its captive market, producer $i$ sets the monopoly price $p_i^M = a_i$ that solves

$$max_{p_i} p_i \left(a_i - \frac{p_i}{2}\right).$$

In market 3, suppliers play a Hotelling equilibrium and set the Hotelling price $p_i = p^H = t$.

Suppose now that suppliers must set the same price in all markets in which they compete and they set their price to maximize profits in both markets:

$$max_{p_i} \left\{ p_i \left(a_i - \frac{p_i}{2}\right) + \left(\frac{1}{2} + \frac{p_j - p_i}{2t}\right) p_i \right\}.$$

The unique equilibrium is characterized by

$$\hat{p}_i = t \times \frac{(3+2t) + 4(1+t)a_i + 2a_j}{(3+2t)(1+2t)}.$$

This then leads to

$$\hat{p}_i - p^H = t \times \frac{4(a_i - t)(1+t) + (a_j - t)}{(3+2t)(1+2t)}.$$

Therefore, for small values of $t$, for example $t < a_2$, $\hat{p}_i > p^H$, preventing cross-market price differentiation increases price in the competitive market.

Similarly,

$$p_i^M - \hat{p}_i = t \times \frac{(3+2t)(a_i - t) + 2t(a_i - a_j)}{(3+2t)(1+2t)}.$$

Thus if $t < a_2$, $\hat{p}_1 < p_1^M$, preventing cross-market price differentiation reduces price in market 1. On the other hand, the impact on market 2's price is ambiguous. Thus preventing price differentiation has an ambiguous impact on consumer surplus and prices.

Crampes and Laffont (2016) also examine the case of perfect competition in market 3 (that is, $t = 0$). They show that preventing cross-market price differentiation also has an ambiguous impact on consumer surplus and prices.

## 4.5 Economic Analysis of Forward Commitments

### 4.5.1 Forward Contracts Lead to Lower Retail Prices

#### 4.5.1.1 The model

The analysis developed by Aid et al. (2011) is similar to the model presented in chapter 2, with two differences. First, demand varies with the state of the world but is completely inelastic: consumers pay a constant retail price, hence they do not react to wholesale spot prices, but their demand does not vary with the constant retail price, either. This assumption is not illegitimate in the short term, which is the focus of the Aid et al. (2011) article. As a result of this assumption, the equilibrium retail price is determined by balancing supply and demand in the retail market and not by maximizing surplus, as was done in chapter 2.

Second, Aid et al. (2011) suppose that frictions in financial market make firms risk averse, which is consistent with empirical evidence and recent advances in corporate-finance theory (see, e.g., Tirole 2006). To simplify the analysis, Aid et al. (2011) assume that this risk aversion translates into a mean-variance utility of profit:

$$U(\pi) = \mathbb{E}[\pi] - \frac{\lambda}{2} Var[\pi].$$

where $\lambda \geq 0$ is the firm's coefficient of risk aversion. The higher the $\lambda$, the higher the disutility the firm derives from the variance. If $\lambda = 0$, the firm is risk neutral. Aid et al. (2011) assume that firms are heterogeneous; hence it allows $\lambda_k$ to vary across firms.

In state $\theta$, demand is $D(\theta)$. Competition in production is perfect; hence the wholesale spot price $p(\theta)$ is determined by the intersection of the aggregate supply curve and the vertical demand. The retail price $p^R$ is unique, since retail competition is also perfect. The aggregate retail profit is $\Pi^R(\theta) = (p^R - p(\theta))D(\theta)$.

Firm $k$ profit in state $\theta$ is

$$\pi_k(\alpha_k, \theta) = \alpha_k \Pi^R(\theta) + \pi_k^G(\theta),$$

where $\alpha_k \geq 0$ is firm $k$'s retail market share and $\pi_k^G(\theta)$ is firm $k$'s profit from generating activities, if applicable. Each firm chooses its retail market share to maximize the utility from its profit:

$$U_k(\pi_k) = \mathbb{E}[\alpha_k \Pi^R(\theta) + \pi_k^G(\theta)] - \frac{\lambda_k}{2}\{\alpha_k^2 Var[\Pi^R(\theta)] + Var[\pi_k^G(\theta)] + 2\alpha_k Cov[\Pi^R(\theta), \pi_k^G(\theta)]\}.$$

#### 4.5.1.2 Determination of each firm's market share

We have

$$\frac{dU}{d\alpha_k} = \mathbb{E}[\Pi^R(\theta)] - \lambda_k\{\alpha_k Var[\Pi^R(\theta)] + Cov[\Pi^R(\theta), \pi_k^G(\theta)]\}$$

and

$$\frac{d^2U}{d\alpha_k^2} = -\lambda_k Var[\Pi^R(\theta)].$$

Assuming $\lambda_k > 0$, firm $k$'s utility-maximizing market share is uniquely defined by

$$\alpha_k = \frac{1}{Var[\Pi^R(\theta)]}\left[\frac{\mathbb{E}[\Pi^R(\theta)]}{\lambda_k} - Cov[\Pi^R(\theta), \pi_k^G(\theta)]\right].$$

To simplify, assume that, for all $k$, $\lambda_k > 0$, and the parameters are such that $\alpha_k \in [0, 1]$. Retail market share is decreasing with the firm's risk aversion.

Since $\Pi^R(\theta)$ decreases with the wholesale price $p(\theta)$, while $\pi_k^G(\theta)$ increases with $p(\theta)$, it is reasonable to assume that $Cov[\Pi^R(\theta), \pi_k^G(\theta)] < 0$ for all $k$. If this holds, the retail market share of a vertically integrated firm is higher than that for a pure retailer: vertical integration provides a natural hedge; hence it encourages a firm to take a larger retail position.

#### 4.5.1.3 Determination of the equilibrium retail price

Equilibrium in the retail market implies that $\sum_k \alpha_k = 1$. Defining $\Pi^G(\theta) = \sum_k \pi_k^G(\theta)$ the aggregate generation profit of vertically integrated retailers, and $\frac{1}{\Lambda} = \sum_k \frac{1}{\lambda_k}$ the aggregate risk aversion, this is equivalent to

$$\frac{\mathbb{E}[\Pi^R(\theta)]}{\Lambda} - Cov[\Pi^R(\theta), \Pi^G(\theta)] = Var[\Pi^R(\theta)] \Leftrightarrow \mathbb{E}[\Pi^R(\theta)]$$
$$= \Lambda(Var[\Pi^R(\theta)] + Cov[\Pi^R(\theta), \Pi^G(\theta)]).$$

Aid et al. (2011) show that the quadratic equation

$$\mathbb{E}[(x - p(\theta))D(\theta)] = \Lambda(Var[(x - p(\theta))D(\theta)] + Cov[(x - p(\theta))D(\theta), \Pi^G_{(\theta)}])$$

admits two roots and that unique retail price $p^R$ is the smallest of these two.

#### 4.5.1.4 Impact of vertical integration on the equilibrium retail price

Suppose no retailer is vertically integrated. The retail price $p^R_{NI}$ is uniquely defined by

$$\mathbb{E}[(p^R_{NI} - p(\theta))D(\theta)] = \Lambda Var[(p^R_{NI} - p(\theta))D(\theta)].$$

The risk-neutral price is $p_0^R$, uniquely defined by

$$\mathbb{E}[(p_0^R - p(\theta))D(\theta)] = 0 \Leftrightarrow p_0^R = \frac{\mathbb{E}[p(\theta)D(\theta)]}{\mathbb{E}[D(\theta)]}.$$

Define $g^{NI}(x) = \mathbb{E}[(x - p(\theta))D(\theta)] - Var[(x - p(\theta))D(\theta)]$. $g^{NI}(.)$ is a quadratic function, $\frac{d^2 g^{NI}}{dx^2}(x) = -2Var[D(\theta)] < 0$, hence $g^{NI}(.)$ is increasing then decreasing. $g^{NI}(p_0^R) = -Var[(p_0^R - p(\theta))D(\theta)] < 0$ and $p^R_{NI} > p_0^R$ is the first root of equation $g^{NI}(x) = 0$.

Suppose now retailer $k$ decides to integrate vertically. Equilibrium retail price $p_k^I$ is the smallest root of equation

$$g_k(x) = \mathbb{E}[(x - p(\theta))D(\theta)] - \Lambda(Var[(x - p(\theta))D(\theta)]$$
$$+ Cov[(x - p(\theta))D(\theta), \pi_k^G(\theta)]) = 0.$$

The function $g_k(x)$ has the same shape as $g^{NI}(x)$. Since $Cov[(x - p(\theta))D(\theta), \pi_k^G(\theta)] < 0$ by assumption, $g_k(x) > g^{NI}(x)$. Thus the smallest root of $g_k(x) = 0$ is smaller than the smallest root $g^{NI}(x) = 0$, that is, $p_k^I < p^{NI}$: vertical integration by retailer $k$ reduces the equilibrium retail price.

### 4.5.2 Forward Contracts Reduce Producers' Exercise of Market Power

There is no uncertainty in the analysis proposed by Allaz and Vila (1993). Hence, there is no need to introduce the state of the world $\theta$. Consider $N$ producers, indexed by $n=1, \ldots, N$, facing inverse demand $P(Q)$. For simplicity, assume that all producers have the same operating constant cost $c$ per unit and that inverse demand is well behaved; hence a unique Cournot equilibrium exists, and it is symmetric. Absent forward market, the equilibrium Cournot output $Q^C$ is uniquely defined by

$$P(Q^C)+\frac{Q^C}{N}\partial_Q P(Q^C)=c.$$

#### 4.5.2.1 Equilibrium in the spot market

Suppose now producer $n$ has sold quantity $h_n$ in the forward market, at price $F$. If $h_n \leq q_n$, his spot market profit is

$$\pi_n=(q_n-h_n)(P(Q)-c)+h_n(F-c).$$

If $h_n \geq q_n$, producer $n$ sells its entire output at price $F$, and purchases $(h_n-q_n)$ on the spot market at price $P(Q)$ to sell it at price $F$:

$$\begin{aligned}\pi_n&=q_n(F-c)+(h_n-q_n)(F-P(Q))=q_n(F-c)+(h_n-q_n)(F-c+c-P(Q))\\&=(q_n-h_n)(P(Q)-c)+h_n(F-c).\end{aligned}$$

Suppose that forward contracts are financial swaps, that is, a forward sale generates (positive or negative) profits $(F-P(Q))$ per unit. In this case, there is no reason to impose $h_n \leq q_n$. Producer $n$ profit is

$$\tilde{\pi}_n=q_n(P(Q)-c)+h_n(F-P(Q)).$$

All expressions are equivalent

$$\begin{aligned}\tilde{\pi}_n&=(q_n-h_n)(P(Q)-c)+h_n(F-P(Q)+P(Q)-c)\\&=(q_n-h_n)(P(Q)-c)+h_n(F-c)=\pi_n,\end{aligned}$$

hence physical and financial forward contracts have the same economic impact in this case.

Given $h_n$, producer $n$ chooses $q_n$ that solves

$$\frac{\partial\pi_n}{\partial q_n}=P(Q)-c+(q_n-h_n)\partial_Q P(Q)=0. \tag{4.1}$$

A producer who reduces output loses operating margin $(P(Q)-c)$ and captures the price increase $(-\partial_Q P(Q))$ on quantity $(q_n-h_n)$. The impact of forward sales on the equilibrium output is summarized below:

**Proposition 4.1.** If producer $n$ increases his forward sales: (a) he increases his output, since he benefits less from a spot price increase, (b) producer $m$ reduces his output, since outputs are strategic substitutes, and (c) total output is increased, since the direct effect exceeds the indirect effect.

***Proof.*** Denote $H = \sum_{n=1}^{N} h_n$. Summing up these first-order conditions, the equilibrium output $Q(H)$ is uniquely determined by

$$P(Q(H)) + \frac{Q(H)-H}{N}\partial_Q P(Q(H)) = c. \tag{4.2}$$

Full differentiation of equation (4.2) with respect to $H$ yields

$$\left[\partial_Q P(Q) + \frac{Q(H)-H}{N}\partial_{QQ}P(Q)\right] Q^{'}(H) + \frac{1}{N}(Q^{'}(H)-1)\partial_Q P(Q) = 0$$

$$\Leftrightarrow Q^{'}(H) = \frac{\partial_Q P(Q)}{\partial_Q P(Q)(N+1) + (Q(H)-H)\partial_{QQ}P(Q)} > 0,$$

which proves (c) since $Q^{'}(H) = \frac{dQ}{dH} = \frac{\partial Q}{\partial h_n}$.

Full differentiation of equation (4.1) with respect to $h_n$ yields

$$[\partial_Q P(Q) + (q_n - h_n)\partial_{QQ}P(Q)]\frac{\partial Q}{\partial h_n} + \left(\frac{\partial q_n}{\partial h_n} - 1\right)\partial_Q P(Q) = 0$$

$$\Leftrightarrow \frac{\partial q_n}{\partial h_n} = 1 + \frac{[\partial_Q P(Q) + (q_n - h_n)\partial_{QQ}P(Q)]Q^{'}(H)}{-\partial_Q P(Q)}.$$

$$= \frac{N\partial_Q P(Q) + [Q(H) - H - (q_n - h_n)]\partial_{QQ}P(Q)}{\partial_Q P(Q)(N+1) + (Q(H)-H)\partial_{QQ}P(Q)} > 0,$$

which proves (a).

Full differentiation of equation (4.1) with respect to $h_m$ for $m \neq n$ yields

$$[\partial_Q P(Q) + (q_n - h_n)\partial_{QQ}P(Q)]\frac{\partial Q}{\partial h_m} + \frac{\partial q_n}{\partial h_m}\partial_Q P(Q) = 0$$

$$\Leftrightarrow \frac{\partial q_n}{\partial h_m} = \frac{[\partial_Q P(Q) + (q_n - h_n)\partial_{QQ}P(Q)]Q^{'}(H)}{-\partial_Q P(Q)} < 0,$$

which proves (b).

Finally, in $\frac{\partial Q}{\partial h_n} = \frac{\partial q_n}{\partial h_n} + \sum_{m \neq n} \frac{\partial q_m}{\partial h_n}$, the first term (the direct effect) is positive, the other terms (the indirect effect) are negative, hence $\frac{\partial q_n}{\partial h_n} > \frac{\partial Q}{\partial h_n} > 0$. Since the aggregate effect is positive, the direct effect is larger than the indirect effect. □

#### 4.5.2.2 Equilibrium in the forward market

Participants in the forward market are producers, who mostly sell, and arbitrageurs, who mostly buy. Assume a fraction of the arbitrageurs perfectly anticipate the impact of forward sales on producers' decisions in the spot market, that is, they know the function $P(Q(H))$. This assumption is not as heroic as it seems: as mentioned in chapter 3, trading firms often employ large teams of quantitative analysts, well versed in microeconomic modeling.

The no-arbitrage condition implies that the equilibrium forward price is equal to the spot price, that is, $F = P(Q)$. The proof proceeds by contradiction. Suppose that producers sell volume $H_0$ and that the price for the last forward contract is $F_0 > P(Q(H_0))$. A smart trader can capture a risk-free arbitrage by selling additional volume $\Delta H$ forward. This will depress the forward price and will last until $F = P(Q(H_0 + \Delta H))$. The same argument applies if $F_0 < P(Q(H_0))$.

Thus producer $n$ profit is

$$\pi_n = q_n(h_n, h_{-n})(P(Q(H)) - c):$$

since there is no uncertainty, forward sales have no direct impact on profit. Since forward sales bring no risk-management value, producers choose them purely for strategic reasons, that is, to impact output decisions. Then,

$$\begin{aligned}\frac{\partial \pi_n}{\partial h_n} &= \frac{\partial q_n}{\partial h_n}(P(Q(H)) - c) + q_n \partial_Q P(Q(H)) \frac{\partial Q}{\partial h_n} \\ &= \partial_Q P(Q) \left[ q_n \frac{\partial Q}{\partial h_n} - \frac{\partial q_n}{\partial h_n}(q_n - h_n) \right].\end{aligned}$$

Suppose $N = 1$, then $\frac{\partial Q}{\partial h_n} = \frac{\partial q_n}{\partial h_n} = \frac{\partial Q}{\partial H}$ and

$$\frac{\partial \pi}{\partial H} = \partial_Q P(Q) \times \frac{\partial Q}{\partial H} \times H < 0:$$

a monopoly does not sell any volume forward.

For $N > 1$, since $\partial_Q P(Q) < 0$, an interior hedging equilibrium (if it exists) satisfies

$$\frac{\partial \pi_n}{\partial h_n} = 0 \Leftrightarrow q_n \frac{\partial Q}{\partial h_n} - \frac{\partial q_n}{\partial h_n}(q_n - h_n) = 0 \Leftrightarrow h_n = q_n \left( 1 - \frac{\frac{\partial Q}{\partial h_n}}{\frac{\partial q_n}{\partial h_n}} \right).$$

Proving existence and unicity of a symmetric equilibrium requires sufficient conditions. One simple case is linear inverse demand with constant slope: $P(Q) = a - bQ$. Then, $\partial_{QQ} P(Q) = 0$, $Q^{'}(H) = \frac{1}{N+1}$, and $\frac{\partial q_n}{\partial h_n} = \frac{N}{N+1}$.

Then,

$$\frac{\partial^2 \pi_n}{\partial h_n^2} = -b\left[\frac{N}{N+1}\frac{1}{N+1} - \frac{N}{N+1}\left(\frac{N}{N+1} - 1\right)\right] = -b\frac{2N}{(N+1)^2} < 0:$$

the profit function is globally concave, hence the first-order condition defines the unique equilibrium. We have

$$h_n = q_n\left(1 - \frac{\frac{N}{N+1}}{N-1+\frac{1}{N+1}}\right) = q_n\left(1 - \frac{N}{N^2-1+1}\right) = q_n\frac{N-1}{N}.$$

#### 4.5.2.3 Resulting equilibrium output

Inserting $h_n = q_n \frac{N-1}{N}$ into equation (4.1) produces a symmetric system of $N$ linear equations with $N$ unknowns, which admits a unique solution. Algebraic manipulations show that the aggregate output $Q^H$ is

$$Q^H = \frac{N^2}{N^2+1}(a-c).$$

As expected, it exceeds the Cournot equilibrium output $Q = \frac{N}{N+1}(a-c)$, since $\frac{N^2}{N^2+1} > \frac{N}{N+1}$ for all $N > 1$.

### 4.5.3 Forward Contracts May Be Used Strategically to Foreclose Entry

#### 4.5.3.1 Set-up

Aghion and Bolton (1987) propose the following model. An incumbent firm produces one unit of the good at cost $c = \frac{1}{2}$. The buyer has value 1 for one unit of the good. Another firm may enter, which produces one unit at cost $c_e$ uniformly distributed on [0, 1]. The main ingredient of the model is that the cost of the potential entrant is unknown at the time the contract is signed.

The timing is as follows:

1. The buyer and the incumbent firm sign a contract or not.
2. The new firm enters or not.
3. If applicable, the buyer exits the contract or not.
4. Production occurs and payments are made.

#### 4.5.3.2 No contract is signed

In this case, the new firm enters if and only if it has a cost advantage, that is, $c_e \leq \frac{1}{2}$. The probability of entry is $\phi = Pr\left(c_e \leq \frac{1}{2}\right) = \frac{1}{2}$. Suppose that, when the buyer faces two offers, she chooses the entrant. When entry occurs, the entrant charges spot price $p = \frac{1}{2}$ and pockets profit $\left(\frac{1}{2} - c_e\right)$.

When no entry occurs, the incumbent firm captures all the surplus and sells at spot price $p = 1$. When entry occurs, competition drives the spot price down to $p = \frac{1}{2}$ and the buyer receives surplus $1 - p = 1 - \frac{1}{2} = \frac{1}{2}$. Her expected surplus is $\phi \times (1 - p) + (1 - \phi) \times 0 = \frac{1}{2} \times \frac{1}{2} = \frac{1}{4}$. By not signing a contract, the buyer receives an average surplus equal to $\frac{1}{4}$. Therefore, she will require at least the same average surplus if she signs a contract.

#### 4.5.3.3 A contract is signed

Aghion and Bolton (1987) prove that we can restrict our attention to contracts that include a sale price $P$ and a breakup fee $P_0$, that is, a contract is a set $\mathcal{C} = \{P, P_0\}$.

Suppose no entry occurs. The buyer purchases the good at the contract price $P$ and receives a surplus $(1 - P)$. Suppose now entry occurs. If the buyer does not break the contract, she pays $P$ and receives a surplus $(1 - P)$. She will exit the contract only if her surplus is (weakly) higher. In equilibrium, the entrant will charge spot price $\tilde{p}$ such that the buyer's surplus is precisely equal to $(1 - P)$, that is, $\tilde{p} + P_0 = P \Leftrightarrow \tilde{p} = P - P_0$, and the buyer breaks the contract.

When a contract is signed, the buyer receives surplus $(1 - P)$, which must exceed (weakly) her expected surplus when no contract is signed. This condition puts an upper bound on $P$: $1 - P \geq \frac{1}{4} \Leftrightarrow P \leq \frac{3}{4}$. The possibility of entry limits the contract price the incumbent firm offers.

#### 4.5.3.4 Incumbent choice of contract

The new firm enters if it can make a profit, that is, if and only if $c_e \leq \tilde{p} = P - P_0$. The probability of entry when a contract is signed is $\phi' = Pr(c_e \leq P - P_0) = P - P_0$. Since he sets $P$ and $P_0$, the incumbent controls the probability of entry.

When entry occurs the incumbent receives the breakup fee $P_0$ and does not produce. When entry does not occur, the incumbent receives the contract price $P$ and produces. His expected profit is $\phi' \times P_0 + (1 - \phi') \times \left(P - \frac{1}{2}\right)$. The incumbent then chooses $\mathcal{C} = \{P, P_0\}$ to maximize his expected profit:

$$max_{\{P, P_0\}} \left((P - P_0) \times P_0 + (1 - P + P_0) \times \left(P - \frac{1}{2}\right)\right).$$
$$st: P \leq \frac{3}{4}.$$

The incumbent selects the highest contract price acceptable to the buyer, $P = \frac{3}{4}$, and selects $P_0$ to control the probability of entry. In this case, the incumbent selects $P - P_0 = \frac{1}{4}$. The contract is therefore $\mathcal{C} = \left\{\frac{3}{4}, \frac{1}{2}\right\}$. This is socially inefficient: entry occurs with probability $\frac{1}{4}$, while it should occur when $c_e \leq c = \frac{1}{2}$, with probability $\frac{1}{2}$. The incumbent firm therefore reduces the net surplus.

By allowing some entry, the incumbent firm increases its profit. Specifically, the incumbent's expected profit $\pi^I$ is

$$\pi^I = \frac{1}{4} \times \frac{1}{2} + \frac{3}{4} \times \frac{1}{4} = \frac{5}{16}.$$

If no contract is signed, incumbent's profit is $\phi \times 0 + (1-\phi) \times \left(1 - \frac{1}{2}\right) = \frac{1}{2} \times \frac{1}{2} = \frac{1}{4}$. Therefore, by offering an acceptable contract, the incumbent strictly improves his position. Similarly, the incumbent could block entry by setting $P = P_0 = \frac{3}{4}$. Its profits would then be $\frac{3}{4} - \frac{1}{2} = \frac{1}{4}$. The incumbent prefers to allow entry and receive $P_0$ when entry occurs.

# 5 A Kuhnian Paradigm Shift: Price-Responsive Electricity Demand

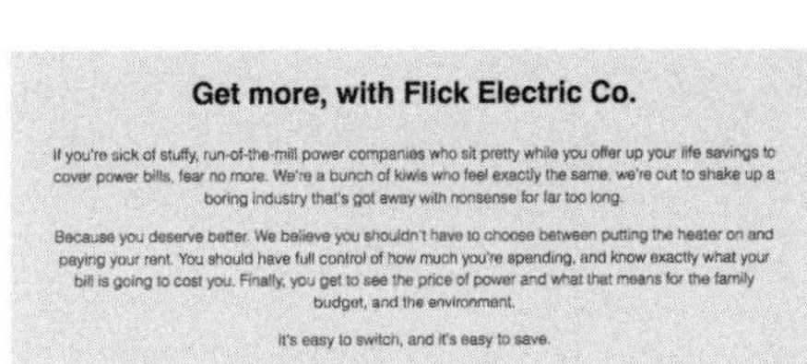

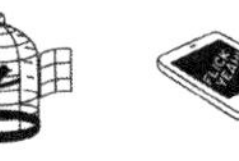

## 5.1 Introduction

Demand response, that is, the ability of customers to modify their consumption (and more generally their behavior) in response to spot prices, is the holy grail of the power industry. Absence of sufficient demand response is responsible for two of the ills besetting power markets: first, administrative rationing may be required to adjust demand to supply during peak hours, as discussed in chapter 2. Second, market power is magnified by the absence of demand response, as discussed in chapter 3.

Historically, demand response was impossible. There was no spot price to which customers could respond. Utilities offered some form of interruptible contracts to large customers: in exchange for a reduced power price, the utility could demand the customer reduce its load. In practice, these interruptions were seldom called upon, and these contracts were thinly veiled subsidies to large electricity users.

Today, the combination of three recent developments makes demand response a technical possibility. First, organized spot markets for electricity generate spot prices at small time intervals, for example, every five minutes. Communication technologies make these spot prices instantly available to all market participants. Second, customers sites are progressively equipped with interval meters, which measure consumption at small time intervals (e.g., every half hour), enabling suppliers to measure and contract on consumers' actual consumption. Most large users already have interval meters, and smart meters, which provide interval reading (among other features), are being rolled out to most consumers in many countries and states. Third, smart appliances are able to automatically adjust to real-time spot prices, without any user's intervention. These developments are discussed in section 5.2.

Thus it is expected that demand response is just around the corner. Households will remotely control their appliances and automatically and continuously adjust their behavior to the fluctuations of spot markets. They will become "consom-actors." This will bring untold benefits to the power industry.

Rigorous economic analysis, summarized in this chapter, tells a slightly different story, which is encouraging, although slightly less uplifting. This chapter's first analytical finding, presented in section 5.3, is that there is a tension between economics and politics in the choice of the contractual arrangements underpinning demand response: economists favor real-time pricing, that is, exposing (larger) users to the wholesale spot price (possibly on a fraction of their consumption), while policy makers favor peak-time resale, that is, allowing users to resell the electricity they do not consume into the wholesale spot market. While these two mechanisms would be equivalent if markets were perfect, implementation challenges make the former economically preferable to the latter.

This tension arises from the different perspectives economists, on the one hand, and citizens and policy makers, on the other, have on electric power. For the former, electricity is an input that contributes to firms' production of goods and services and delivers to residential consumers services such as heating, lighting, and accessing the Internet. Electricity usage is by nature flexible: a Megawatt-hour should be consumed only when its cost is lower than the value it generates. Adapting consumption to the spot price is natural. For the latter, electricity is an essential good, almost a right. Providing (or arranging for the provision of) universal and reliable access to electricity falls within a government's responsibility. Electricity consumption should be based on needs and not be dictated by prices. Users have the right to consume and the right to resell.

This chapter's second analytical finding, presented in section 5.4, is that, in the short-term at least, enabling demand response is more economic for large users (e.g., industrial sites, commercial buildings, and communities) than for small residential ones. The intuition is simple: the benefits of demand response are roughly proportional to the size of an equipment (measured in kW of peak demand), while the costs of instrumentation are roughly equal for all equipments. Therefore facility managers of large commercial/and

**Box 5.1**
Inflexible-demand models

In composing my PhD dissertation in 1997, I used inelastic demand, as did all of the academic literature I reviewed at the time. Subsequent work (e.g., Fabra, von der Fehr, and de Frutos 2006, reviewed in chapter 3; Garcia and Shen 2010) also relies on inelastic demand. This choice somehow simplifies the economic analysis, as the impact of prices on demand is neglected. The surplus maximization problem simplifies to a cost minimization one. As discussed in chapters 2 and 3, analysts must introduce a cap on price, often taken to be the VoLL, to compute equilibria.

office buildings will adjust temperature to respond to surges in electricity spot prices before small households will.

Taken together, these two findings lead to policy recommendations, presented in section 5.5. Underlying analytics are presented in sections 5.6 and 5.7.

## 5.2 An Economic and Technical Reality

For most of its first century of existence, the electricity industry operated under the "fixed demand to serve" paradigm. Engineers and economists structured their analysis assuming that demand depends only on the state of the world (weather condition, hour of the day, day of the week, and week of the year). Countless economic and mathematical models were written under this paradigm, as illustrated in box 5.1

This paradigm is still very much in the minds of policy makers and many observers of the industry. Even an economically literate publication such as the *Financial Times* comments on the safety-margin numbers published by National Grid, Great Britain's transmission-system operator, thus implicitly considering that electricity demand is fixed and independent of price. This is all the more surprising that the same *Financial Times* often points out the "lump of labor fallacy," which consists in assuming that labor demand is fixed and independent of price.

### 5.2.1 Existence of a Wholesale Spot Price

The availability of a reliable price signal is the first necessary condition for the emergence of price-responsive demand. Since the early 2000s, this is a reality in most of North America, Europe, Australia, and New Zealand.

This does not mean that the wholesale spot price always measures exactly the opportunity cost of power. As mentioned in chapter 2, some elements such as start-up cost may not be perfectly included in spot prices. In addition, as mentioned in chapter 3, market power may distort energy prices. I do not believe these distortions are significant enough to jettison the wholesale spot price as an indicator of scarcity, however, other parties disagree. For example, public utilities commissions in the United States consider that wholesale spot prices are not "just and reasonable" (Wolak, 2013, page 4).

Existence of a capacity mechanism is a more challenging issue. Many countries (or markets, in the United States) are implementing capacity markets or capacity mechanisms in which a large fraction of capacity costs is recovered. The analysis presented in chapter 9, as well as empirical evidence, indicates that price spikes in the wholesale spot market will be less frequent. In the limit case, if the entire fixed-capacity costs are recovered by a capacity mechanism, the wholesale spot-market price is simply the variable production cost; hence it no longer represents the true opportunity cost of electricity.

Demand response is not incompatible with the presence of a capacity market: users can sell their demand-response "capacity" in the capacity mechanism. This approach, while conceptually simple, is extremely difficult to implement.

## 5.2.2 Technical Feasibility

Today, demand response is a technical reality, even in real time. The most natural demand-response market is the day-ahead market. For example, expecting prices to be high tomorrow at 5:00 p.m., a user can plan to reduce his consumption at that hour, hence increase his net surplus. For example, an industrial user can move his production process forward or backward to avoid the peak-price period, if the gains from reduced consumption exceed the cost of such a change. A residential user can decide to reduce heating for one

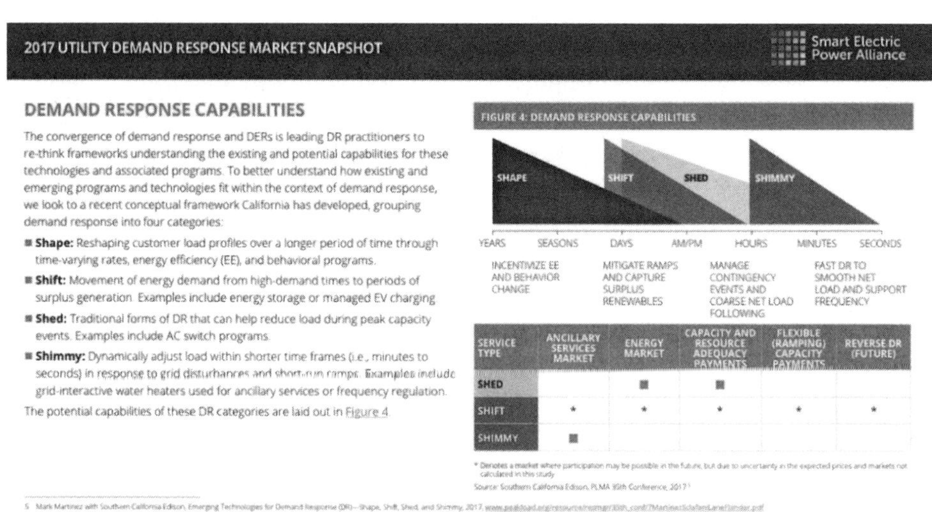
2017 UTILITY DEMAND RESPONSE MARKET SNAPSHOT

Smart Electric Power Alliance

DEMAND RESPONSE CAPABILITIES

The convergence of demand response and DERs is leading DR practitioners to re-think frameworks understanding the existing and potential capabilities for these technologies and associated programs. To better understand how existing and emerging programs and technologies fit within the context of demand response, we look to a recent conceptual framework California has developed, grouping demand response into four categories:

- **Shape:** Reshaping customer load profiles over a longer period of time through time-varying rates, energy efficiency (EE), and behavioral programs.
- **Shift:** Movement of energy demand from high-demand times to periods of surplus generation. Examples include energy storage or managed EV charging.
- **Shed:** Traditional forms of DR that can help reduce load during peak capacity events. Examples include AC switch programs.
- **Shimmy:** Dynamically adjust load within shorter time frames (i.e., minutes to seconds) in response to grid disturbances and short-run ramps. Examples include grid-interactive water heaters used for ancillary services or frequency regulation.

The potential capabilities of these DR categories are laid out in Figure 4.

FIGURE 4: DEMAND RESPONSE CAPABILITIES



| SERVICE TYPE | ANCILLARY SERVICES MARKET | ENERGY MARKET | CAPACITY AND RESOURCE ADEQUACY PAYMENTS | FLEXIBLE (RAMPING) CAPACITY PAYMENTS | REVERSE DR (FUTURE) |
|---|---|---|---|---|---|
| SHED | | ■ | ■ | | |
| SHIFT | * | * | * | * | * |
| SHIMMY | ■ | | | | |

* Denotes a market where participation may be possible in the future, but due to uncertainty in the expected prices and markets not calculated in this study

Source: Southern California Edison, PLMA 36th Conference, 2017[5]

5 Mark Martinez with Southern California Edison, Emerging Technologies for Demand Response (DR)—Shape, Shift, Shed, and Shimmy, 2017,

**Figure 5.1**
The scope of demand-response capabilities.
*Source*: 2017 Utility Demand Response Market Snapshot, Smart Electric Power Alliance and Navigant.

hour tomorrow, if the gains from reduced consumption exceed the disutility of wearing a sweater indoors for one hour.

Demand response also occurs in real-time markets. Suppose the price for the next five minutes spikes, owing to a generation outage unforeseen in the day-ahead market. Automated demand-reduction solutions are feasible for systems that have some inertia, for example, heating and cooling: industrial fridges and air-conditioning systems can be turned off momentarily before temperature changes notably. The extent of demand response depends on each technology. But as demand for demand response increases, new technologies are tested, and the cost of providing demand response decreases.

Figure 5.1, taken from a survey conducted by the Smart Electric Power Alliance, provides a good summary of the various types of demand-response capabilities and their use.

## 5.3 Which Contractual Approach for Demand Response?

Two families of contractual approaches are used to make demand respond to price, often called passive and active demand response. A user reducing her consumption when the power price is high, hence saving on high power prices, would be an example of passive demand response. A user reducing her consumption and reselling her unused power into the market when the power price is high, would be a case of active demand response. For the reader's convenience, I use these terms, although I find them misleading: in both cases, users actively decide to reduce their usage.

In reality, all demand response is active. Even if demand reduction is automated, users actively decide to install the equipment that will automatically reduce their consumption under prespecified circumstances.

### 5.3.1 Passive Demand Response

#### 5.3.1.1 Real-time pricing

The most natural approach is real-time pricing (RTP). As suggested by its name, RTP means that customers purchase power at the spot price, in practice the day-ahead price, for every hour (or half hour). If competition is perfect, and no frictions are present, real-time pricing has consumers pay the true opportunity cost of every kilowatt-hour they consume; hence it leads to the first-best.

Real-time pricing, while being immensely popular with economists, is viewed with suspicion by consumers and policy makers, who fear that customers would be exposed to unjustified swing in their supply costs. In addition, the transaction costs for consumers to react to real-time prices are perceived to be high. For these reasons, alternative contractual forms are proposed.

#### 5.3.1.2 Time-of-use pricing

Time-of-use (TOU) pricing has the consumer pay different predetermined prices for different predetermined periods. For example, customers would pay \$30/MWh on weekends and at nights, \$60/MWh during weekdays except super peak, and \$200/MWh on super peak (e.g., summer afternoon in the South of United States). The certainty provided by TOU pricing is liked by consumers and policy makers. Unfortunately, the efficiency gains from TOU pricing are limited: there is no guarantee that the higher retail price periods determined *ex ante* correspond to the high spot price periods arising ex post. For example, many power plants schedule their maintenance during the "shoulder" period, that is, away from the winter and summer peaks. If a major outage occurs during this period, it may lead to capacity being close to demand for a few hours, hence a price spike. TOU prices usually do not cover this possibility.

#### 5.3.1.3 Critical peak pricing

An improvement on TOU pricing is critical peak pricing (CPP). As in TOU pricing, customers pay different predetermined prices for different periods. The main difference is that under CPP the critical hours are determined dynamically by the SO as a function of system conditions. The number of critical hours per year is predetermined.

Critical peak pricing is an excellent compromise. It reproduces the underlying micro economics exposed in chapter 2: prices are low for the vast majority of the hours, and rise for a few hours only. The scheme is much more predictable and acceptable for consumers than real-time pricing. It has been applied successfully in France since the 1980s. However, it is still viewed suspiciously by some consumers and policy makers.

### 5.3.2 Active Demand Response

The main model is peak-time rebate, which mirrors the interruptible contracts that utilities used to offer to their industrial clients. A consumer who reduces his or her demand when called up by the system operator (SO) during peak hours receives a rebate, usually the reduced quantity times the spot price during the corresponding hours. Alternatively, a consumer may resell to the spot market the electricity he has purchased. Since consumers resell only during on-peak hours, this is called peak-time resale (PTR).

Peak-time resale is hugely popular with consumers, policy makers, and demand-response operators, since consumers cannot lose: either they consume as expected, or they resell power to the market and get compensated for their virtuous behavior. Both families of approaches would lead to the same outcome if competition were perfect. Unfortunately, PTR implementation is fraught with complexities.

#### 5.3.2.1 On-peak peak-time resale and real-time pricing are equivalent

Consider a market with multiple consumers and multiple retailers. To simplify, suppose all transactions go through a single market operator, and nothing changes until between the day-ahead and real-time spot markets.

A month ago, a consumer purchased 100 MWh from her retailer, to be consumed tomorrow between 5:00 p.m. and 6:00 p.m. The first 80 MWh are used for production, hence she values them at €220/MWh. The last 20 MWh are used to heat the building and are valued at €80/MWh, since the building retains heat for one hour.

Before tomorrow's market closes, say today at 8:00 a.m., our consumer learns that the wholesale spot price for the hour starting at 5:00 p.m. is expected to be €200/MWh, higher than the value she places on the last 20 MWh, which she then resells in the day-ahead wholesale spot market. Suppose the equilibrium price when demand reduction is taken into account is €180/MWh. Since this is lower than the value she places on the next block of energy, she does not attempt to sell any more.

If different customers participate in the day-ahead market, the process continues until the wholesale spot price is lower than or equal to the value of the last Megawatt-hour consumed. Thus consumers on PTR consume only the Megawatt-hours they value more than the wholesale spot price. On-peak, PTR produces the same consumption incentives as real-time pricing; hence it is optimal.

#### 5.3.2.2 Customers must purchase the electricity they resell

The first difficulty with PTR implementation is that consumers must purchase the kilowatt-hours from their supplier before reselling them to the market. It may sound obvious, but it turns out not to be. In the United States, no less an authority than the Federal Energy Regulatory Commission has suggested that consumers (or the demand-response aggregators representing them) should receive the entire spot price for every kilowatt-hour they do not consume and should not be liable to their supplier for these kilowatt-hours.

This practice is, of course, contrary to economic reasoning: one cannot sell what one has not produced or purchased. This holds true for any good. There is no reason electricity should be different. To be allowed to sell electricity to the market and receive the wholesale spot price, consumers must have somehow procured this energy. In the previous example, our consumer purchases energy from their supplier (100 Megawatts for an entire hour) and decides to consume a fraction of it (80 Megawatt-hours) and resell the remainder (20 Megawatt-hours) to the market; hence she receives the market price for these 20 Megawatt-hours.

Suppose our consumer is the only one reselling power and limit the analysis to our consumer. The physical energy flows are as follows: from 5:00 to 6:00 p.m. on the day, producers produce 80 MWh, and our consumer consumes 80 MWh. What are the associated financial flows? The market operator (MO) purchases 80 MWh at €180/MWh from producers and 20 MWh from our consumer, also at €180/MWh. Thus the MO purchases 100 MWh at €180/MWh, even though only 80 MWh are actually consumed. To balance his accounts, the market operator must therefore sell to the retailers 100 MWh at €180/MWh. If he sold only the energy actually consumed, he would record a loss of €180/MWh on 20 MWh.

This point is so important, and so misunderstood or misrepresented, that it is worth repeating it: the market operator purchases 80 MWh actually consumed from the producers but also 20 MWh not consumed from the consumers. Therefore, to balance his account, he must sell 100 MWh to the retailers, his only buyers.

The same argument applies to the retailers. Suppose they sell only 80 MWh. They end up with a systematic loss: they buy 100 MWh from the market operator and sell only 80 MWh. This is not sustainable for them, hence they must sell 100 MWh to their customers, who then physically consume 80 MWh and resell 20 MWh.

Promoters of demand response have argued that consumers should not pay for the 20 MWh they do not actually consume. Why should one pay for a good one does not consume? This argument is incomplete, hence misleading. One should pay for a good one resells.

The very lively debate surrounding this issue in the United States and in Europe is a perfect illustration of the interplay between microeconomics, politics, and lobbying. Policy makers are eager to see demand responses to price. They are wary of real-time pricing and are concerned that incumbent utilities are too conservative in their demand-response offers. Therefore, they design policies to facilitate entry by new demand-response operators. The latter make a superficially compelling argument (one should not pay for Megawatt-hours one does not consume), and policy makers are happy to believe them.

#### 5.3.2.3 How are kWh-that-would-have-been-consumed-but-were-not measured?

Even if the first difficulty is solved, another one looms: How can we measure the kilowatt-hours that are not consumed? They are the difference between the consumption that would have occurred (but has not), called the baseline, and the actual consumption (measured). The baseline is impossible to measure, since it does not occur. The problem is challenging for two reasons: first, consumers (or their demand-response aggregators) have a strong incentive to inflate their baseline, since every kilowatt-hour of baseline generates an arbitrage profit equal to the difference, when positive, between the on-peak price and the retail contract price. Second, customers have superior and proprietary information about what their consumption would have been. For example, as a factory manager, I know that tomorrow only one production line will operate, since the other one is maintained, or as a residential customer I know that I am visiting my mother next weekend, hence I will not be consuming power.

The situation can be described using options. Retail contracts are full-requirements: at any hour, a customer can purchase all the electricity he requires at an agreed-upon price (usually constant). Customers thus have a volume option, sometimes called a swing option. This option may be correlated with the spot price, for example, consumers use more electricity for air-conditioning when the weather is very hot, which drives up the spot price for electricity. However, a specific consumer's usage may not be correlated with the electricity spot price.

When customers have the ability to resell, they are granted a put option on the electricity price: resell the electricity when the spot price exceeds the strike price of the option, consume it otherwise. If the baseline were observable, the strike price would be the value customers place on electricity (technically, customers hold a portfolio of put options, at strike prices the different values of the marginal Megawatt-hours they consume). However, since the baseline is by construction unobservable, the strike price is the retail contract price.

Market designers can choose one of two approaches. First, they may require retailers to estimate the baseline using statistics. Since this approach is not perfect, market monitors must monitor demand-response volumes to attempt to detect fraudulent behavior. Policy makers (and consumers) must accept residual cheating as the price to pay to encourage demand response.

Determination of the baseline is even more challenging if it has to be conducted multiple months or years in advance to enable demand response to effectively participate in a capacity mechanism (see chapter 9). This constitutes an important area of research, with important policy implications.

Alternatively, market designers may suggest that retailers offer incentive-compatible contracts, that is, contracts that force consumers to truthfully reveal their baseline. Astier and Léautier (2016) examine this problem, using the tools of contract theory under asymmetric information. The results from their analysis are not encouraging. First, incentive-compatible contracts are, in fact, exceedingly simple: customers need to purchase their future baseline consumption at the expected spot price, then resell or purchase the difference between the baseline and their actual consumption at the spot price. This contract is equivalent to a variable critical-peak pricing contract: pay the contract price off-peak and the spot price during on-peak hours.

Second, consumers who have the possibility to pay the contract price all the time will never accept a contract that forces them to pay more on-peak. As Astier and Léautier (2016) put it: "either policy-makers give customers implicit subsidies which incentivize them to cheat, or no one will enroll in PTR programs. It underlines the fact that PTR is not a 'free carrot': the high political acceptability of the mechanism comes at the cost of implicit subsidies that compromise incentive compatibility." Retailers must then propose retail contracts more favorable off-peak to induce (a fraction of) users to opt in on peak-time resale contracts. The resulting opt-in dynamics are complex since customers are heterogeneous. Consumers whose usage occurs mostly on-peak are subsidized by consumers whose usage occurs mostly off-peak. The opt-in equilibrium depends on the details of the regulatory framework, in particular, whether policy makers decide to maintain crosssubsidies among consumers, and on the competitiveness of retailing. If no cross subsidies are maintained, Astier and Léautier (2016) prove that all customers opt in to PTR contracts if retailing is perfectly competitive, but that only a fraction may opt in if a monopoly retailer exists.

## 5.4 Which Customers Should Be Incentivized to Respond to Spot Prices?

Most stakeholders argue that all customer classes should become price responsive: the more elastic the demand, the better. This argument ignores the costs of making users price responsive. When these costs are taken into account, priority should be given to large users. The economic analysis of demand response presented here is largely inspired by Léautier (2014a). It first discusses the underlying economics, then it uses the model presented in chapter 2 to quantify the marginal value of switching customers to RTP; finally, it discusses some of the attendant costs.

### 5.4.1 Marginal Impact of Real-Time Pricing

This subsection first presents—and proves using images—a few results. Formal proofs are presented in section 5.7.

#### 5.4.1.1 Impact on average price

**Result** The time-weighted average wholesale spot price remains constant, irrespective of the share of price-reactive customers.

As discussed in chapter 2, since the first technology (the baseload technology) produces in all states of the world, the free-entry condition implies that the expected spot price is simply the long-run marginal cost of the baseload technology. It is therefore independent of the share of price-sensitive customers. This contradicts commonly held wisdom that real-time pricing lowers time-weighted average power price. Demand-weighted average price, however, may decrease.

#### 5.4.1.2 Impact on net surplus

**General case**

**Result** If retailers offer an optimal two-part contract (i.e., the optimal constant price per Megawatt-hour derived in chapter 2 and a subscription charge set to balance their accounts), increasing the share of price-reactive customers increases the net surplus. This increase is equal to the net surplus from price-reactive customers minus the surplus from constant-price customers.

Consider first the situation off-peak. The wholesale spot price $p$ is the variable production cost $c$, which is lower than the retail price $p^R$. The first panel in figure 5.2 represents the net surplus if consumers face retail price $p^R$. The gross surplus $S\left(p^R\right)$ is the surface under the inverse demand curve $P\left(Q\right)$ up to $D\left(p^R\right)$, the demand for price $p^R$. The net surplus is the gross surplus $S\left(p^R\right)$ minus the demand $D\left(p^R\right)$ valued at wholesale spot price $p$. The second panel represents the net surplus if consumers face the real-time price $p$ that matches supply and demand, computed as the surface under the demand curve up to $D\left(p\right)$ minus $pD\left(p\right)$. Since customers face the wholesale spot price $p$, the net surplus is

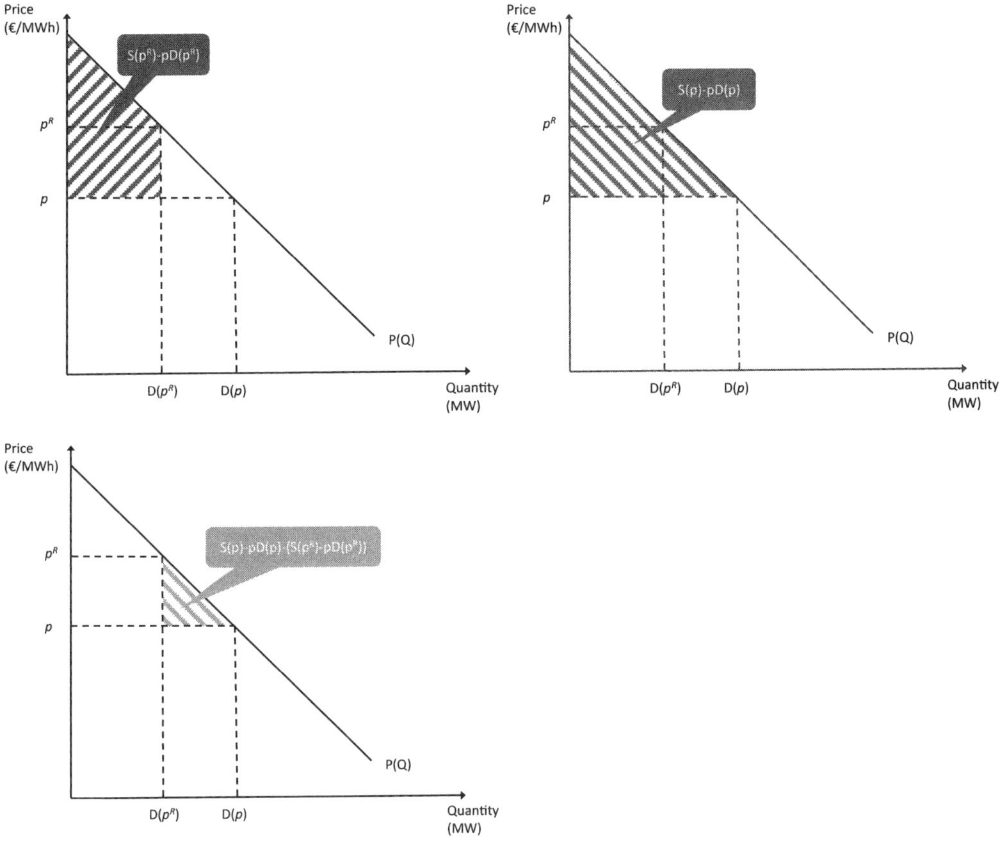


**Figure 5.2**
Net surplus gain from a marginal increase in the share of customers on RTP off-peak.

a triangle. Since $p^R > p$ off-peak, consumers consume too little. Therefore, if consumers are switched to facing the real-time price, consumption, hence net surplus, increases. This is illustrated in the third panel: the gain in net surplus is the positive difference between the net surpluses. It is a triangle.

Consider now the situation on-peak. To simplify, assume the share of price-reactive customers is high enough that constant-price customers are not rationed. The wholesale spot price $p$ is higher than the retail price $p^R$. The first panel in figure 5.3 represents the net surplus if consumers face retail price $p^R$. The gross surplus $S\left(p^R\right)$ is the surface under the inverse demand curve $P\left(Q\right)$ up to $D\left(p^R\right)$. The net surplus is the gross surplus $S\left(p^R\right)$ minus the demand $D\left(p^R\right)$ valued at wholesale spot price $p$. The net surplus is the sum of a triangle under the inverse demand curve and another triangle above the inverse demand curve: consumers face the price $p^R < p$, hence they consume more than is economically

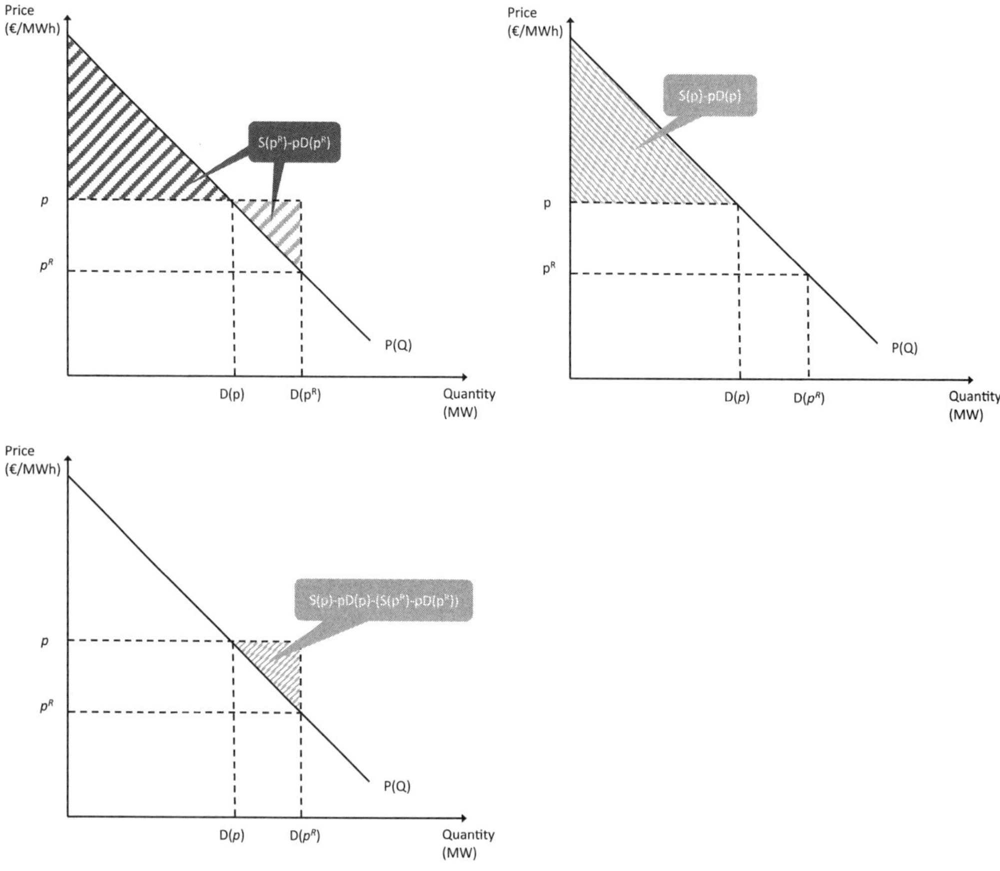


**Figure 5.3**
Net surplus gain from a marginal increase in the share of customers on RTP on-peak, if no rationing occurs.

efficient. The second panel represents the net surplus if consumers face the real-time price $p$ that matches supply and demand, computed as the surface under the demand curve up to $D(p)$ minus $pD(p)$. Since customers face the wholesale spot price $p$, the net surplus is a triangle. If consumers are switched to facing the real-time price, noneconomic consumption is eliminated, hence net surplus increase. This is illustrated in the third panel: the gain in net surplus is the triangle above the inverse demand curve.

Therefore, if retailers offer optimal two-part contracts, increasing the share of price-reactive customers increases net surplus. The assumption that retailers offer the optimal two-part contracts is essential. Borenstein and Holland (2005) consider that retailers are limited to linear contracts (i.e., a constant price per unit consumed expressed in \$/MWh). Then, if retail competition is perfect, the retail price is such that the retail profit is equal to zero. Borenstein and Holland (2005) propose a counterexample where increasing the share of price-reactive consumers reduces net surplus.

However, there exists no empirical nor theoretical reason to limit retail offers to linear contracts. Policy makers in many U.S. states require retailers to offer multiple-part contracts, with different unit prices for different blocks of consumption. In other countries, retailers routinely offer two-part contracts. The analysis presented above follows that approach: retailers optimally choose the unit price, and their budget balance is achieved by the fixed part of their offer.

Figures 5.2 and 5.3 illustrate another important point. Numerous analyses of the value of demand response (e.g., Faruqui et al. 2009) include reduced consumption, lower investment, and lower emissions. This approach is incorrect, as it ignores the loss of profit or utility from not consuming, hence it overstates the value of demand response. The correct approach is the one presented in figures 5.2 and 5.3. As is well known from public economics, net surplus change is measured with triangles, and not rectangles.

### Linear inverse demand

**Result** If inverse demand is linear with constant slope and rationing does not occur at the long-term equilibrium, the marginal value of switching RTP is proportional to the variance of the wholesale spot price and is decreasing as the share of customers already on RTP increases.

This result is illustrated in figures 5.2 and 5.3. Inverse demand is $P(Q)=a-bQ$, hence $D(p)=\frac{a-p}{b}$. The marginal value of RTP in each state of the world is the surface of the triangle, $(p-p^R)\times(D(p)-D(p^R))=\frac{(p-p^R)^2}{b}$. Thus the expected net marginal value of RTP is proportional to $\mathbb{E}[(p-p^R)^2]$. If no rationing occurs, $p^R=\mathbb{E}[p]$, hence $\mathbb{E}[(p-p^R)^2]=\mathbb{E}[(p-\mathbb{E}[p])^2]=var[p]$: the marginal value of RTP is proportional to the variance of the wholesale spot price.

As the share of customers already on RTP increases, the elasticity of demand increases, hence the variance of wholesale spot prices decreases.

This result may appear technical. However, it has two practical implications. First, it confirms the intuition that, as more customers switch to RTP, the variance of prices decreases. Second, assuming the marginal cost of switching customers to RTP is constant, it makes it possible that the optimal share of customers on RTP lies within (0, 1), that is, that the deployment of smart meters is not a 0 or 1 decision.

How realistic are these conditions? Linearity of demand is, of course, a very strong assumption. However, it has some empirical justification: Patrick and Wolak (1997), who measure real-time elasticities in five industrial sectors in the United Kingdom, find the absolute value of the elasticity to be relatively high at peak hours, which is consistent with a linear demand relationship. Furthermore, the result also holds by continuity if demand is "not too far" from being linear.

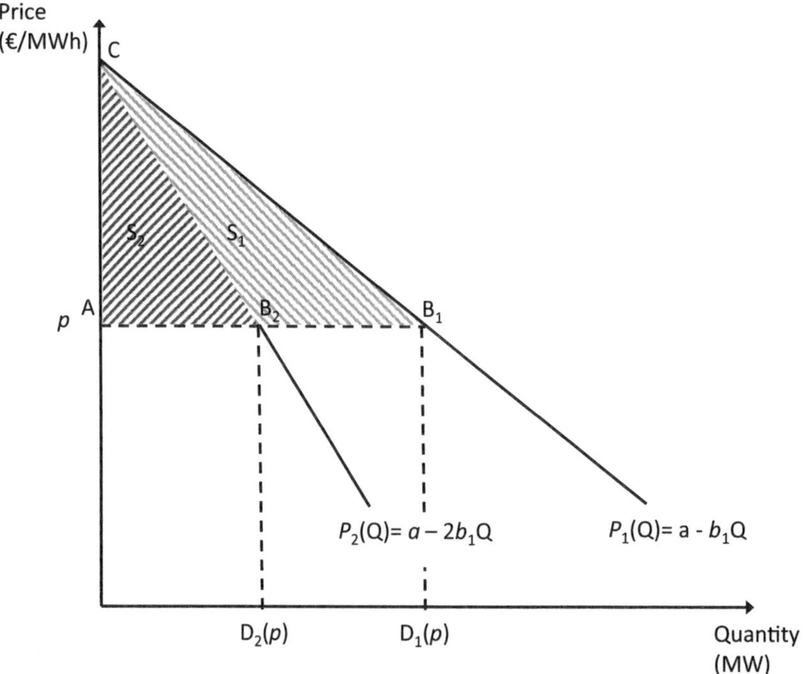


**Figure 5.4**
Two customers with different sizes.

No curtailment at the optimum is not as restrictive at it seems. For the model estimated in section 2.7, it is met as soon as the share of price-responsive demand exceeds 13.7 percent, which is reasonable.

### Different customers' sizes

**Result** If demand is linear with constant slope, and no rationing occurs, the marginal value of RTP is proportional to a customer size (measured in MW of peak demand).

Consider two consumers, indexed by $i = 1, 2$. For any price level, consumer 1 consumes twice as much as consumer 2. As illustrated in figure 5.4, this occurs if and only if the slope of the inverse demand curve of consumer 1 is half that of consumer 2. Therefore, consumer 1's net surplus is twice that of consumer 2.

For algebraically inclined readers, the proof is as follows: $D_i(p) = \frac{a_i - p}{b_i}$ is the quantity consumed by consumer $i$. For any price $p$,

$$D_1(p) = 2D_2(p) \Leftrightarrow \frac{a_1 - p}{b_1} = 2\frac{a_2 - p}{b_2} \Leftrightarrow a_1 b_2 - 2a_2 b_1 + (2b_1 - b_2)\, p = 0.$$

The only solution is $b_2 = 2b_1$ and $a_1 = a_2 = a$.

Using the notation of figure 5.4, the net surplus for consumer $i$ is the surface of the triangle $AB_iC$ and denoted $S_i$. Since (a) the triangles are rectangle, (b) they have the same height $AC$, and (iii) $AB_1 = D(p_1) = 2D(p_2) = AB_2$, we have $S_1 = 2S_2$. The argument can be repeated for any size ratio, hence the net surplus is proportional to a customer's size. This intuitive result matters enormously for policy making, as it justifies switching larger customers to RTP first.

### 5.4.2 Marginal Value of RTP

This subsection provides numerical estimate of the marginal value of switching to RTP, using the simple model presented in section 2.7. First, it examines the marginal value of switching 1 percent of customers to RTP, assuming all customers are identical. As indicated earlier, this marginal value decreases as the share of customers already on RTP increases. Second it examines the marginal value of switching one customer to RTP, recognizing customers have different sizes, but assuming identical load profiles.

#### 5.4.2.1 Marginal value of switching 1 percent of customers to RTP

The model and parameters are that of section 2.7. To maintain comparability with section 2.7, the technology is chosen to be a gas turbine, although in practice no power market would use simply gas turbines. Léautier (2014a) extends the results to multiple technologies.

Suppose first demand elasticity is $\eta = -0.01$. Rationing of constant-price customers does not occur if the share of customers on RTP, denoted $\alpha$, exceeds $\alpha_{\min} = 3.9\%$. Figure 5.5 presents the marginal value of switching 1 percent of customers to RTP, measured in € millions per year per percent, as share of customers on RTP increases from $\alpha_{\min}$ to 100 percent.

This marginal value decreases sharply. A 1 percent increase in the fraction of customers on RTP increases annual surplus annually by €6.4 million for $\alpha = \alpha_{\min}$ and by only €0.8 million for $\alpha = 1$.

Léautier (2014a) shows that similar results hold for different configurations of demand and supply. If the price elasticity of demand is higher, that is, $\eta = -0.1$, the marginal surplus has the same shape, although it is higher for every $\alpha$: since customers react more to prices, increasing their share on RTP has a higher impact.

Léautier (2014a) also considers three production technologies mix: a single production technology, combined-cycle gas turbine; two technologies, nuclear and gas turbine; and a richer mix, including nuclear, CCGT, and gas turbines. Changing the mix has only limited impact on the marginal value of price responsiveness. For every $\alpha$, the marginal value of switching to RTP decreases as the number of technologies increases. This result has a nice intuitive explanation: as the number of technologies increases, supply flexibility increases, hence the marginal value of demand flexibility decreases. This result holds for the technologies selected and their cost structure. It may or may not hold for any technology mix.

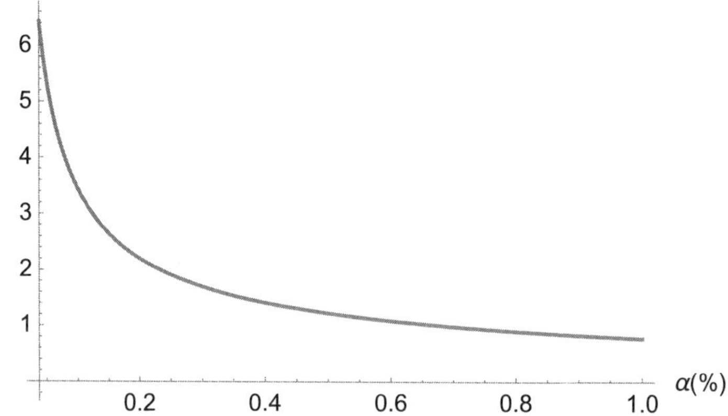


**Figure 5.5**
Marginal surplus of switching 1 percent of customers to RTP, as a function of the share of customers already on RTP.

### 5.4.2.2 Marginal value to RTP per site

The marginal value of switching per site is approximatively the marginal value of switching 1 percent of customers to RTP times the incremental increase in $\alpha$ from a single site, denoted $\delta\alpha$.

Additional data are required to perform this analysis. The French Energy Regulatory Commission (Commission de Régulation de l'Energie) divides the French electricity market in four main categories. The total number of sites and the total consumption (MWh) for each category are as follows:[1]

1. *Large nonresidential.* Around 36,000 sites have peak demand higher than 250 kW: large industrials users, hospitals, shopping malls, large buildings. They represent 0.1 percent of the total number of sites and 42 percent of total demand.
2. *Medium nonresidential.* Around 360,000 sites have peak demand between 36 and 250 kW, mostly small companies. They represent 1 percent of the total number of sites and 15 percent of total demand.
3. *Small nonresidential.* Around 4.6 million sites have peak demand smaller than 36 kilo Volt Ampere: professional offices, small workshops. They represent 13 percent of the total number of sites, 10 percent of total demand.
4. *Residential sites.* Around 30.7 million sites have peak demand lower than 36 kilo Volt Ampere. They represent 86 percent of the total number of sites, and only 32 percent of total demand.

1. The segmentation is based on net power (kW) for large users and apparent power (kilo Volt Ampere) for small users. The former is slightly smaller than the latter. This does not alter the results.

Upon further request, the French commission provided me with the distribution of meter size for the 30.7 million residential users.[2] The overwhelming majority of residential users have meter size of less than 6 kilo Volt Ampere (18.6 million, 60.6 percent of residential sites) or 9 kVA (6.3 million, 20.5 percent of residential sites). Since we assume customers have identical load profiles, the distribution of meter sizes measured in kilo Volt Ampere can be translated into the distribution of energy usage expressed in MWh.

To simplify, assume all sites in each class are assumed to have the same size, thus $\delta\alpha$ is constant for each class, computed as the "size" of each class in percentages divided by the number of sites. For example, for large nonresidential users, $\delta\alpha = \frac{0.42}{36{,}000} = 1.18 \times 10^{-5}$. For small residential customers (meter size between 3 and 6 kVA), analysis presented in section 5.7 shows that $\delta\alpha = 8.50 \times 10^{-9}$.

If $\alpha = \alpha_{min}$, the marginal value of switching one large non residential site is €7,585 per site per year, while it is €5.46 per site per year for a 6 kilo Volt Ampere residential customers. This value drops to €1.89 per site per customer if 20 percent of demand is already on RTP, and €0.76 per site per year if 50 percent of demand is already on RTP. These estimates are in line with other analyses (e.g., Allcott 2011).

The analysis presented above suggests that the value of RTP lies overwhelmingly with high-consuming customers. The analysis suffers from three limitations. First, and most important, users are assumed to have an identical demand profile. Since residential customers consume mostly during on-peak hours, the marginal value of switching them to RTP should be higher than the estimate presented here. Second, it ignores the impact of RTP on reducing exercise of market power, which could be significant. Third, it relies on a linear representation of inverse demand, fitted on French data. Other representations of demand and data sets could provide different results.

All these points are valid. Estimating the full value of switching a customer to RTP remains an important empirical research issue. Still, I believe this result, mostly driven by the differential in size between large industrial and small residential users, will prove robust.

### 5.4.3 Cost of Demand Response

To switch to RTP, a site must be fitted with an interval meter. Estimates of the cost (including installation) vary widely. A median estimate is €250, higher than in Italy (€70) and lower than in the United Kingdom. Assuming a cost of capital at 10 percent, the annualized cost of each meter is €25/meter/year.

The second cost is commercial. Suppliers have to design demand-response offers and market them to consumers. Again, this varies across countries and across marketing

2. I am grateful to Jean-Yves Ollier and Christophe Leininger from the Commission de Régulation de l'Energie for their invaluable help in obtaining these data.

strategies. Some suppliers contact clients directly, which is, of course, very costly. Others interface with them via their Internet platform, which is cheaper, provided enough customers join. A reasonable estimate of the cost of convince a residential customer to switch pricing structure is around €50. In addition, suppliers need to be able to bill customers based on their real-time usage, without too many errors. This requires very heavy investment in information and billing systems.

The final cost is the investment at the customer's site to respond to high prices. For example, a user may purchase a battery or install automatic control devices on appliances and control them remotely.

## 5.5 Policy Implications

This analysis has three main policy implications. First, it is essential that high-consuming customers face real-time prices, as is the case in most restructured power markets. Ensuring this captures most of the benefits of demand response.

Second, the economic case for exposing all residential customers to RTP appears weak. Policy makers have always been hesitant to do so, for fear that customers will find the exposure to volatility of spot prices unbearable. The analysis presented here suggests that the economic benefit is small. Critical-peak pricing appears to be a robust compromise for commercial and small industrial consumers and also possibly for residential consumers. This conclusion is predicated the small economic gain (less than few euros per year and per site) and the high cost (tens of euros per year per site) associated with switching a residential customer to RTP. These may change in the future. In particular, the cost of demand response may plummet as new information and control technologies are developed.

However, all is not lost for demand response. On the contrary, a small fraction of price-reactive customers is sufficient to ensure that administrative curtailment is not required on-peak.

Third, since demand-response benefits lie mostly with large users, the case for complete rollout of smart meters is weakened. This analysis does not constitute a full-blown marginal cost-benefit analysis. First, it does not include operational benefits of smart meters, such as reduction in metering costs and other optimization for the distribution network owner or operator. Second, the marginal value of RTP represents the marginal value of customers modifying their demand for energy according to energy prices but does not include the value of the reliability customers may provide to the system by reducing their load for small durations. Since reliability is often valued more highly than energy, this would increase the value from smart meters. Third, the marginal value of RTP does not include the procompetitive impact of more-elastic demand.

The analysis does not discuss the marginal cost of installing smart meters, in particular, potential (dis)economies of scale. I am not aware of any empirical evidence on the subject.

On the one hand, the marginal cost of producing and installing a smart meter is likely to decrease as their number increases, owing to economies of scale in production and learning by doing in installation. On the other hand, the per unit data gathering, storage, and processing costs may well be significantly higher for 30 million customers than for 30,000. Additionally, the cost of installing a meter is the marginal cost of enabling the switch to real-time price usage. The analysis does not include the cost of informing consumers and inducing them to switch. Nor does it factor in the fact that, for a variety of reasons, not all consumers equipped with smart meters will switch.

When properly accounting for all these, it may be the case that the benefits still exceed the costs. However, the analysis above suggests that a large share of the energy management benefits can be obtained with a much more limited smart meters rollout.

Finally, the above analysis has a commercial implication. Given the small value of price responsiveness compared with the cost of convincing clients to adopt it, developing a profitable residential energy management offer will prove challenging. Energy management firms will target developers and owners and managers of commercial and industrial buildings.

## 5.6 Peak-Time Resale: An Economic Representation

Consider a single representative consumer. Her utility when she consumes energy $q$ is $U(q)$. Her marginal utility is $p(q)$ as in the rest of this text. Wholesale spot price in the state of the world $\theta$ is $p(\theta)$. Marginal cost of production is $c(q)$.

Our customer has purchased from her retailer quantity $\bar{q}(\theta)$ at retail price $p^R$, constant across states of the world. She consumes $q(\theta)$ and resells quantity $(\bar{q}(\theta)-q(\theta))\geq 0$ to the market operator. If she decides not the resell any volume, $q(\theta)=\bar{q}(\theta)$.

All transactions (purchases and sales) go through a single-market operator and are settled at the wholesale spot price $p(\theta)$. This section is derived from Crampes and Léautier (2015), and Astier and Léautier (2016).

### 5.6.1 Accounting

Actual production is equal to actual consumption, which is $q(\theta)$. The market operator purchases $q(\theta)$ from the producers and pays them $p(\theta)q(\theta)$. He also purchases $(\bar{q}(\theta)-q(\theta))$ from the consumer; hence he pays her $p(\theta)(\bar{q}(\theta)-q(\theta))$. Suppose first that the market operator sells to the retailer $q(\theta)$ the quantity actually consumed by his client. His profit is

$$\pi^M(\theta)=-p(\theta)\,q(\theta)-p(\theta)\,(\bar{q}(\theta)-q(\theta))+p(\theta)\,q(\theta)=-p(\theta)\,(\bar{q}(\theta)-q(\theta)).$$

The market operator records a loss whenever the customer actually resells energy she does not consume. Why? Because the market operator actually purchases

$\bar{q}(\theta) = q(\theta) + \bar{q}(\theta) - q(\theta)$. Therefore, he needs to also sell $\bar{q}(\theta)$ to the retailer, which is his only buyer. Then, his profit becomes

$$\pi^M(\theta) = -p(\theta)\, q(\theta) - p(\theta)\left(\bar{q}(\theta) - q(\theta)\right) + p(\theta)\, \bar{q}(\theta)$$
$$= p(\theta)\left(-q(\theta) - \bar{q}(\theta) + q(\theta) + \bar{q}(\theta)\right) = p(\theta) \times 0 = 0.$$

The market operator realizes zero profit in every state of the world, which is the objective. Therefore, to balance the market, the retailer must purchase the consumed and resold quantities.

Suppose the retailer charges the customer only for $q(\theta)$, the energy she actually consumes. His profit is

$$\pi^R(\theta) = p^R q(\theta) - p(\theta)\, \bar{q}(\theta) = \left(p^R - p(\theta)\right) \bar{q}(\theta) - p^R\left(\bar{q}(\theta) - q(\theta)\right).$$

The retailer records loss $p^R\left(\bar{q}(\theta) - q(\theta)\right)$ whenever the customer actually resells energy. Why? Because he has to purchase $\bar{q}(\theta)$ and resells only $q(\theta)$. The retailer must actually sell $\bar{q}(\theta)$. His profit then becomes

$$\pi^R(\theta) = \left(p^R - p(\theta)\right) \bar{q}(\theta).$$

Profit in state $\theta$ may be positive or negative, but the retailer should break even on average (possibly by adding a fixed fee).

### 5.6.2 Relation between Peak-Time Resale and Real-Time Pricing

The consumer first purchases maximum consumption $\bar{q}(\theta)$, then chooses $q(\theta)$ to maximize her net surplus:

$$max_{q(\theta) \leq \bar{q}(\theta)} U\left(q(\theta)\right) + p(\theta)\left(\bar{q}(\theta) - q(\theta)\right) - p^R \bar{q}(\theta).$$

The Lagrangian is

$$\mathcal{L} = U\left(q(\theta)\right) + p(\theta)\left(\bar{q}(\theta) - q(\theta)\right) - p^R \bar{q}(\theta) + \lambda(\theta)\left(\bar{q}(\theta) - q(\theta)\right),$$

hence the first-order condition is

$$p\left(q(\theta)\right) - p(\theta) + \lambda(\theta) = 0.$$

Two situations are possible. First, if $\tilde{q}(\theta)$ uniquely defined by

$$p\left(\tilde{q}(\theta)\right) = p(\theta),$$

satisfies $\tilde{q}(\theta) < \bar{q}(\theta)$. Then, $\lambda(\theta) = 0$, and the consumer chooses $q(\theta) = \tilde{q}(\theta)$. This occurs when $\tilde{q}(\theta)$ is small, hence $p(\theta)$ is high. Second, for low prices, we may have

$\tilde{q}(\theta) \geq \bar{q}(\theta)$. Then, no resale occurs, and the consumer chooses $q(\theta) = \bar{q}(\theta)$. Peak-time resale, indeed, leads to resale at peak time.

Peak-time resale leads to optimal consumption on-peak but does not increase off-peak consumption. It generates lower surplus than does real- time pricing.

Suppose the customer pays only for the energy she consumes. Her program would be

$$max_{q(\theta) \leq \bar{q}(\theta)} U(q(\theta)) + p(\theta)(\bar{q}(\theta) - q(\theta)) - p^R q(\theta).$$

On-peak, the consumer chooses

$$p(\check{q}(\theta)) = p(\theta) + p^R > p(q(\theta)) \Leftrightarrow \check{q}(\theta) < q(\theta).$$

The consumer receives $(p(\theta) + p^R)$ for every unit she resells: the market operator pays her $p(\theta)$, and she saves $p^R$ on her retail purchases. Therefore, she oversells or underconsume so.

## 5.7 An Economic Model of Switching to Real-Time Pricing

This section examines the long-term impact of a marginal increase in the share of price-reactive customers. The model used is that of chapter 2, in which we increase $\alpha$ by a small quantity. The notation is the same, although the dependency on $\alpha$ is made explicit, for example, the price in state $\theta$ is $p(\alpha, \theta)$, and cumulative capacity is $K_N(\alpha)$. As in Borenstein (2005) and Allcott (2012), all values, in particular, generation capacity and mix and retail prices, are optimal, that is, we consider the long-term equilibrium of the industry. As mentioned in chapter 1, this is unrealistic, since installed- generation mix and retail prices are rarely optimal. However, it enables us to isolate the impact of switching to RTP.

### 5.7.1 Impact on Average Price

**Proposition 5.1.** Increasing the share of price-reactive customers has no impact on the expected price.

***Proof.*** As shown in chapter 2, for every $\alpha$, the expected price is equal to the long-term marginal cost of the baseload technology:

$$\mathbb{E}[p(\alpha, \theta)] = r_1 + c_1.$$

Thus

$$\frac{d\mathbb{E}[p(\alpha, \theta)]}{d\alpha} = 0.$$

□

### 5.7.2 Impact on Net Surplus

**Proposition 5.2.** The expected hourly marginal value of switching to RTP is

$$W'(\alpha) = \mathbb{E}\left[S(p,\theta) - pD(p,\theta) - \gamma\left(S\left(p^R,\theta\right) - pD\left(p^R,\theta\right)\right)\right] \geq 0.$$

Furthermore, if inverse demand is linear with constant slope $P(Q) = a(\theta) - bQ$, and rationing does not occur at the optimal capacity, $W'(\alpha)$ is proportional to the variance of the wholesale price, and decreasing with the share of customers on RTP:

$$W'(\alpha) = \frac{1}{2b} var[p(\alpha,\theta)] \text{ and } W''(\alpha) < 0.$$

***Proof.*** From chapter 2, the expected hourly surplus is:

$$\begin{aligned} W(\alpha) = \mathbb{E}\big[&\alpha\left(S\left(p\left(\alpha,\theta\right),\theta\right) - p\left(\alpha,\theta\right)D\left(p\left(\alpha,\theta\right),\theta\right)\right) \\ &+ (1-\alpha)\gamma\left(S\left(p^R,\theta\right) - pD\left(p^R,\theta\right)\right)\big]. \end{aligned}$$

Since all decisions (i.e., installed capacity, wholesale spot prices, serving ratios, and retail price) are optimal for every $\alpha$, we apply the envelope theorem, which yields

$$W'(\alpha) = \frac{\partial W}{\partial \alpha} = \mathbb{E}\left[S\left(p^*,\theta\right) - p^*D\left(p^*,\theta\right) - \gamma\left(S\left(p^{R*},\theta\right) - p^*D\left(p^{R*},\theta\right)\right)\right].$$

Then, since $\gamma \leq 1$ and $p(\theta) = argmax_x\left\{S(x,\theta) - pD(x,\theta)\right\}$,

$$\gamma\left(S\left(p^{R*},\theta\right) - p^*D\left(p^{R*},\theta\right)\right) \leq S\left(p^R,\theta\right) - pD\left(p^R,\theta\right) \leq S(p,\theta) - pD(p,\theta).$$

Thus

$$\begin{aligned} W(\alpha) \geq \mathbb{E}\Big[&\left(S(p^*,\theta) - p^*(\alpha,\theta)D\left(p^*,\theta\right)\right) - \Big(S\left(p^{R*},\theta\right) \\ &- p^*(\alpha,\theta)D\left(p^{R*},\theta\right)\Big)\Big] \geq 0. \end{aligned}$$

Suppose demand is linear with constant slope $P(q,\theta) = a(\theta) - bq$, and rationing does not occur at the optimal capacity. We prove formally below the result obtained by observing figures 5.2 and 5.3. Since demand is linear $D(p,\theta) = \frac{a(\theta)-p}{b}$,

$$\begin{aligned} S(p,\theta) &= \int_0^{D(p,\theta)} \left(a(\theta) - bq\right)dq \\ &= \left(a(\theta) - \frac{b}{2}D(p,\theta)\right)D(p,\theta) = \left(\frac{a(\theta)+p}{2}\right)D(p,\theta). \end{aligned}$$

The net surplus, represented as triangles in figures 5.2 and 5.3, is

$$S(p,\theta)-pD(p,\theta)=\left(\frac{a(\theta)+p}{2}-p\right)D(p,\theta)=\left(\frac{a(\theta)-p}{2}\right)D(p,\theta)=\frac{b}{2}D^2(p,\theta).$$

Since no rationing occurs,

$$\begin{aligned}S\left(p^R,\theta\right)-pD\left(p^R,\theta\right)&=\left(a(\theta)-\frac{bD\left(p^R,\theta\right)}{2}-\left(a(\theta)-bD(p,\theta)\right)\right)D\left(p^R,\theta\right)\\&=\frac{b}{2}\left(2D(p,\theta)-D\left(p^R,\theta\right)\right)D\left(p^R,\theta\right).\end{aligned}$$

Thus

$$\begin{aligned}W'(\alpha)&=\frac{b}{2}\times\mathbb{E}\left[D^2(p^*,\theta)-\left(2D(p^*,\theta)-D\left(p^{R*},\theta\right)\right)D\left(p^{R*},\theta\right)\right]\\&=\frac{1}{2b}\mathbb{E}\left[\left(D(p^*,\theta)-D\left(p^{R*},\theta\right)\right)^2\right]=\frac{1}{2b}\mathbb{E}\left[\left(p^*-p^{R*}\right)^2\right].\end{aligned}$$

Then, since (a) the fixed retail price is chosen optimally, (b) demand is linear with constant slope, and (c) no rationing occurs, $p^{R*}=\mathbb{E}\left[p^*\right]$, thus $W'(\alpha)=\frac{1}{2b}var\left[p^*\right]$.

Full differentiation of $W'(\alpha)$ with respect to $\alpha$ yields

$$W''(\alpha)=$$

$$\mathbb{E}\left[\begin{array}{c}\left(\frac{\partial S(p^*,\theta)}{\partial p}-p^*\frac{\partial D(p^*,\theta)}{\partial p}-D(p^*,\theta)\right)\frac{\partial p^*}{\partial\alpha}-\gamma\left(\frac{\partial S(p^{R*},\theta)}{\partial p}-p^*\frac{\partial D(p^{R*},\theta)}{\partial p}\right)\frac{\partial p^{R*}}{\partial\alpha}\\-\left(S(p^*,\theta)-p^*D(p^*,\theta)\right)\frac{\partial\gamma^*}{\partial\alpha}+\gamma D\left(p^{R*},\theta\right)\frac{\partial p^*}{\partial\alpha}\end{array}\right].$$

Most terms are equal to zero. First, $\frac{\partial S}{\partial p}=p\frac{\partial D}{\partial p}$. Second, $\mathbb{E}\left[\gamma\left(\frac{\partial S(p^{R*},\theta)}{\partial p}-p^*\frac{\partial D(p^{R*},\theta)}{\partial p}\right)\right]$ $=0$ by construction of $p^{R*}$. Finally, $S(p^*,\theta)-p^*D(p^*,\theta)=0$ when rationing occurs, while $\frac{\partial\gamma^*}{\partial\alpha}=0$ when it does not. Thus

$$W''(\alpha)=\mathbb{E}\left[\left(\gamma D\left(p^{R*},\theta\right)-D(p^*,\theta)\right)\frac{\partial p^*}{\partial\alpha}\right].$$

Léautier (2014a) proves that, if demand is linear with constant slope and no rationing occurs, $W''(\alpha)<0$. □

Since the wholesale spot price is equal to the VoLL, constant-price customers derive no surplus when they are rationed. The marginal value of switching to RTP is therefore $(S(p^*,\theta)-p^*D(p^*,\theta))$. It is much higher when customers are rationed than when they are not. This effect should be even stronger if rationing is not perfectly anticipated. This observation confirms the intuition that, ceteris paribus, the marginal value of RTP is higher when customers are curtailed.

### 5.7.3 A Closed-Form Solution

This section derives a simple expression of the marginal value of switching to RTP for a single production technology. As indicated earlier, Léautier (2014a) extends the results to multiple technologies.

**Proposition 5.3.** If only production technology is present, the expected hourly marginal value of switching to *RTP* is:

$$W'(\alpha) = \frac{1}{b}\left\{\frac{1}{2+\lambda}\left[\left(\frac{a_1}{\alpha}\right)^{\lambda}(1+\lambda)\,r^{\lambda+2}\right]^{\frac{1}{\lambda+1}} - \frac{r^2}{2}\right\}, \tag{5.1}$$

where $\lambda = \frac{\lambda_1}{\lambda_2}$.

***Proof.*** We have

$$\begin{aligned} var\left[p(\theta)\right] &= var\left[p^*(\theta) - c\right] = \mathbb{E}\left[\left(p^*(\theta)-c\right)^2\right] - \left(\mathbb{E}\left[p^*(\theta)-c\right]\right)^2 \\ &= \int_{\hat{\theta}_0(K^*,c)}^{+\infty} \left(\rho\left(K^*,\theta\right)-c\right)^2 f(\theta)\,d\theta - r^2, \end{aligned}$$

since $\mathbb{E}\left[p^*(\theta)\right] = c + r$. Then, integrating by parts twice,

$$\begin{aligned} \int_{\hat{\theta}_0(K,c)}^{+\infty} (\rho(K,\theta)-c)^2 f(\theta)\,d\theta &= 2\int_{\hat{\theta}_0(K,c)}^{+\infty} (\rho(K,\theta)-c_1)\frac{\partial\rho}{\partial\theta}(1-F(\theta))\,d\theta \\ &= 2\frac{a_1\lambda_2}{\alpha}\int_{\hat{\theta}_0(K,c)}^{+\infty} (\rho(K,\theta)-c)\,e^{-(\lambda_1+\lambda_2)\theta}\,d\theta \\ &= 2\frac{a_1^2}{\alpha^2}\frac{\left(e^{-\lambda_2\hat{\theta}_0(K,c)}\right)^{2+\lambda}}{(1+\lambda)(2+\lambda)} = 2\frac{a_1^2}{\alpha^2}\frac{\left(\frac{\alpha(1+\lambda)r}{a_1}\right)^{\frac{2+\lambda}{1+\lambda}}}{(1+\lambda)(2+\lambda)}, \end{aligned}$$

since $e^{-\lambda_1\hat{\theta}_0(K^*,c)} = \left(\frac{\alpha(1+\lambda)r}{a_1}\right)^{\frac{\lambda}{1+\lambda}}$. Algebra yields

$$var\left[p^*(\theta)\right] = \frac{2}{2+\lambda}\left[\left(\frac{a_1}{\alpha}\right)^{\lambda}(1+\lambda)\,r^{\lambda+2}\right]^{\frac{1}{\lambda+1}} - r^2,$$

which proves the result. □

The variance of the price is equal to the variance of the operating margin $(p^*(\theta) - c_1)$, which is positive only for $\theta \geq \hat{\theta}_0(K,c)$. The variance is the expectation of the square of the operating margin minus the square of the expected operating margin. At the optimum, these terms can be expressed as a function of the cost of capital $r$.

For $N > 1$, the logic is similar. Additional terms corresponding to inframarginal technologies are added, and

$$W'(\alpha) =$$

$$\frac{1}{b}\left(\begin{array}{c} \frac{y(K_N^*, c_N, \alpha)^{2+\lambda}}{\alpha^2 a_1^\lambda (1+\lambda)(2+\lambda)} - \frac{r_1^2}{2} \\ + \sum_{n=1}^{N-1}\left((c_{n+1} - c_n)\left(r_{n+1} - \frac{y(K_n^*, c_{n+1}, \alpha)^{1+\lambda}}{\alpha a_1^\lambda (1+\lambda)}\right) + \frac{y(K_n^*, c_n, \alpha)^{2+\lambda} - y(K_n^*, c_{n+1}, \alpha)^{2+\lambda}}{\alpha^2 a_1^\lambda (1+\lambda)(1+2\lambda)}\right)\mathbb{I}_{(N>1)} \end{array}\right),$$

where $\mathbb{I}_{(N>1)} = 1$ if $N > 1$, and 0 otherwise, and

$$y\,(K, c, \alpha) = a_1 e^{-\lambda_2 \hat{\theta}_0(K,c)} = a_0 - bK - \left(\alpha c + (1-\alpha)\,p^R\right).$$

### 5.7.4 Annual Marginal Surplus

**Annual marginal surplus per percent** $W_a'(\alpha)$ the expected annual value of switching to RTP is 8,760 times the hourly value: $W_a'(\alpha) = \frac{8{,}760}{2b} var\left[p^*\right] = \frac{8{,}760}{2bQ^\infty} Q^\infty var\left[p^*\right]$.

The associated units are

$$hours/year \times GW \times (\text{€/MWh})^2 \,/\, (\text{€/MWh}) = hours/year \times GW \times \text{€}/MW/hours$$
$$= 10^3\text{€} \times \text{year}^{-1}.$$

Thus $\frac{W_a'(\alpha)}{10^3} = \frac{8.76}{2b} var\left[p^*\right]$ is the marginal value of switching to RTP, expressed in millions per year.

Easier to understand is the marginal value for a 1 percent increase in $\alpha$. Formally, observing that $W_a'\left(\alpha\right) \simeq \frac{\delta W_a}{\delta \alpha}$, we have $\delta W_a\left(\alpha\right) = W_a'\left(\alpha\right) \times \delta\alpha$. Since a 1 percent increase corresponds to $\delta\alpha = 10^{-2}$, the annual marginal value of switching 1 percent of customers to RTP, expressed in € millions per year per percent, is $\delta W_a\left(\alpha\right) = \frac{W_a'(\alpha) \times 10^{-2}}{10^3} = \frac{8.76 \times 10^{-2}}{2b} var\left[p^*\right]$.

**Annual marginal surplus per site** Suppose customers can be grouped in $I$ classes, indexed by $i = 1, \ldots, I$. All customers have identical demand profiles. Customers are identical within each class, and sizes vary across classes. Denote $N_i$ the number of class $i$ consumers, and $\Delta\alpha_i$ their share of total demand. Since consumers are identical within a class, the share of total demand of a single class $i$ consumer is $\delta_i \alpha = \frac{\Delta\alpha_i}{N_i}$. Then the marginal value of switching one class $i$ consumer to RTP is $\delta W_{ai}\left(\alpha\right) = W_a'\left(\alpha\right) \times \delta_i \alpha$.

In the main text, the share of total demand of class $i$ consumers is given for nonresidential consumers. For residential consumers, $\Delta_i \alpha$ must be computed using the data provided. Denote $I_0$ the set of classes corresponding to residential customers, that is, a class $i$ consumer is residential if and only if $i \in I_0$, and $s_i$ the size of meters of all class $i$ consumers. The share of total demand of class $i$ consumers is $\Delta_i\alpha = \frac{N_i s_i}{\sum_{j\in I_0} N_j s_j} \sum_{j\in I_0} \Delta\alpha_j$; hence the share of total demand of a single class $i$ consumer is $\delta_i\alpha = \frac{s_i}{\sum_{j\in I_0} N_j s_j} \sum_{j\in I_0} \Delta\alpha_j$.

For example, data provided by the French Energy Regulatory Commission indicate that the total share of demand of residential customers is $\sum_{j\in I_0} \Delta\alpha_j = 32\%$, the sum of the meters size for all residential customers is $\sum_{j\in I_0} N_j s_j = 225.9$ million kVA, and the number of small residential customers, that is, customers whose meter size lies between 3 and 6 kVA is $N_i = 18.6$ million. Therefore, the share of total demand of small residential customers is $\Delta_i\alpha = \frac{18.6\times 6}{225.8} \times 32\% = 15.8\%$, and the share of total demand of a single small residential customer is $\delta_i\alpha = \frac{6}{225.8} \times 32\% = 8.5 \times 10^{-9}$.

# III NETWORK ISSUES

# 6 Fred Schweppe Meets Marcel Boiteux: Transmission Pricing

Everybody talks about loop flow, but nobody does anything about it. Most prevailing firm transmission rights are specified in terms of "contract paths" or "interface transfer capabilities" that do not address the special conditions in electric networks. The present paper suggests the use of a "contract network" as a basic building block of a market in power transmission. A contract network and the associated rights can accommodate a system for short-term efficient pricing and long-term firm use of a transmission network.
—William Hogan (1992)

## 6.1 Introduction

### 6.1.1 A New Set of Problems

When the power industry was restructured, policy makers had to develop rules for transmission access, pricing, and expansion. This was a new set of problems. Historically, access and pricing were nonissues for vertically integrated regional monopolies. Expansion followed engineering and economic analyses conducted by the utility: the transmission grid was built to transport electricity from the production centers (usually in the countryside or near mines) to the consumption centers (usually towns). Its cost was considered part of the cost of developing new generation facilities and was included in the bundled regulated rate. This introductory section exposes how these problems have been solved. The rest of this chapter presents the underlying analytics.

### 6.1.2 Access

Ensuring fair access to the transmission grid for all market participants was, of course, essential. Vertical separation emerged as the most pragmatic solution. Ownership and operations of transmission assets were handed over to an independent company, separate from all other market participants, called a Transco, or a transmission system operator (TSO) in Europe. In the United States, utilities retained ownership of the transmission assets while operations of transmission systems were handed over to independent non profit entities called independent system operators (ISOs). In France and Germany, transmission assets ownership and operations were entrusted to independent separate companies, yet still fully owned by the utility, sometimes called legally separated transmission system operators (LTSOs). The limited number of complaints by market participants suggests that, even without full vertical separation, fair access to the transmission grid has been secured.

### 6.1.3 Pricing

#### 6.1.3.1 The challenge

Prices users pay to access and use the transmission grid aim to achieve three objectives: to elicit the right production and consumption decisions in real time, to elicit the right investment decisions, and to cover the full costs of the network.

Transmission pricing is a complex issue. The difficulty arises from the laws of physics that rule power flows on a grid, transmission-capacity limits, and the cost structure of a transmission grid.

**Laws of physics** One does not really "move electric power" as one moves a crate of tomatoes. Rather, one creates an electric current (technically an electromagnetic wave), which moves on the power grid following the laws of physics, not the law of economics. These have three consequence:

First, energy losses: electric current heats up conductors. This heat is dissipated in the atmosphere, hence it is lost. A producer needs to produce around 105 Megawatt-hours for a client to consume 100 Megawatt-hours.

Second, loop flows: energy sold by a French producer to a German consumers travels from France to Germany but also travels through Belgium, the Netherlands, Switzerland, and other countries.

Finally, congestion: when the flow from France to Great Britain is equal to the capacity of the interconnection, the latter is congested, and it is impossible to increase exports from France into Great Britain.

**Transmission-capacity limits** Power flows on transmission lines are limited. Thus a power market can be viewed as a series of power islands linked by bridges of limited capacity. When the traffic is low, it flows freely. When traffic is high, congestion sets in.

Transmission-capacity limits arise for two reasons. First, there are thermal limits: if power flowing on a line is too high, the line will heat up and may break. Alternatively, the line will sag, and may touch the trees, which would produce a short circuit.

Second, there are operating limits. If a power plant or another line on the network fails, power flows are instantaneously rearranged, following the laws of physics. The operating limit on each line is such that, in the event of one (or more) failure on the system, the resulting flow on this line does not exceed the physical limit. This is called the $(N-1)$ criterion, or the single-contingency rule: the system is operated to withstand the loss of one major component but no more. Some system operators use a $(N-2)$ criterion and operate their system to withstand the loss of two major components.

Operating limits are often much lower than thermal limits. For example, in the early 2000s, the thermal capacity of the Washington, D.C., interconnection linking Quebec to the center of New England was 2,000 MW, while its operating limit was hovering around 1,200 to 1,300 MW. This was economically costly for Hydro-Québec, which used the interconnection to export cheap electricity into New England, and for New England load serving entities, who bought it. More recently, in Europe, the Agency for Cooperation of Energy Regulators (ACER) reports that, on average over 2015, the operating limit on the Belgium to Netherlands interconnection was around 25 percent of the thermal limit.

Operating limits need not be identical in both directions, since the resulting flows in case of a failure are one-directional. For example, in 2015, the interconnection between Switzerland and France was operated at less than 20 percent of thermal capacity from Switzerland to France and at more than 40 percent in the reverse direction.

This operating practice was legitimate in the 1950s, but its cost is prohibitive today: on the specific example of the Belgium to Netherlands interconnection, only a quarter of the invested capital is used (on average). Transmission system operators and asset owners are installing measurement devices and developing algorithms to manage the operating limits dynamically, that is, to meet dynamically the $(N-1)$ criterion. In addition, recent analysis (Ovaere 2017) suggests that the $(N-1)$ criterion itself could be made dynamic. This is an exciting development that will increase the usage of the grid. It does not modify the economic analysis presented in this chapter.

**Cost structure** As was seen in chapter 2, the variable cost of producing electric power is approximately proportional to output, and the fixed cost is approximately proportional to its installed capacity. The cost of moving electric power on one transmission line is not so simple.

The variable cost of transmitting energy when the line is not congested is the cost of transmission losses. It is not proportional to the energy flow. Under a reasonable approximation, transmission losses are proportional to the square of the energy flow.

The fixed cost of transmitting energy is the cost of developing, building, and operating a transmission line. It is not proportional to the capacity of the line. It can be approximated as a fixed part, independent of the capacity of the line, and a part proportional to the capacity of the line. A transmission network therefore exhibits strong returns to scale: it is much cheaper to build a transmission line of capacity 200 MW than to build two parallel transmission lines of capacity 100 MW.

#### 6.1.3.2 The solution: Locational marginal prices

Fortunately (and somewhat surprisingly) engineers and economists have developed a simple and elegant solution to the problem of transmission pricing, called locational marginal prices (LMPs).

The first seminal contribution was produced by Fred Schweppe and his colleagues (1998), who derive the optimal spot prices for electricity, including energy losses, loop flows, and congestion. This work generalizes peak-load pricing presented in chapter 2 to include spatial differentiation. At every point of the grid (called a node), electricity price is determined by the balance of local supply and demand, hence it is equal to the marginal cost of the last Megawatt-hour produced or to the value of the last Megawatt-hour consumed at this node (or both), as in chapter 2. In addition, electricity prices at different nodes are related: the price consumers pay is the price at a reference node plus their consumption's marginal contribution to losses and to congestion. Similarly, the price producers receive is the price at a reference node plus their production's marginal contribution to losses and to congestion. Under a reasonable approximation of the equations governing power flows, these last terms have a simple expression.

Like Marcel Boiteux in 1949, Fred Schweppe and his colleagues had a vertically integrated utility in mind when they derived the optimal sport prices. In 1992 William Hogan (1992) made another seminal contribution by showing how Schweppe's analysis can be used to solve the transmission-pricing problem in restructured electricity markets. This article is perhaps the most influential academic contribution to the design of electricity markets.

Hogan's intuition is that transmission does not need to be explicitly priced, rather, the cost of moving power from node A to node B is implicitly defined as the difference between Schweppe's prices at nodes B and A, called nodal prices, or LMPs. If the network is not congested, the cost of transmission is simply the marginal cost of losses. Otherwise, the cost of transmission also includes the marginal contribution to congestion. Hogan found a simple and elegant solution to an apparently intractable problem. This transmission-pricing approach is called nodal pricing or locational marginal pricing.

Implicitly pricing transmission leaves market participants exposed to the difference in nodal prices, which is known only ex post. Hogan's second contribution was to propose the creation of financial transmission rights (FTRs), which grant market participants the difference between nodal prices at two points of the grid. If a producer located in upstate New York wants to sell its power to a consumer in Long Island, he can purchase an FTR between these two points and hence lock in his profit margin.

As usual in economics, if competition is perfect, setting transmission price at the correct marginal cost for a given network generates incentives for optimal consumption and production in the short term and optimal investment in generation assets and consumption centers in the long term.

#### 6.1.3.3 Transmission pricing in practice

In the late 1990s, some markets in the United States experimented with zonal pricing, that is, the independent system operators defined "zones" on the network, declared a single price for each zone, and "redispatched" plants that were effectively constrained on or off to manage congestion within each zone. Eventually, all ISOs in the United States adopted nodal pricing.

By contrast, at the end of 2017, most European markets are "coupled": day-ahead congestion between countries is managed using zonal pricing (i.e., nodal pricing taking each country as a "node"), and congestion within each country is managed by countertrading.

The European approach to transmission pricing is economically inefficient. Nodal pricing being optimal, it generates a higher net surplus than other approaches, in particular, countertrading. Two studies have quantified the magnitude of these gains. Green (2007) finds that nodal pricing would increase net surplus by 1.3 percent in Great Britain compared with countertrading (which is the current method) if markets are perfectly competitive, and 3.1 percent if producers exercise their market power. More recently, Neuhoff et al. (2013) quantify the impact of nodal pricing in Europe, for different scenarios of renewable penetration. The study estimates gains compared with the status quo (market coupling, previously described). It finds that nodal pricing increases cross border trade (in MW) by up to 34 percent, reduces operating costs by €0.8 to 2.0 billion per year (1.1 to 3.3%), depending on the renewable penetration, and reduces average price in 60 to 75 percent of countries.

Since nodal pricing increases net surplus, why is it not adopted in Europe as it is in North America? One possible argument is that nodal pricing is too complex and difficult to implement. However, while nodal pricing is complex, it has been successfully implemented in numerous markets in North America, Australia, and New Zealand, which suggests the technology is now mature, hence this argument is not compelling.

Another argument is that nodal pricing requires a centralized market architecture (i.e., a central market operator runs a double auction for all producers and all consumers and retailers), as in the United States, and is incompatible with the European decentralized market architecture, where each country's TSO modifies the balanced schedules it receives from market participants to respect transmission constraints. While nodal pricing was designed for centralized markets, it can be made to work in the European context. Within each market (national or a grouping of national markets), the market operator would compute nodal prices and use them to settle transactions. Then, the algorithm currently used for coupling these markets would be modified to reflect nodal pricing in each market. Again, this argument is not sufficient to explain the divergence in approaches.

I believe the difference between Europe and North American can be explained by the history and political economy of both systems. Until very recently, congestion was much less severe within each European country than in U.S. markets (Léautier and Thelen 2009). In the United States, the transmission grids were designed to eliminate congestion within each utility's service territory, which was much smaller than current markets. When markets were created in the United States in the late 1990s, congestion was significant, hence market designers had to find an approach to manage them effectively. Credit for the use of nodal pricing should go to Bill Hogan, who was a tireless—and very effective—advocate.

In Europe, by contrast, the transmission grids were designed to eliminate congestion within each country. Many TSOs argue that adopting nodal pricing will generate significant information-system costs, which would most likely exceed the congestion cost in each country. While this argument was probably true in 2016, it ignores the main benefit of nodal pricing: managing congestion effectively between countries. The construction of a European common market has been slow. For example, market coupling was implemented only in 2015, almost twenty years after the first directive restructuring the electricity industry. The cost of congestion, which occurs mostly at the boundaries between countries, has been less visible than in the United States, hence the impetus to manage it less strong. In addition, Europe still does not have a single pan-European TSO responsible to optimize the use of the transmission grid. Multiple national TSOs cooperate to perform this function; hence the responsibility to manage congestion cost cannot be attributed to a single agency, which dilutes incentives. Finally, Europe does not have a nodal-pricing advocate with Hogan's persuasion skills.

I am, however, cautiously optimistic that Europe will adopt nodal pricing. As discussed in chapter 8, the production-technology mix will be transformed by 2050: Renewable energy sources (RES) are progressively replacing fossil-fuel plants. These new RES will be located in different places. In addition, their production is variable. This will most likely increase congestion on the transmission grid. This can already be observed in Germany, where electricity produced by wind turbines located in the North must be transported to factories located in the South, thus congesting the common grid—and neighboring countries as well. Since building new lines is extremely difficult, the priority will be to manage congestion most effectively. Nodal pricing will offer an efficient approach.

### 6.1.4 Transmission Expansion

#### 6.1.4.1 Underlying economics

Locational marginal prices signal the marginal value of transmission capacity. Formally, analysis of transmission grid expansion is similar to the analysis of generation expansion presented in chapter 2. In the latter, generation capacity has value only at peak, when demand is equal to generation capacity. Optimal generation capacity equalizes the expected operating margin of the plant during on-peak hours with the fixed cost of capacity. Since entry in generation is (reasonably) easy, and the cost of generation capacity is approximately proportional to the size of an asset, the competitive equilibrium reaches the optimum.

Similarly, capacity on a transmission line has value only when the line is congested. In the simplest case of two markets linked by an interconnection, the marginal value of transmission capacity is the expected price difference between the two markets. In a more complex network, the marginal value of capacity on a transmission line is a linear combination of LMPs. Optimal transmission capacity equalizes this marginal value with the marginal cost of capacity.

Unfortunately, the parallel with generation expansion stops here. First, there is no free entry in transmission: the number of suitable transmission sites is physically limited; hence competition among transmission providers is necessarily imperfect. Second, since the fixed costs of transmission capacity are significant, the congestion revenues at the optimal transmission grid do not cover the fixed capacity costs. These create policy challenges, as discussed next.

#### 6.1.4.2 Not In My BackYard (NIMBY)

Building new transmission corridors or significantly expanding existing corridors is extremely difficult. Many transmission projects run into significant opposition by local communities. Consumers gladly use electricity, but citizens' view of transmission lines is best summarized by the acronym NIMBY. In the best case (for their sponsor), this leads to delays and cost increases: all or portions of the line need to be buried underground. In the worst case, they must be canceled altogether.

This opposition accelerates the transition to smart grids. Since transmission engineers can install only very limited new hardware, they will develop software to improve the usage of the transmission grid. We have already mentioned two avenues: dynamic line rating to replace the static $(N-1)$ criterion and widespread use of nodal pricing. Transmission system engineers will explore others as well.

#### 6.1.4.3 Regulated transmission expansion

In OECD countries, increasing the capacity of existing grids to accommodate RES will be the large majority of transmission-expansion projects. These will include a mix of "smarter" operating practices, software that will increase the rated capacity of existing assets, and new physical assets. Economic analysis delivers a good piece of news: if, as part of the incentives included in the regulatory contract, a for-profit transmission-asset owner and operator is made responsible for the congestion cost (suitably defined), she faces incentives to operate and expand the grid optimally. A version of this incentive mechanism was successfully implemented in England and Wales from 1990 to 2006.

Such an incentive mechanism cannot be implemented in the United States, since ISOs are separate from asset owners. Since the former are not for-profit entities, it is almost impossible to subject them to incentive regulation. Since the latter do not operate their assets, it is almost impossible to make them responsible for an outcome (at least partly) outside of their control. Regulators in the United States have devised other measures to encourage transmission expansion, such as "augmented" allowed rates of return. I am unsure whether this will prove sufficient to deliver the transmission super highway required to accommodate new power flows resulting from the inclusion of RES.

#### 6.1.4.4 Merchant transmission expansion

**The case for merchant transmission** Merchant generation investors finance an unregulated-generation asset and remunerate their investment through the wholesale electricity prices. Similarly, merchant-transmission investors finance an unregulated transmission interconnection and remunerate their investment through the value of the FTRs generated by this expansion, as revealed in wholesale electricity markets. Merchant generation and transmission companies can enter into forward contracts for part of all of their capacity or opt to be fully exposed to the spot market. The forward contract prices are related to the expected wholesale spot prices.

Many observers consider that merchant transmission is an essential ingredient to well-functioning power markets. In particular, merchant developers are likely to invest at the seams between two systems (e.g., two countries in Europe, two ISOs in the United States, or two provinces in Australia), a situation not always well covered by incumbent transmission-asset owners and national or statewide regulators. Since they are taking the price risk, merchant developers do not require a full-fledged rate case; hence they can be faster than regulated companies in bringing projects to market. However, two issues limit the scope of merchant transmission.

**A technical issue** In the simple case of an interconnection linking two markets, increasing interconnection capacity increases one-for-one the volume of available FTR. This is not true in general. In the case of the three-node network, increasing transmission capacity on the congested interconnection by one Megawatt creates three and a half Megawatt additional FTRs. Furthermore, increasing capacity on interconnection A may affect the volume of FTRs between points B and C. If this volume is increased, it seems normal that the merchant investor in interconnection A receives the newly generated FTRs. If the volume is decreased, the merchant investor in interconnection A will have to compensate holders of FTRs between points B and C, or owners of the underlying transmission assets generating these FTRs.

This computation may prove difficult for complex networks. For this reason, most merchant projects (all that I am aware of) rely on a direct current (DC) interconnection, which is isolated from the surrounding alternating current (AC) system. While a DC interconnection requires one costly transformer at each extremity of the line, merchant developers find it more effective than negotiating with incumbent TSOs or ISOs the amount of FTRs generated by their interconnection.

**Economic issues** By construction, merchant-transmission asset owners receive the congestion rent on their line. Since the lines are often DC interconnections between two markets, they receive the price difference times the capacity of the line. Joskow and Tirole (2005) examine in detail the issues associated with merchant transmission. The analysis boils down to two ingredients previously discussed: imperfect competition and coverage of fixed costs.

In practice, the number of potential sites for merchant lines linking two markets is limited. Competition among merchant developers is therefore imperfect. They will quite likely behave as Cournot oligopolists and invest less than optimal or will practice limit pricing.

Furthermore, if merchant developers invested optimally, they would cover their variable cost but not their fixed costs. Again, this will lead to lower-than-optimal expansion.

**A policy dilemma** Merchant transmission poses a dilemma for policy makers. On the one hand, incumbent-transmission asset owners have been perceived as reluctant to invest in new lines, in particular, interconnections between different systems. This suggests that allowing new entrants, such as merchant-transmission providers, would increase investment and net surplus. On the other hand, merchant providers structurally invest less than would be optimal. The interaction between merchant and regulated transmission remains an unsettled issue, both theoretically and practically.

### 6.1.5 Generation Expansion

#### 6.1.5.1 Expansion under nodal pricing

Chapter 2 has shown that, if competition is perfect, peak-load pricing in a single market leads to optimal generation investment, that is, the long-term equilibrium-generation mix is optimal. If congestion is managed using nodal prices, this result extends to multiple markets on a transmission grid: nodal prices produce optimal location incentives for producers. This is an application of a general result in economics: if competition is perfect, short-term equilibrium prices lead to efficient production and consumption decisions in the short term but also to efficient investment decisions (absent economies of scale).

#### 6.1.5.2 Expansion under countertrading

If congestion is managed by countertrading, spot prices and constrained payments do not produce optimal location incentives for producers. These prices can be complemented by incentives included in the transmission rate, which guide producers to locate in the "right" regions. However, the resulting incentives are usually not optimal.

#### 6.1.5.3 Coordination between generation and transmission expansion

Historically, vertically integrated utilities used an integrated resource planning process, that is, were coordinating generation and transmission investment. Vertical separation has broken this coordination, hence creating risks that lead to inefficient investment. For example, in the United States and in Europe, investment in the transmission grid has not kept pace with the growth of RES, and in some regions, the latter have to be curtailed for many hours for lack of transmission capacity.

While lack of coordination is undeniable, reality is more nuanced. First, as previously discussed, NIMBY is the main challenge facing transmission expansion, not coordination with generation expansion. Grid companies know which new lines to build to accommodate changing supply and demand patterns; they are often not able to do it.

Second, ISOs and TSOs in North American and in Europe attempt to coordinate investment by compiling and publishing forward projections for generation, demand, and transmission.

Third, for any economic activity, when moving from planning to market-based investment, one moves from a static optimization to a dynamic process mediated by prices, which, under reasonable conditions, converges toward the long-term optimum. For a given transmission grid, nodal prices send optimal locational signals for generation investment. As the generation mix evolves, the appropriate regulatory contract can induce a for-profit transmission-asset owner and operator to expand the transmission grid optimally. Supply and demand conditions, and so the long-term optimum, constantly change. Adapting to these new conditions is the superiority of the dynamic optimization.

Unfortunately, investors do not receive the correct price signals. I am not aware of regulatory mechanisms providing congestion-based incentives for transmission expansion. As mentioned earlier, such provisions were in place in England and Wales from 1990 to 2006, but have since been terminated, as the National Grid company, the transmission-asset owner and operator, was not able to meet his target for a few years in a row. If transmission-asset owners in the United States and in Europe received congestion-based incentives, they would be more likely to increase the transfer capacity of key corridors (through hardware of software). Similarly, if RES producers in Europe received nodal prices for the electricity they produce (plus possibly a premium, as discussed in chapter 8), ceteris paribus, they would be more likely to include their impact on the grid when selecting a location.

In summary, the priority for policy makers is to design and implement congestion-based incentives, leveraging the theoretical arguments presented later in this chapter, and use LMPs in Europe, rather than design mechanisms to coordinate generation and transmission investments.

### 6.1.6 Fixed-Cost Recovery

Finally, if the variable cost of transmission capacity is approximately proportional to the capacity, the congestion rent at the optimal capacity covers exactly the variable cost of transmission capacity but not the fixed cost of the grid. Therefore, additional revenues must be raised to cover the full cost of the grid.

Historically, the total cost to be covered was spread across consumers and producers, through a mix of usage charge (expressed in $/MWh) and access charges (expressed in $/MW of peak demand or peak injection). These different approaches had different incentives properties, for example, an access charge should lead to reduction in peak use of the grid, while a usage charge should lead to a reduction in average use. However, distortions were considered, rightly or wrongly, to be of secondary importance.

Matters are different today, owing to the deployment of decentralized generation technologies and the future deployment of localized storage technologies. Consider a residential user, paying \$60/MWh for network charges (transmission and distribution) and \$80/MWh for energy (supposing that retailer's cost and margin are zero). Our user can install a solar panel on his roof, which produces at \$100 /MWh. Since the electricity produced by the panel costs \$100/MWh, more than the wholesale spot price of \$80/MWh, installing this solar panel is economically inefficient. Our user, however, compares the cost of consuming the electricity he produces (\$100/MWh) to the cost of electricity he purchases from the grid (\$140/MWh). He then rationally installs the panel.

The story does not stop here. The total network cost is unchanged, but the volume drawn from the network is reduced. Thus the network charge per unit increases. This then encourages other users to go off grid. The cycle continues.

This outcome is economically inefficient. Redesigning transmission and distribution rate structures is a priority for regulators and policy makers in multiple countries and a promising area of research.

### 6.1.7 This Chapter's Structure

This chapter presents the economic analysis underlying the arguments presented above. The first sections use the simplest example: two markets linked by an interconnection. This enables readers to appreciate the main insights in a simple and analytically tractable environment. Furthermore, since we assume that a single production technology with constant marginal cost is located in each market, it illustrates the relationship between congestion and peak-load pricing. Section 6.2 presents and compares two commonly used approaches to managing short-term congestion: nodal pricing and redispatching/or countertrading. Section 6.3 discusses long-term issues: optimal expansion of the transmission grid, and optimal generation mix.

Sections 6.4 and 6.5 discuss implementation of the optimum. Section 6.4 discusses incentives for optimal generation investment under the two congestion-pricing regimes introduced earlier. It proves that, while nodal pricing leads to optimal generation investment, redispatch and countertrading, even if a differentiated connection cost is included, do not lead to optimal generation investment. Section 6.5 discusses transmission expansion for two institutional settings: a monopoly and merchant-transmission developers.

Section 6.6 extends the previous analysis to a three-node network, introducing loop flows, one of the most surprising features of power markets. Loop flows have a significant impact on market design, but the economic intuitions remain unchanged.

Finally, section 6.7, intended for researchers, derives the previous results for a general $N$-node power network.

## 6.2 Short-Term Congestion Management

Two approaches exist to manage congestion: nodal pricing and redispatching, the latter sometimes called countertrading. They are presented successively in this section. First, we set up the problem.

### 6.2.1 Set-Up

#### 6.2.1.1 Demand and supply

Consider two markets indexed by $m = 1, 2$. To simplify the analysis, suppose that (a) each market is a single point on the network, called a node, and (b) thermal losses are negligible. For $m = 1, 2$, denote $Q^s_m(\theta)$ and $Q^d_m(\theta)$ the aggregate quantities produced and consumed in market $m$. The net export from market $m$ is $Q_m(\theta) = Q^s_m(\theta) - Q^d_m(\theta)$. Since losses are negligible, net exports from market 1 must be equal to net imports into market 2: $Q_1(\theta) = -Q_2(\theta)$. Electrical engineers call $Q_m(\theta)$ the net injection at node $m$ into the grid, and the set $\{Q^s_m(\theta), Q^d_m(\theta)\}_{m=1,2}$ a dispatch.

We continue to assume that customers have the same underlying demand $D(p, \theta)$, where $p$ is the wholesale spot electricity price, and the total mass of customers is normalized to 1. Denote $\alpha_m \in (0, 1)$ the fraction of customers located in market $m$, hence $Q^d_m(\theta) = \alpha_m D(p_m(\theta), \theta)$, where $p_m(\theta)$ is the wholesale spot price for electricity in market $m$.

A single technology is installed in each market, with constant variable cost $c_m$. Suppose the variable cost of production is lower in market 1 than in market 2: $c_1 < c_2$. Technology 1 located in market 1 is the baseload technology, and technology 2 located in market 2 is the peaking technology. An interconnection links markets 1 and 2. Since in the short term only variable costs matter, we expect market 1 to export into market 2: $Q_1(\theta) = -Q_2(\theta) > 0$.[1] The power flow from market 1 to market 2 is

$$\begin{aligned}\varphi(\theta) &= Q_1(\theta) = Q^s_1(\theta) - Q^d_1(\theta) = Q^s_1(\theta) - \alpha_1 D(p_1(\theta), \theta) \\ &= -Q_2(\theta) = \alpha_2 D(p_2(\theta), \theta) - Q^s_2(\theta).\end{aligned}$$

This situation is represented in Figure 6.1

#### 6.2.1.2 Transmission capacity limit

The transfer capacity of the interconnection is limited. In our simple example, the interconnection's capacity from market 1 to market 2 is denoted $\Phi^+$.

### 6.2.2 Efficient Production and Consumption

We first suppose generation capacity always exceeds demand, to focus on the impact of transmission constraints.

1. An interconnection is sometimes understood to be a transmission line between two separate countries. This is not the case here.

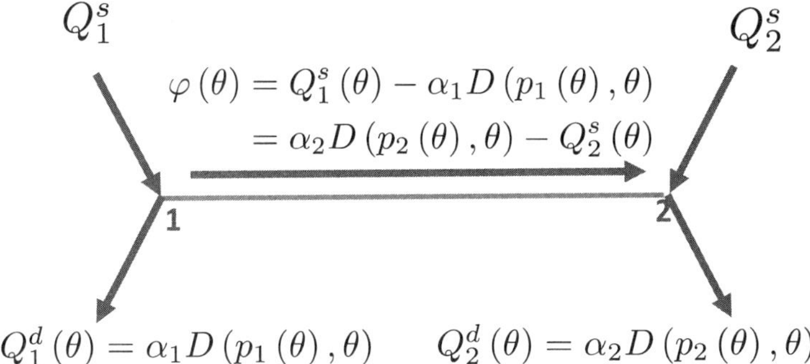


**Figure 6.1**
Production, consumption, and power flows for the two-market network.

**Interconnection not congested** If the interconnection capacity $\Phi^+$ exceeds demand, the interconnection is never congested. The cheapest generation located in market 1 serves the entire demand. The power price is $p^U(\theta) = c_1$, identical in both markets, hence we have

$Q_1^d(\theta) = \alpha_1 D(c_1, \theta),\ Q_2^d(\theta) = \alpha_2 D(c_1, \theta),\ and\ Q_1^s(\theta) = D(c_1, \theta).$

The power flow from market 1 to market 2 is

$\varphi(\theta) = Q_1(\theta) = Q_1^s(\theta) - Q_1^d(\theta) = \alpha_2 D(c_1, \theta).$

The situation is presented in figure 6.2. The interconnection is not congested as long as $\varphi(\theta) = \alpha_2 D(c_1, \theta) \leq \Phi^+$.

**Interconnection congested** Suppose demand is large, specifically $\alpha_2 D(c_1, \theta) > \Phi^+$. Using notation introduced in section 2.3, $\alpha_2 D(c_1, \theta) > \Phi^+ \Leftrightarrow \theta > \hat{\theta}_0\left(\frac{\Phi^+}{\alpha_2}, c_1\right)$. The line is congested, the flow is limited to $\Phi^+$, and the values of the marginal Megawatt-hour in markets 1 and 2 are no longer equal. In market 1, the value of a marginal Megawatt-hour is the marginal cost of production $c_1$, demand is $Q_1^d(\theta) = \alpha_1 D(c_1, \theta)$, and production is $Q_1^s(\theta) = \alpha_1 D(c_1, \theta) + \Phi^+$.

In market 2, consumption is determined by the intersection of inverse demand with vertical supply $\Phi^+ : Q_2^d(\theta) = \Phi^+$. The value of a marginal Megawatt-hour is therefore determined by consumers and equal to $P\left(\frac{\Phi^+}{\alpha_2}, \theta\right)$. As long as this value is lower than the variable cost of technology 2, that is, as long as $P\left(\frac{\Phi^+}{\alpha_2}, \theta\right) < c_2$, technology 2 does not produce. This situation is represented in figure 6.3.

If demand increases such that $P\left(\frac{\Phi^+}{\alpha_2}, \theta\right) \geq c_2 \Leftrightarrow \theta > \hat{\theta}_0\left(\frac{\Phi^+}{\alpha_2}, c_2\right)$, then technology 2 produces. The value of a marginal Megawatt-hour in market 1 remains at the marginal cost $c_1$. In market 2, the value of a marginal Megawatt-hour is the marginal cost $c_2$; hence

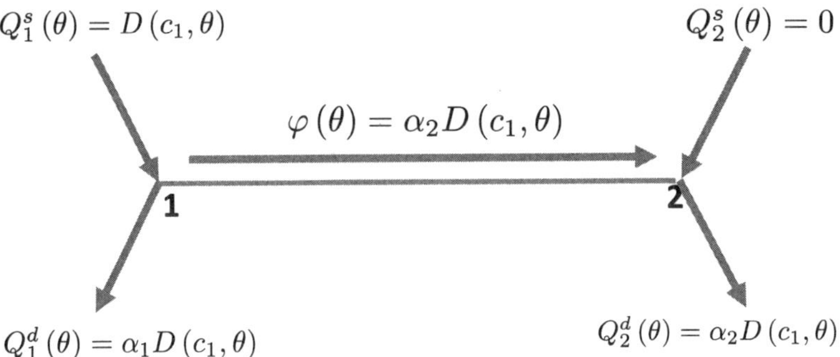


**Figure 6.2**
Production, consumption, and power flows absent congestion.

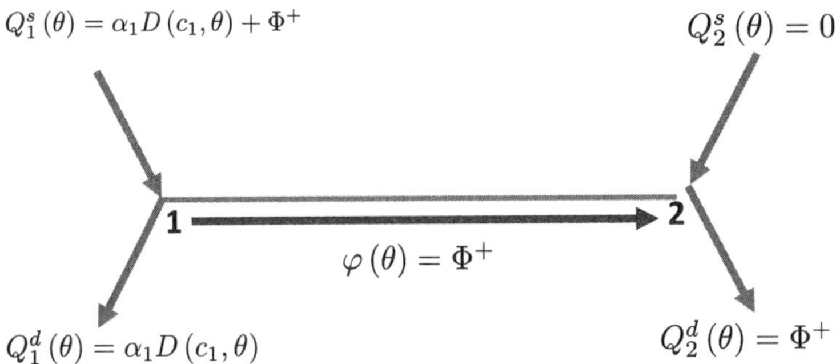


**Figure 6.3**
Production, consumption, and power flows when the line is congested, and technology 2 does not produce.

consumption is $Q_2^d(\theta) = \alpha_2 D(c_2, \theta)$ and production is $Q_2^s(\theta) = \alpha_2 D(c_2, \theta) - \Phi^+$. Power flows are represented in figure 6.4.

The resulting supply curves in both market are presented in figure 6.5.

The supply curve in market 2 presented in figure 6.5 is a staircase, similar to the supply curve for a single market with two technologies. In chapter 2, the vertical portions of the supply curve are caused by the limited capacity of each technology. In this chapter, they are caused by the limited capacity of the interconnection. This analogy is not fortuitous. On the contrary, it captures an essential intuition: the economics of transmission and generation capacity are closely related. This relation is discussed throughout this chapter.

**Marginal value of transmission capacity** Increasing the interconnection capacity by one Megawatt has no value when the interconnection is not congested. This is the equivalent of chapter 2's observation that the marginal value of generation capacity is zero when demand is lower than capacity.

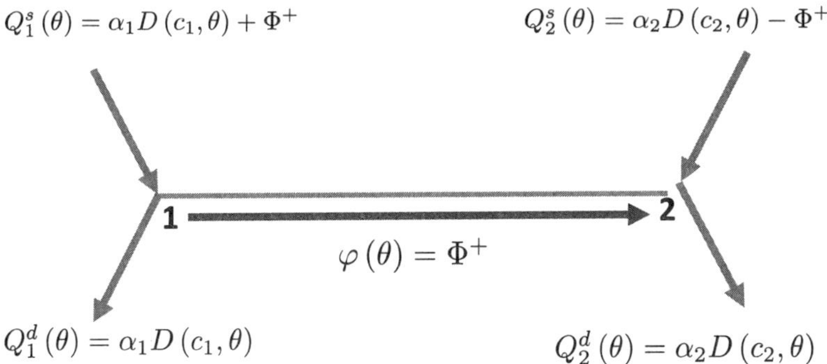


**Figure 6.4**
Production, consumption, and power flows when the line is congested, and technology 2 produces.

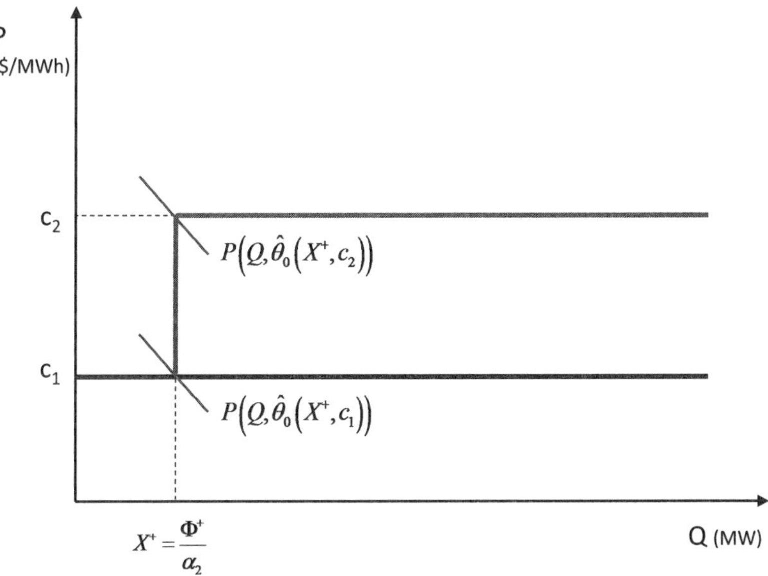


**Figure 6.5**
Value of a marginal MWh in both markets. In market 1, the value is constant, equal to the marginal cost of production $c_1$. In market 2, when the interconnection is congested, the value increases. It is first set by the value of the last MWh consumed, $P\left(\frac{\Phi^+}{\alpha_2}, \theta\right)$, then by the constant marginal cost of production $c_2$.

When the interconnection is congested and technology 2 is not yet producing, it enables the production of one additional Megawatt-hour using technology 1 in market 1 and its consumption in market 2, generating net surplus $\left(P\left(\frac{\Phi^+}{\alpha_2},\theta\right)-c_1\right)$. When technology 2 is producing, it enables the substitution of one Megawatt-hour produced with technology 1 for one Megawatt-hour produced with technology 2, thus saving $(c_2-c_1)$. The marginal value of interconnection capacity is therefore

$$\Lambda(\Phi^+)=\mathbb{E}\left[\left(P\left(\frac{\Phi^+}{\alpha_2},\theta\right)-c_1\right)\mathbb{I}_{\left\{\hat{\theta}_0\left(\frac{\Phi^+}{\alpha_2},c_1\right)<\theta<\hat{\theta}_0\left(\frac{\Phi^+}{\alpha_2},c_2\right)\right\}}\right]$$
$$+\mathbb{E}\left[(c_2-c_1)\mathbb{I}_{\left\{\theta\geq\hat{\theta}_0\left(\frac{\Phi^+}{\alpha_2},c_2\right)\right\}}\right].$$

The marginal value of interconnection capacity has a similar structure as the marginal value of the baseload technology. It is decreasing as the capacity of the interconnection increases: if the interconnection is congested and the capacity increases, the probability of congestion decreases, and the prices on the vertical segment of the supply curve also decrease. If the interconnection capacity is extremely large, the interconnection is never congested, and $\Lambda(\Phi^+)=0$.

### 6.2.3 Nodal Pricing

**Locational marginal prices** Locational marginal prices in each market are equal to the value of a marginal Megawatt-hour:

$$p_1(\theta)=c_1 \textit{ and } p_2(\theta)=\begin{cases} c_1 & \textit{if } \theta\leq\hat{\theta}_0\left(\frac{\Phi^+}{\alpha_2},c_1\right) \\ P\left(\frac{\Phi^+}{\alpha_2},\theta\right) & \textit{if } \hat{\theta}_0\left(\frac{\Phi^+}{\alpha_2},c_1\right)<\theta<\hat{\theta}_0\left(\frac{\Phi^+}{\alpha_2},c_2\right) \\ c_2 & \textit{if } \theta\geq\hat{\theta}_0\left(\frac{\Phi^+}{\alpha_2},c_2\right) \end{cases}.$$

**Marginal value of transmission capacity** The marginal value of transmission capacity can be expressed using LMPs:

$$\Lambda(\Phi^+)=\mathbb{E}[(p_2(\theta)-p_1(\theta))].$$

Therefore, LMPs signal the marginal value of transmission capacity.

**Merchandizing surplus** In a centralized market, producers in each market receive, and consumers in each market pay, the nodal price for all Megawatt-hours produced and sold. The central market operator (i.e., the independent system operator in the United States, the transmission system owner in Europe) buys and sells power at the nodal prices at each node. He collects a positive merchandizing surplus:

$$MS=\mathbb{E}[p_1(\theta)\times(Q_1^d(\theta)-Q_1^s(\theta))+p_2(\theta)\times(Q_2^d(\theta)-Q_2^s(\theta))].$$

Since

$$Q_2^d(\theta) - Q_2^s(\theta) = -(Q_1^d(\theta) - Q_1^s(\theta)) = \varphi(\theta),$$

we have

$$MS = \mathbb{E}[(p_2(\theta) - p_1(\theta)) \times \varphi(\theta)].$$

Since prices differ only when the interconnection is congested, that is, $\varphi(\theta) = \Phi^+$, we have

$$MS = \mathbb{E}[(p_2(\theta) - p_1(\theta))] \times \Phi^+ = \Lambda(\Phi^+) \times \Phi^+.$$

The merchandizing surplus is always positive. It is the marginal value of capacity on the interconnection $\Lambda(\Phi^+)$ times the interconnection capacity $\Phi^+$. This merchandizing surplus can be used to provide market participants with insurance against congestion costs.

**Financial transmission rights (FTRs)** Consider a producer in market 1, selling to a customer in market 2, at price $p$. If the line is congested, the producer cannot sell directly in market 2. Instead, he produces at cost $c$ and sells to the market operator at $p_1(\theta)$ in market 1, and he purchases from the market operator at $p_2(\theta)$ and sells to its customer at $p$ in market 2. His profit per unit is thus

$$\pi(\theta) = p_1(\theta) - c + p - p_2(\theta) = p - c + (p_1(\theta) - p_2(\theta)).$$

The producer is thus exposed to the difference in nodal prices, called the basis risk by traders.

In his seminal 1992 article, Bill Hogan suggests market participants can hedge this uncertainty by selling (or buying) financial forward products that pay the difference between nodal prices for every unit of hedge purchased. These financial transmission rights are auctioned by the SO, and perfectly transferable.

If he owns an FTR that pays $(p_2(\theta) - p_1(\theta))$, our producer's unit profit becomes

$$\pi(\theta) = p - c + (p_1(\theta) - p_2(\theta)) + (p_2(\theta) - p_1(\theta)) = p - c.$$

Profit is thus insured against fluctuations in nodal prices.

The market operator pays FTR owners the price difference associated with their FTR. As previously mentioned, he receives the merchandizing surplus from purchasing and selling power at different prices, equal to the difference in nodal prices times the interconnection capacity. Thus in the simple two-node case, the market operator can pay exactly as many FTRs as there is capacity on the line. Hogan (1992) and Bushnell and Stoft (1996) generalize the result and show that if the sum of FTRs auctioned is feasible (i.e., if it is consistent with the grid configuration), the latter always exceeds the former: the market operator can always cover FTR payments.

Thus the market operator auctions off FTRs forward, and market participants purchase FTRs at their expected value. When the market is run, congestion appears (or not), and the market operator pays exactly the congestion amount.

The auction proceeds go to the transmission-asset owners, as part of their regulated revenues. The incentive properties of FTR payments are discussed later.

For interested readers, Rosellon and Kristiansen (2013), which gathers contributions from leading academics, provide a recent and in-depth coverage of FTRs.

Nodal pricing provides a simple and elegant solution to two problems: congestion management and pricing of transmission services. It does so by applying to transmission pricing the same insight that gave rise to peak-load pricing, presented in chapter 2: when a facility is used below its capacity, the price is the variable cost of production, the cost of fuel in the case of a power plant, and the marginal losses in the case of an interconnection. In our example, the latter is neglected, hence the price is zero. If the interconnection is not congested, nodal prices are equal, hence the price of transmission service is zero. On the other hand, when a facility is used at capacity, the price exceeds the variable cost. When the interconnection is congested, nodal prices are different, and the price of transmission service is no longer zero.

Power markets in the United States have all converged toward nodal pricing, which has helped them manage congestion efficiently.

### 6.2.4 Redispatching and Countertrading

Redispatching, sometimes called countertrading, is another approach to manage congestion. It yields the same production and consumption as nodal prices but has different transfers.

The market operator receives all offers (supply and demand). She first computes the unconstrained dispatch, ignoring transmission constraints. The resulting price is $p^U$. In our example, since generation capacity in market 1 is assumed to exceed total demand, the unconstrained price is $p^U(\theta) = c_1$ in all states of the world.

Second, the market operator compares this dispatch with the interconnection capacity. If the resulting flows are lower than the interconnection capacity, that is, as long as $\alpha_2 D(c_1, \theta) \leq \Phi^+ \Leftrightarrow \theta \leq \hat{\theta}_0\left(\frac{\Phi^+}{\alpha_2}, c_1\right)$, all offers are accepted and no congestion occurs, hence no redispatching is required.

For $\theta > \hat{\theta}_0\left(\frac{\Phi^+}{\alpha_2}, c_1\right)$, exports from market 1 into market 2 are limited to $\Phi^+$; however, the price in both markets remains the unconstrained price $p^U(\theta) = c_1$. The market maker must constrain off a fraction of producers in market 1, hence $Q_1^s(\Phi^+, \theta) = \alpha_1 D(c_1, \theta) + \Phi^+ < D(c_1, \theta)$, since $\Phi^+ < \alpha_2 D(c_1, \theta)$. The system operator compensates constrained-off producers for the operating profit they have lost. Since operating profit is equal to zero, the system operator does not pay any constrained-off payment.

The market maker must also constrain off units of demand in market 2, to maintain $Q_2^d(\theta) = \Phi^+$. For simplicity, we assume she is able to choose the ones with the lowest valuation, that is, to ration demand efficiently. She pays them a constrained-off payment, equal to their valuation minus the unconstrained price.

As long as the highest constrained-off valuation is lower than $c_2$, technology 2 is not turned on. Production, demand, and flows are identical to nodal pricing, as represented in figure 6.3.

When $P\left(\frac{\Phi^+}{\alpha_2}, \theta\right) \geq c_2 \Leftrightarrow \theta \geq \hat{\theta}_0\left(\frac{\Phi^+}{\alpha_2}, c_2\right)$, the market maker calls on producers in market 2 and gives them a constrained-on payment equal to $c_2$. Since the price in all markets remains $p^U(\theta) = c_1$, there is excess demand in market 2. All units with valuation higher than $c_2$ are served, while the others are constrained off. Production, demand, and flows are again identical to nodal pricing, as represented in figure 6.4.

### 6.2.5 Nodal Pricing versus Redispatching with Increasing Marginal Costs

To maintain consistency with the rest of this text, the previous section compares nodal pricing and redispatching/countertrading if marginal production costs are constant. This section conducts the same analysis if the supply curve in each market is increasing, since this constitutes the "standard" presentation. A smoothly increasing supply curve can be seen as the limit of the staircase supply curve presented in section 2.5 when each power plant is considered a separate technology, and assumed to be very small.

Consider two markets indexed by $m = 1, 2$. The marginal cost of production in market $m$ is increasing and denoted $c_m(Q)$. The value of consuming a marginal Megawatt-hour is $P_m(Q)$. Market 1 is "cheaper" than market 2. For example, the marginal cost and marginal value of any quantity for any $Q \geq 0$ are higher in market 2: $c_2(Q) > c_1(Q)$ and $P_2(Q) > P_1(Q)$.

#### 6.2.5.1 Interconnection unconstrained

Suppose first no congestion occurs. The superscript U refers to the unconstrained situation. At the equilibrium, prices in each market are equal, and equal to the marginal cost of the last Megawatt-hour produced, and the value of the last Megawatt-hour consumed. This situation is presented in figure 6.6, which also illustrates energy balance on a power network: demand is not equal to supply in each market. Rather, total demand is equal to total supply, and exports for market 1 are equal to imports into market 2.

#### 6.2.5.2 Nodal pricing

Suppose now the interconnection is congested. Exports from cheaper producers located in market 1 are constrained by the interconnection capacity $\Phi^+$. More expensive producers located in market 2 are therefore called on to produce. Price in each market is equal to the marginal cost of the last Megawatt-hour produced and the value of the last Megawatt-hour consumed in this market. The size of the interconnection determines the difference between nodal prices. The congestion rent received by the market operator is

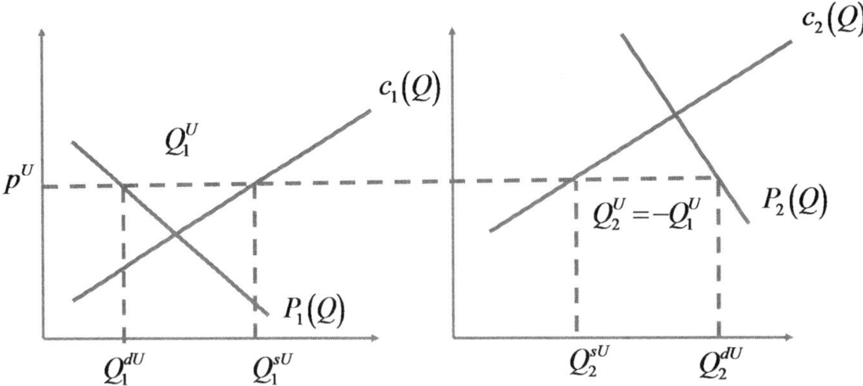


**Figure 6.6**
Optimal production and consumption if marginal costs are continuously increasing and the interconnection is unconstrained. The unconstrained price is equal to the marginal cost of the last MWh produced and the value of the last MWh consumed in each market.

the surface of a rectangle: the interconnection capacity $\Phi^+$ times the difference in nodal prices. This situation is presented in figure 6.7.

#### 6.2.5.3 Redispatching and countertrading

If the power flows exceed the interconnection capacity, the market operator must modify the transactions to adjust the flows. He purchases power from the producers at node 2 (and the consumers, if possible), to increase production at node 2 (and reduce demand) and thereby reduce imports. At the same time, he reduces production in market 1 (and increases demand, if possible), to reduce exports by the same amount as imports are increased. The market operator adjusts until the power flow is exactly equal to the capacity of the line. The dispatch is thus exactly identical to the nodal pricing one, as represented in figure 6.7.

**Compensation to constrained-off and -on producers** However, transfers are different. Under nodal pricing, all producers sell and all customers and retailers buy at $p_1$ in market 1, while consumers and retailers purchase at $p_2 > p_1$ in market 2. As illustrated in figure 6.7, the operator receives a positive surplus, equal to the congestion rent.

Under countertrading, the market operator purchases at price $p^U$ actual production $Q_1^s(\Phi^+, \theta)$ and $Q_2^s(\Phi^+, \theta)$, sells at price $p^U$ the volume $(Q_1^d(\Phi^+, \theta) + Q_2^d(\Phi^+, \theta))$, and compensates constrained-on and -off producers and consumers for their adjustments.

Figure 6.8 illustrates the compensation to constrained-off and -on producers. Market 1 is export constrained, that is, $Q_1^s(\Phi^+, \theta) < Q_1^{sU}(\theta)$. For $x \in [Q_1^s(\Phi^+, \theta), Q_1^{sU}(\theta)]$, constrained-off producers are paid the net operating profit they would have received: $(p^U - c_1(x))$. Market 2 is import constrained, that is, $Q_2^s(\Phi^+, \theta) > Q_2^{sU}(\theta)$. For $x \in [Q_2^{sU}(\theta), Q_2^s(\Phi^+, \theta)]$, constrained-on producers are paid their cost $c_2(x)$, hence their constrained-on payment is $(c_2(x) - p^U)$. Thus compensation to constrained-off and constrained-on producers in state $\theta$ is

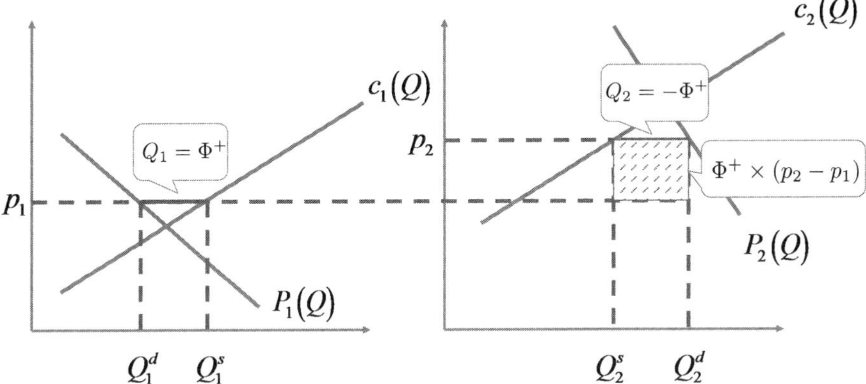


**Figure 6.7**
Optimal production and consumption if marginal costs are continuously increasing and congestion is managed through nodal pricing. Price in each market is equal to the marginal cost of the last Megawatt-hour produced and the value of the last Megawatt-hour consumed in this market. The size of the interconnection determines the difference between nodal prices. The congestion rent received by the market operator is the surface of a rectangle: the interconnection capacity $K$ times the difference in nodal prices.

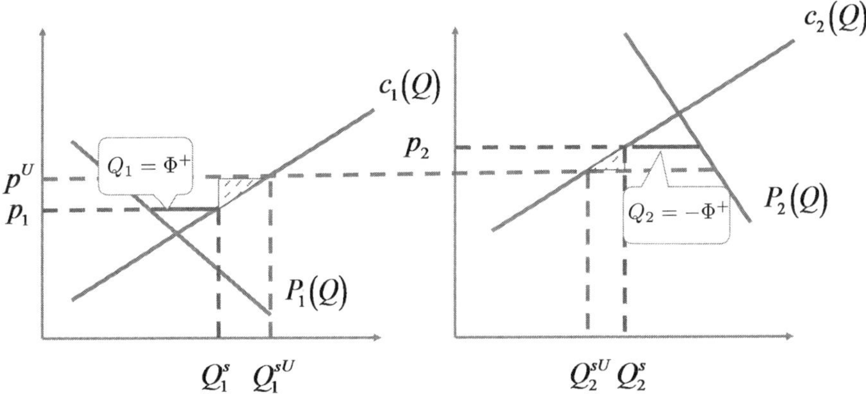


**Figure 6.8**
Payments to constrained-on and constrained-off producers if marginal costs are continuously increasing and congestion is managed through redispatching/countertrading.

$$\int_{Q_1^s(\Phi^+,\theta)}^{Q_1^{sU}(\theta)} (p^U - c_1(x))dx + \int_{Q_2^{sU}(\theta)}^{Q_2^s(\Phi^+,\theta)} (c_2(x) - p^U)dx$$

$$= \sum_{m=1}^{2} \int_{Q_m^{sU}(\theta)}^{Q_m^s(\Phi^+,\theta)} (c_m(x) - p^U)dx.$$

It is equal to the surface of the two triangles in figure 6.8.

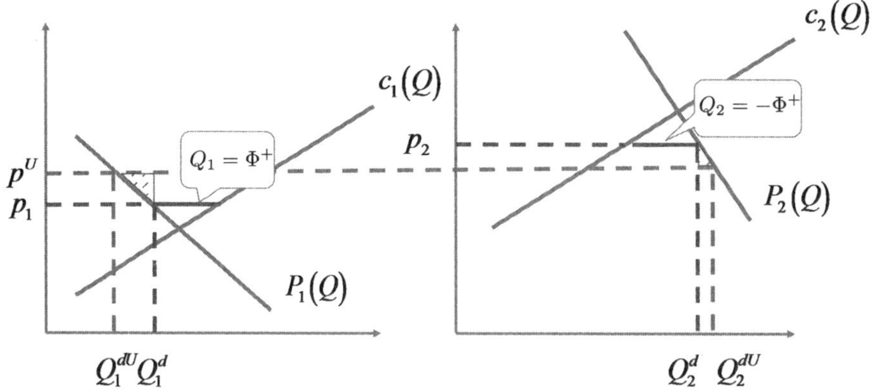


**Figure 6.9**
Compensation to constrained-on and constrained-off consumers if congestion is managed through redispatching/countertrading.

**Compensation to constrained-on and -off consumers** Figure 6.9 illustrates the compensation to constrained-on and constrained-off consumers. Market 1 is export-constrained, that is, $Q_1^d(\Phi^+, \theta) > Q_1^{dU}(\theta)$. For $x \in [Q_1^{dU}(\theta), Q_1^d(\Phi^+, \theta)]$, constrained-on consumers are compensated for their surplus loss from consuming, hence they are paid $(p^U - P_1(x, \theta))$. Market 2 is import constrained, that is, $Q_2^d(\Phi^+, \theta) < Q_2^{dU}(\theta)$. For $x \in [Q_2^d(\Phi^+, \theta), Q_2^{dU}(\theta)]$, constrained-off consumers receive the net surplus they would have derived from consuming: $(P_2(x, \theta) - p^U)$. Thus compensation to constrained-on and constrained-off consumers in state $\theta$ is

$$\int_{Q_1^{dU}(\theta)}^{Q_1^d(\Phi^+,\theta)} (p^U - P_1(x, \theta))dx + \int_{Q_2^d(\Phi^+,\theta)}^{Q_2^{dU}(\theta)} (P_2(x, \theta) - p^U)dx$$

$$= \sum_{m=1}^{2} \int_{Q_m^d(\Phi^+,\theta)}^{Q_m^{dU}(\theta)} (P_m(x, \theta) - p^U)dx.$$

It is equal to the surface of the two triangles in figure 6.9.

**Market operator net profit** In state $\theta$, the market operator's payment to producers, including compensation to constrained-off and constrained-on producers is

$$C(\Phi^+, \theta) = p^U(Q_1^s(\Phi^+, \theta) + Q_2^s(\Phi^+, \theta)) + \sum_{m=1}^{2} \int_{Q_m^{sU}(\theta)}^{Q_m^s(\Phi^+,\theta)} (c_m(x) - p^U)dx.$$

The market operator's revenues from electricity sale net of the compensation to constrained-on and constrained-off consumers are

$$R(\Phi^+,\theta)=p^U(Q_1^d(\Phi^+,\theta)+Q_2^d(\Phi^+,\theta))-\sum_{m=1}^{2}\int_{Q_m^d(\Phi^+,\theta)}^{Q_m^{dU}(\theta)}(P_m(x,\theta)-p^U)dx.$$

The operator's expected net profit is

$$\mathbb{E}[R(\Phi^+,\theta)-C(\Phi^+,\theta)]$$

$$=-\mathbb{E}\left[\begin{array}{c} p^U(Q_1(\Phi^+,\theta)+Q_2(\Phi^+,\theta))+ \\ \sum_{m=1}^{2}[\int_{Q_m^{sU}(\theta)}^{Q_m^s(\Phi^+,\theta)}(c_m(x)-p^U)dx+\int_{Q_m^d(\Phi^+,\theta)}^{Q_m^{dU}(\theta)}(P_m(x,\theta)-p^U)dx] \end{array}\right]$$

$$=-\mathbb{E}\left[\sum_{m=1}^{2}\left[\int_{Q_m^{sU}(\theta)}^{Q_m^s(\Phi^+,\theta)}(c_m(x)-p^U)dx+\int_{Q_m^d(\Phi^+,\theta)}^{Q_m^{dU}(\theta)}(P_m(x,\theta)-p^U)dx\right]\right],$$

since $Q_1(\Phi^+,\theta)+Q_2(\Phi^+,\theta)=0$. The market operator's expected profit is negative and equal to minus the redispatching cost. It is covered by all consumers, through an uplift charge.

## 6.3 Optimal Generation and Transmission Expansion

The first question when considering investment is timing: Does investment in generation arise before or after investment in the transmission grid?

To simplify the exposition, we consider each situation separately. First, we assume that the transmission grid is fixed and examine optimal generation expansion. Second, we conduct the opposite exercise: we assume generation is given and characterize the optimal grid. Third, we characterize the optimal network and generation mix. Finally, we discuss the thorny issue of actual coordination between generation and transmission expansion.

### 6.3.1 Optimal Generation Capacity

We determine the optimal generation mix for our two-market example, combining the nodal pricing presented in section 6.2 with the peak-load pricing analysis presented in chapter 2. In this particular case, we are able to compute closed-form solutions, which provide simple illustrations for the main results and intuitions.

**Prices in different states of the world** For $m=1,2$, denote $k_m$ the installed capacity located in market $m$, and $K_m$ the cumulative capacity up to $m$. The superscript $^U$ refers to the unconstrained situation, for example $K_2^U$ is the cumulative capacity if the network is never congested. Analysis presented in section 6.2 shows that the flow from market 1 to market 2 is equal to $\varphi(\theta)=\alpha_2 D(c_1,\theta)$. Maximum flow occurs when baseload technology produces at capacity, that is, $D(c_1, \theta) = K_1$, and peaking technology is not yet turned on. Then, it is equal to $\varphi(\theta) = \alpha_2 K_1$.

If $\Phi^+ \geq \alpha_2 K_1^U$, where the interconnection is never congested. To make matters interesting, assume $\Phi^+ < \alpha_2 K_1^U$, hence the interconnection is congested from market 1 to market 2 if demand is high enough. To simplify the exposition, define $X^+ = \frac{\Phi^+}{\alpha_2}$ and $X_1 = \frac{K_1 - \Phi^+}{\alpha_1}$.

Two additional conditions on the values of the parameters are required: (a) $\alpha_1 K_2^U \leq \alpha_2 K_1^U$, and (b) $(\Phi^+ + \Phi^-) \geq \alpha_1 k_2^U$. If these conditions do not hold, other cases need to be considered, although the economic intuition is unchanged. See Léautier (2013), from which this section is inspired, for details. To simplify the exposition, I assume $\Phi^+ = \Phi^-$, hence the analysis presented assumes $\Phi^+ \geq \frac{\alpha_1 k_2^U}{2}$.

As presented in section 6.2 and illustrated in figure 6.5, as long as $\theta \leq \hat{\theta}_0(X^+, c_1)$, no congestion occurs, and the value of a marginal Megawatt-hour is $p_1(\theta) = p_2(\theta) = c_1$. For $\hat{\theta}_0(X^+, c_1) < \theta < \hat{\theta}_0(X^+, c_2)$, exports from market 1 into market 2 are limited to $\Phi^+$, the value of a marginal Megawatt-hour is $p_1(\theta) = c_1$ in market 1, and $p_2(\theta) = P(X^+, \theta)$ in market 2. For $\theta \geq \hat{\theta}_0(X^+, c_2)$, the value of a marginal Megawatt-hour is $p_1(\theta) = c_1$ in market 1, and $p_2(\theta) = c_2$ in market 2.

We now introduce $K_1$, the baseload capacity located in market 1, and $K_2$, the cumulative capacity. Consider states of the world such that $\alpha_1 D(c_1, \theta) + \Phi^+ > K_1 \Leftrightarrow D(c_1, \theta) > \frac{K_1 - \Phi^+}{\alpha_1} = X_1 \Leftrightarrow \theta > \hat{\theta}_0(X_1, c_1)$. In market 1, production is limited to $Q_1^s(\theta) = K_1$; hence available supply is $(K_1 - \Phi^+)$ and the price is the value consumers place on a marginal Megawatt-hour: $p_1(\theta) = P(X_1, \theta)$. As long as demand in market 2 is lower than peaking capacity, $p_2(\theta) = c_2$. The situation is represented in figure 6.10.

When the price in market 1 reaches the price in market 2, that is, when $P(X_1, \theta) \geq c_2 \Leftrightarrow \theta \geq \hat{\theta}_0(X_1, c_2)$, the interconnection is no longer constrained: the value of a marginal

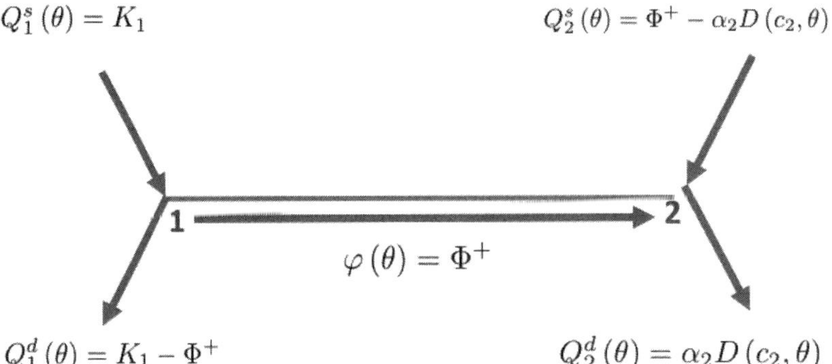


**Figure 6.10**
Production, consumption, and power flows when the line is congested, technology 2 produces, and technology 1 produces at capacity.

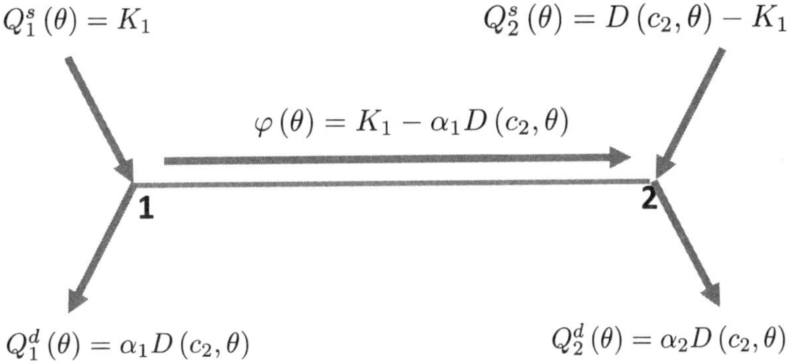


**Figure 6.11**
Production, consumption, and power flows when technology 1 produces at capacity, technology 2 produces, and the line is no longer congested.

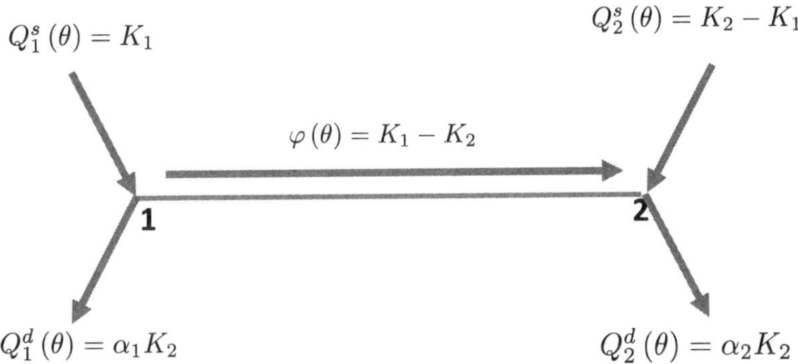


**Figure 6.12**
Production, consumption, and power flows when technologies 1 and 2 produce at capacity, and the line is no longer congested.

Megawatt-hour is the same in both markets. Until cumulative capacity is reached $p_1(\theta) = p_2(\theta) = c_2$, hence demand in each market is $Q_m^d(\theta) = \alpha_m D(c_2, \theta)$ and production in market 1 remains $Q_1^s(\theta) = K_1$, hence production in market 2 is $Q_2^s(\theta) = D(c_2, \theta) - K_1$. The flow on the interconnection is $\varphi(\theta) = K_1 - \alpha_1 D(c_2, \theta)$. We verify directly that the interconnection is not congested:

$$\varphi(\theta) \leq \Phi^+ \Leftrightarrow K_1 - \alpha_1 D(c_2, \theta) \leq \Phi^+ \Leftrightarrow \frac{K_1 - \Phi^+}{\alpha_1} = X_1 \leq D(c_2, \theta) \Leftrightarrow \theta \geq \hat{\theta}_0(X_1, c_2).$$

This situation is illustrated in figure 6.11.

When $D(c_2, \theta) \geq K_2 \Leftrightarrow \theta \geq \hat{\theta}_0(K_2, c_2)$, cumulative capacity is reached, and $p_1(\theta) = p_2(\theta) = P(K_2, \theta)$. This situation is illustrated in figure 6.12.

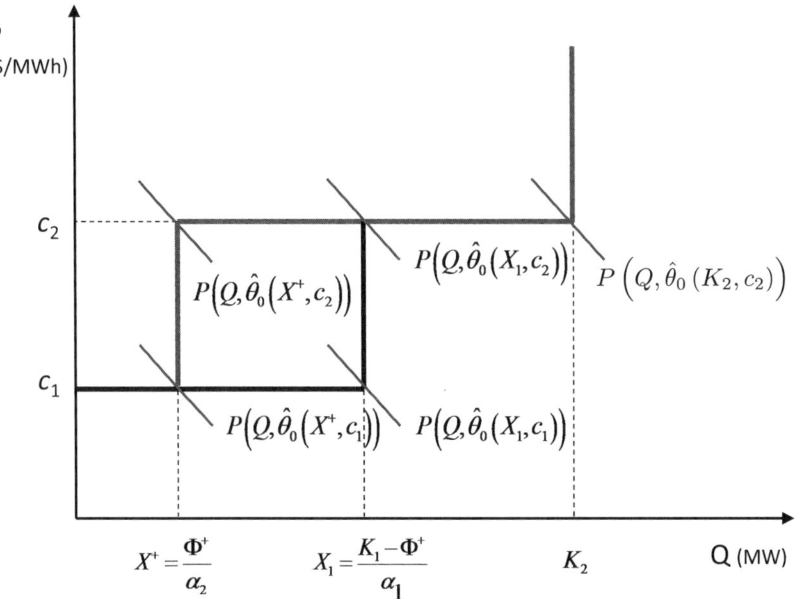


**Figure 6.13**
Supply curves, taking generation capacity into account. Price rises above variable cost of the peaking technology to cover capacity cost. On-peak, congestion disappears.

The resulting supply curves are presented in figure 6.13. The left portion of the supply curves is identical to figure 6.5.

## Cumulative and baseload capacity

**Proposition 6.1.** Optimal capacity.

1. The cumulative optimal capacity $K_2$ is equal to the cumulative uncongested capacity:

$$K_2 = K_2^U. \tag{6.1}$$

2. Adjusted baseload capacity $X_1$ is equal to the uncongested baseload capacity:

$$X_1 = K_1^U \Leftrightarrow K_1(\Phi^+) = \alpha_1 K_1^U + \Phi^+. \tag{6.2}$$

***Proof.*** The optimality condition defining cumulative capacity $K_2$ is

$$\mathbb{E}[(p_2(\theta) - c_2)^+] = \mathbb{E}[(P(K_2, \theta) - c_2)^+] = \Psi_0(K_2, c_2) = r_2,$$

which proves the first result.

The optimality condition defining baseload technology $K_1$ is

$$\mathbb{E}[(p_1(\theta) - c_1)^+] = r_1.$$

As shown in figure 6.13, $p_1(\theta) \geq c_1$ for all $\theta \geq \hat{\theta}_0(X_1, c_1)$, and we successively have $p_1(\theta) = P(X_1, \theta)$ until $\theta = \hat{\theta}_0(X_1, c_2)$, then $p_1(\theta) = c_2$ until $\theta = \hat{\theta}_0(K_2, c_2)$, and finally $p_1(\theta) = P(K_2, \theta)$ for all $\theta \geq \hat{\theta}_0(K_2, c_2)$. Following the approach presented in chapter 2, this condition can be written as

$$\Psi_0(X_1, c_1) - \Psi_0(X_1, c_2) = r_1 - r_2,$$

which proves the second result. □

Proposition 6.1 calls for a few observations. First, congestion stops on peak. This appears counterintuitive. One would argue that, since peaking technology (located in market 2) has higher marginal cost than the baseload technology (located in market 1), once the interconnection becomes congested, it always remains so. This intuition turns out to be invalid, since it ignores the necessary recovery of investment cost: when the baseload technology produces at capacity, price in market 1 increases, and eventually reaches the marginal cost of the peaking technology.

Second, as a consequence of the previous observation, congestion has no impact on the cumulative installed capacity. This may again appear surprising. The intuition is that cumulative capacity is determined by its marginal value when total capacity is constrained. In these states of the world, the interconnection is no longer congested, and the peaking technology is price setting. Thus congestion no longer matters.

Third, equation (6.2) shows that congestion on the transmission line reduces the capacity installed at market 1 and increases the capacity installed at market 2. Since $\Phi^+ < \alpha_2 K_1^U$ (otherwise, the interconnection would not be congested), $K_1(\Phi^+) < \alpha_1 K_1^U + \alpha_2 K_1^U = K_1^U$; hence since cumulative capacity $K_2 = K_2^U$, $k_2(\Phi^+) > k_2^U$. In addition, increasing transmission capacity increases optimal generation capacity in market 1 and reduces generation capacity in market 2, as one would expect.

### 6.3.2 Optimal Transmission Capacity

#### 6.3.2.1 Marginal cost of grid expansion

For a single interconnection, the cost of capacity can be approximated as $\Gamma(\Phi^+) = \Gamma_0 + \gamma \times \Phi^+$. $\Gamma_0$ represents the fixed cost of building the interconnection, which does not depend on the capacity, and $\gamma$ represents the variable cost of building the interconnection, which increases approximately linearly with the capacity of the interconnection.

In a large network, matters are much more complex. First, increasing the size of a line (or building a parallel new line) increases the transfer capacity of that line (or interface between two regions), but it also impacts other lines. Therefore, different solutions are available to increase capacity on a given interface, including increasing capacity on other interfaces. The resulting cost function is the minimum over all feasible solutions (Boyer et al. 2006 provide a very clear example of a natural gas network).

Second, as indicated by Hogan et al. (2010), the resulting cost function does not always possess "nice" mathematical properties. In some instances, increasing capacity on one interconnection reduces capacity on another. Suppose, for example, line $A$ is the largest contingency setting the operating limit of line $B$. Increasing the capacity of line $A$ increases the flow on line B if line $A$ were to fail. Thus the operating limit on line $B$ has to be reduced. In this case, there may be no optimal grid, or multiple optima.

Since this book's focus is the incentives generated by different remuneration mechanisms, it assumes that the cost function is such that a unique optimal grid exists. It is the case for a single interconnection linking two markets. Readers should be aware, however, that this property may not always hold for meshed networks.

#### 6.3.2.2 Optimal interconnection capacity

As previously discussed, the marginal value of the interconnection is the expected difference in LMPs, and is decreasing as the capacity of the line increases. At the optimum, this marginal value is equal to the marginal cost of capacity on this interconnection:

$$\mathbb{E}[p_2(\theta) - p_1(\theta)] = \gamma.$$

For given capacities $K_1$ and $K_2$, not necessarily optimal, using the marginal values presented in figure 6.13, we have

$$\Lambda(\Phi^+, K_1) = \int_{\hat{\theta}_0(X^+,c_1)}^{\hat{\theta}_0(X^+,c_2)} (P(X^+, \theta) - c_1) f(\theta) d\theta + \int_{\hat{\theta}_0(X^+,c_2)}^{\hat{\theta}_0(X_1,c_1)} (c_2 - c_1) f(\theta) d\theta$$
$$+ \int_{\hat{\theta}_0(X_1,c_1)}^{\hat{\theta}_0(X_1,c_2)} (c_2 - P(X_1, \theta)) f(\theta) d\theta.$$

As in chapter 2, adding and subtracting $\int_{\hat{\theta}_0(X^+,c_2)}^{+\infty} (c_2 - c_1) f(\theta) d\theta$ yields a simple expression:

$$\Lambda(\Phi^+, K_1) = \Psi_0(X^+, c_1) - \Psi_0(X^+, c_2) - (\Psi_0(X_1, c_1) - \Psi_0(X_1, c_2)).$$

The marginal value of transmission capacity combines two effects. The term $(\Psi_0(X^+, c_1) - \Psi_0(X^+, c_2))$ is the expected difference in LMPs, assuming that generation capacity always exceeds demand, as presented in figure 6.5. It captures the value of increasing electricity production at short-term marginal cost $c_1$. The term $(\Psi_0(X_1, c_1) - \Psi_0(X_1, c_2))$ therefore captures the value of the long-term impact of transmission capacity. $\Lambda(\Phi^+, K_1)$ is decreasing in $\Phi^+$ and is increasing in $K_1$ by inspection.

If it exists, the optimal interconnection capacity $\Phi^+(K_1)$ is therefore defined by

$$\Psi_0(X^+, c_1) - \Psi_0(X^+, c_2) - (\Psi_0(X_1, c_1) - \Psi_0(X_1, c_2)) = \gamma. \tag{6.3}$$

If the interconnection is no longer congested, that is, $\Phi^+ = \alpha_2 K_1$, its marginal value is equal to $0$[2], hence $\Phi^+(K_1) < \alpha_2 K_1$. We have imposed $\Phi^+ \geq \frac{\alpha_1 k_2^U}{2}$. A sufficient condition

2. This result can be verified directly. If $X^+ = \frac{\Phi^+}{\alpha_2} = K_1$, then $X_1 = \frac{K_1 - \Phi^+}{\alpha_1} = K_1$, thus $\Lambda(\alpha_2 K_1) = \Psi_0(K_1, c_1) - \Psi_0(K_1, c_2) - (\Psi_0(K_1, c_1) - \Psi_0(K_1, c_2)) = 0$.

for the existence of a unique solution $\Phi^{+}(K_1) \in \left(\frac{\alpha_1 k_2^U}{2}, \alpha_2 K_1\right)$ is $\Lambda\left(\frac{\alpha_1 k_2^U}{2}, K_1\right) > \gamma$, which is assumed to hold.

Since the marginal cost of capacity is positive, equation (6.3) implies that interconnection must be congested in some states of the world. An interconnection that could never be congested in any state of the world would be too large. This does not mean that every line must be congested a few hours every year. It means that an interconnection that would never be congested in all the scenarios studied would be oversized.

This result differs from current planning practice in two aspects. First, transmission grids within a utility service territory were often built never to be congested. In many cases, the transmission grid was built to transport electricity from the production centers to the consumption centers. The power flows were highly predictable, and often one-directional. Since the cost of transmission was (in most instances) a fraction of the cost of generation, interconnection size was simply the size of the power plant, and the cost of building the interconnection was included in the total cost of generation.

In the future, matters will most likely be different. We expect renewables and decentralized production to progressively represent a large share of the generation mix. Power flows will thus become more variable, and interconnections will be used in both directions. Furthermore, building transmission lines is proving extremely challenging, hence costly. Thus we will have to accept congested lines for some period of time and use the relationship in equation (6.3) to optimize the network.

Second, transmission planners often compute separately economic benefits and reliability benefits of an interconnection. Equation (6.3) provides a unified treatment of these two sources of value. Economic benefit is the substitution of cheap for dear power, as discussed earlier. Reliability benefit is the reduction in the probability of having to curtail customers in case of tension on the system. As seen in chapter 2, curtailment (e.g., in market 2) would result in the price being set at the VoLL in market 2. Thus equation (6.3) captures this benefit by recognizing that the price $p_2(\theta)$ may sometimes be equal to the VoLL.

Finally, since the marginal value of transmission capacity $\Lambda(\Phi^{+}, K_1)$ is increasing in $K_1$, $\Phi^{+}(K_1)$ is also increasing. We have previously observed that $K_1(\Phi^{+})$ is increasing in $\Phi^{+}$. Taken together, these two observations indicate that baseload capacity and interconnection capacity are complements: increasing one increases the other's profitability.

The relationship between peaking and transmission capacity is more subtle. Increasing peaking capacity has no impact on the marginal value of transmission capacity since $k_2$ does not appear in equation (6.3). On the other hand, increasing transmission capacity reduces the value of peaking capacity. Thus transmission capacity is a substitute for peaking capacity, but the substitution does not go in the reverse direction.

#### 6.3.2.3 Fixed-costs recovery

Electricity generation exhibits approximately constant returns to scale: if prices are set to cover marginal costs, they also cover average cost. This is not the case for networks, which exhibit increasing returns to scale; hence covering marginal cost does not guarantee average cost coverage (e.g., Bell et al. 2011). In the two-market example, the congestion rent at the optimal capacity covers exactly the variable cost of the grid:

$$\Lambda(\Phi^{+}, K_1) = \gamma \Leftrightarrow \Lambda(\Phi^{+}, K_1) \times \Phi^{+} = \gamma \times \Phi^{+}.$$

This leaves the entire fixed cost $\Gamma_0$ uncovered. This result has long been confirmed empirically: Pérez-Arriaga, Rubio, and Puerto Gutierrez (1995) show that the congestion rents recover only approximately 25 percent of total costs of a representative transmission grid. An additional charge must be levied to cover the remaining costs.

### 6.3.3 Optimal Generation Mix and Transmission Grid

Suppose generation mix is optimal, given interconnection capacity $\Phi^{+}$. Then, equation (6.3) yields

$$\Lambda(\Phi^{+}, K_1(\Phi^{+})) = \Psi_0(X^{+}, c_1) - \Psi_0(X^{+}, c_2) - (r_1 - r_2) = \gamma. \tag{6.4}$$

If $\Lambda\left(\frac{\alpha_1 k_2^U}{2}, K_1\left(\frac{\alpha_1 k_2^U}{2}\right)\right) > \gamma$, there exists a unique solution $\Phi^{+*} \in \left(\frac{\alpha_1 k_2^U}{2}, \alpha_2 K_1\right)$.

The optimal transmission capacity is such that the reduction in operating costs is precisely equal to the increase in generation capital cost and the capital cost of transmission.

## 6.4 Generation Investment under Different Congestion Management Regimes

Chapter 2 shows that if competition is perfect, peak-load pricing in a single market leads to optimal generation investment; that is, the long-term equilibrium generation mix is optimal. This section proves that this result extends to two markets linked by an interconnection if congestion is managed using nodal prices. It also provides a sufficient condition for redispatching (countertrading) to lead to the optimal generation investment.

### 6.4.1 Congestion Managed Using LMPs

**Proposition 6.2.** If producers face LMPs, competitive equilibrium leads to optimal generation investment.

***Proof.*** Competitive equilibrium capacity in market $m$ is determined by the free-entry condition $\mathbb{E}[(p_m(\theta) - c_m)^{+}] = r_m$, which leads to the optimal generation investment, given the constraints on the transmission grid. □

This is simply an application of a general result in economics: if competition is perfect, equilibrium prices lead to efficient production and consumption decisions in the short term, but also to efficient investment decisions.

### 6.4.2 Congestion Management Using Redispatching and Differentiated Access Charge

European countries use redispatching and not nodal pricing to manage congestion. Mindful that this approach is unlikely to lead to optimal investment, power market designers attempt to guide investment decisions by sending location signals through differentiated transmission access charges. Variants of this approach are known as "deep" connection charge, or long-run incremental cost (LRIC) transmission pricing. It is described and evaluated for example (along with other transmission pricing approaches) by Bell et al. (2011), and has been advocated by different academics (e.g., Brunekreeft, Neuhoff, and Newbery 2005, and more recently, Newbery 2011) when nodal pricing is politically infeasible. Economic intuition suggests that having producers pay for their (incremental) contribution to network costs will induce them to choose the location that minimizes generation and transmission costs.

However, this intuition is not quite correct: LRIC does not necessarily induce optimal generation investment. Our two-market network provides a simple counterexample.

#### 6.4.2.1 Constrained-on and constrained-off payments

By construction, the unconstrained dispatch and prices are those described in section 2.5 and summarized in table 6.1.

We first compute payments to constrained-off producers in market 1. To do so, we must order the relevant states of the world. Since the line is congested, $\Phi^+ < \alpha_2 K_1 \Leftrightarrow X^+ < K_1 \Leftrightarrow \hat{\theta}_0\left(X^+, c_1\right) < \hat{\theta}_0\left(K_1, c_1\right)$. In addition,

$$\Phi^+ < \alpha_2 K_1 \Leftrightarrow \frac{K_1 - \Phi^+}{\alpha_1} > \frac{K_1 - \alpha_2 K_1}{\alpha_1} \Leftrightarrow X_1 > K_1 \Leftrightarrow \hat{\theta}_0\left(X_1, c_1\right) > \hat{\theta}_0\left(K_1, c_1\right).$$

For $\hat{\theta}_0\left(X^+, c_1\right) \le \theta \le \hat{\theta}_0\left(X_1, c_1\right)$, a fraction of producers in market 1 are constrained-off. For $\hat{\theta}_0\left(X^+, c_1\right) \le \theta \le \hat{\theta}_0\left(K_1, c_1\right)$, the unconstrained price is $p^U(\theta) = c_1$, hence their unconstrained operating profit is equal to 0, hence they receive no constrained-off payment. For $\hat{\theta}_0\left(K_1, c_1\right) \le \theta \le \hat{\theta}_0\left(X_1, c_1\right)$, constrained-off producers receive payment $p^U(\theta) - c_1 = P\left(K_1, \theta\right) - c_1$, while dispatched producers receive payment $p^U(\theta)$, hence their operating profit is $P\left(K_1, \theta\right) - c_1$: since they have the same marginal cost, all producers capture the same operating profit, whether they are dispatched or constrained-off. For

**Table 6.1**
Unconstrained price and dispatch.

| State | $p^U$ | $Q_1^{sU}$ | $Q_2^{sU}$ |
|---|---|---|---|
| $\theta \le \hat{\theta}_0(K_1, c_1)$ | $c_1$ | $D(c_1, \theta)$ | 0 |
| $\hat{\theta}_0(K_1, c_1) \le \theta < \hat{\theta}_0(K_1, c_2)$ | $P(K_1, \theta)$ | $K_1$ | 0 |
| $\hat{\theta}_d(K_1, c_2) \le \theta < \hat{\theta}_0(K_2, c_2)$ | $c_2$ | $K_1$ | $D(c_2, \theta) - K_1$ |
| $\hat{\theta}_0(K_2, c_2) \le \theta$ | $P(K_2, \theta)$ | $K_1$ | $K_2$ |

$\theta \geq \hat{\theta}_0\,(X_1, c_1)$, producers in market 1 produce at capacity, hence none is constrained-off. They all receive the unconstrained price $p^U\,(\theta)$.

For $\hat{\theta}_0\left(X^+, c_2\right) \leq \theta \leq \hat{\theta}_0\,(X_1, c_2)$, a fraction of producers in market 2 are constrained-on. They receive a constrained-on payment equal to their variable cost $c_2$, hence their operating profit is equal to 0. For $\theta \geq \hat{\theta}_0\,(X_1, c_2)$, congestion stops, and no constrained-on payments are paid.

#### 6.4.2.2 Generators' free entry under redispatching

The market maker charges $\delta_m$ for locating in market $m$, expressed in €/MWh (i.e., annual charge expressed in €/MW divided by the 8,760 hours per year) to induce producers to locate at the "right" location. $\delta_m > 0$ means producers are discouraged from locating in market $m$, while $\delta_m < 0$ means producers are encouraged to locate in market $m$. We expect $\delta_1 > 0$ and $\delta_2 \leq 0$.

The market maker computes $\delta_m$ as the incremental network cost caused by a producer's location in market $m$. Since variable production costs are constant, the previous discussion shows that producers derive no profits when they are constrained on: the constrained-on payments cover exactly their variable production costs. Free-entry condition in market $m$ is therefore

$$\mathbb{E}[(p^U(\theta) - c_m)^+] = r_m + \delta_m.$$

We first compute $\delta_m$ for $m = 1, 2$, then we derive the impact of LRIC transmission pricing on investment.

#### 6.4.2.3 Computation of $\delta_m$

The most appropriate way to compute $\delta_m$ is to compute the incremental network cost for the optimal capacity $\Phi^+(K_1)$ defined by equation (6.3).

First, increasing generation capacity in market 2 has no impact on the optimal interconnection, hence $\delta_2 = 0$. Second, as was shown previously, the optimal transmission capacity increases with generation capacity in market 1, thus $\delta_1 > 0$. We compute $\frac{d\Phi^+}{dK_1}$ by applying the implicit function theorem to equation (6.3). A bit of algebra leads to the exact expression of $\delta_1$ :

$$\delta_1 = \gamma \frac{d\Phi^+(K_1)}{dK_1} = \frac{\gamma}{1 + \frac{\alpha_1 J(X^+)}{\alpha_2 J(X_1)}} > 0,$$

where $J(x) = \int_{\hat{\theta}_0(x,c_1)}^{\hat{\theta}_0(x,c_2)} \partial_Q P(x, \theta) f(\theta) d\theta$.

#### 6.4.2.4 Equilibrium investment under LRIC transmission pricing

**Proposition 6.3.** On the simple two-market network, if $P(Q, \theta) = a_0 - a_1 e^{-\theta} - Q$ and $f(\theta) = e^{-\theta}$, LRIC transmission pricing leads to optimal generation investment if and only if the transmission grid is optimal.

***Proof.*** Since $\delta_2 = 0$, the free-entry condition in market 2 is

$$\mathbb{E}[(p^U(\theta) - c_2)^+] = \mathbb{E}[(P(K_2, \theta) - c_2)^+] = r_2;$$

hence cumulative generation capacity is optimal.

Since $\Psi_0(K_2, c_2) = r_2$, the free-entry condition in market 1 is

$$\begin{aligned} &\mathbb{E}[(p^U(\theta) - c_1)^+] = \Psi_0(K_1, c_1) - \Psi_0(K_1, c_2) - r_2 = r_1 + \delta_1 \\ &\Leftrightarrow \Psi_0(K_1, c_1) - \Psi_0(K_1, c_2) = r_1 - r_2 + \delta_1 > r_1 - r_2. \end{aligned} \tag{6.5}$$

Investment in market 1 is reduced by LRIC transmission pricing, which is directionally correct. It is not clear whether it leads to the optimal transmission capacity, that is, that $K_1 = \alpha_1 K_1^U + \Phi^+$ is the unique solution of equation (6.5).

Suppose $P(Q, \theta) = a_0 - a_1 e^{-\theta} - Q$ and $f(\theta) = e^{-\theta}$. Then,

$$\Psi_0(K, c) = \int_{\hat{\theta}_0(K,c)}^{+\infty} (a_0 - c - a_1 e^{-\theta} - Q) e^{-\theta} d\theta = a_1 e^{-2\hat{\theta}_0(K,c)}.$$

Since $\hat{\theta}_0(K, c)$ is uniquely defined by $a_0 - a_1 e^{-\hat{\theta}_0(K,c)} - K = c$, we have

$$\Psi_0(K, c) = \frac{(a_0 - c - K)^2}{a_1}.$$

Then, equation (6.3) defining optimal interconnection capacity $\Phi^+$ for a given baseload generation capacity $K_1$ becomes

$$\begin{aligned} &\frac{(a_0 - c_1 - X^+)^2 - (a_0 - c_2 - X^+)^2}{a_1} - \frac{(a_0 - c_1 - X_1)^2 - (a_0 - c_2 - X_1)^2}{a_1} = \gamma \\ &\Leftrightarrow \frac{c_2 - c_1}{a_1} \left[ \left( a_0 - \frac{c_1 + c_2}{2} - X^+ \right) - \left( a_0 - \frac{c_1 + c_2}{2} - X_1 \right) \right] = \gamma \\ &\Leftrightarrow X_1 - X^+ = \frac{K_1 - \Phi^+}{\alpha_1} - \frac{\Phi^+}{\alpha_2} = \frac{K_1 - \frac{\Phi^+}{\alpha_2}}{\alpha_1} = \frac{a_1}{c_2 - c_1}\gamma \Leftrightarrow \Phi^+ = \alpha_2 \left( K_1 - \frac{\alpha_1 a_1}{c_2 - c_1}\gamma \right). \end{aligned}$$

Since the optimal baseload capacity for a given interconnection capacity is $K_1 = \alpha_1 K_1^U + \Phi^+$, the long-term optimal interconnection capacity is defined by

$$\Phi^{+*} = \alpha_2 \left( \alpha_1 K_1^U + \Phi^{+*} - \frac{\alpha_1 a_1}{c_2 - c_1}\gamma \right) \Leftrightarrow \Phi^{+*} = \alpha_2 \left( K_1^U - \frac{a_1}{c_2 - c_1}\gamma \right).$$

We then compute $\delta_1$: $\frac{d\Phi^+(K_1)}{dK_1} = \alpha_2 \Rightarrow \delta_1 = \gamma \frac{d\Phi^+(K_1)}{dK_1} = \alpha_2 \gamma$. Observing that $\Psi_0(K_1^U, c_1) - \Psi_0(K_1^U, c_2) = r_1 - r_2$, equation (6.5) yields

$$K_1 = K_1^U - \frac{\alpha_2 a_1}{c_2 - c_1}\gamma.$$

Finally,

$$K_1^U - \frac{\alpha_2 a_1}{c_2 - c_1}\gamma = \alpha_1 K_1^U + \Phi^+ \Leftrightarrow \Phi^+ = \alpha_2 \left( K_1^U - \frac{a_1}{c_2 - c_1}\gamma \right) \Leftrightarrow \Phi^+ = \Phi^{+*}.$$

The competitive-equilibrium cumulative capacity $K_2$ under LRIC transmission pricing is optimal since (i) $\delta_2 = 0$, and the unconstrained price is modified compared with the nodal price only when it is lower than $c_2$; hence it does not impact the recovery of the capacity cost.

The competitive-equilibrium baseload capacity $K_1$ under LRIC transmission pricing is optimal if and only if interconnection capacity is at the long-term optimum.

This result can be interpreted in different ways. The good piece of news is that, if the interconnection is optimal, LRIC transmission pricing leads to optimal generation investment. I am not certain this result holds for a more general specification. Confirming or disproving this result on a general network for different specifications of demand is an interesting avenue for further research.

The bad piece of news is that if the interconnection capacity is not the long-term optimum, even in the simplest case, LRIC transmission pricing leads to overinvestment or under investment. Since transmission networks are rarely at the long-term optimum, in practice, LRIC transmission pricing does not provide correct generation-investment incentives.

Furthermore, the "deep-connection charges" are rarely optimally computed, which creates an additional distortion. The method employed in Great Britain relies on an estimation of the number of miles traveled by the electric current, which ignores differences in other characteristics. In Scandinavia, the geographical latitude is used as a proxy for the impact of the transmission grid.

European policy makers sometimes argue that, since full LMP implementation is highly unlikely, LRIC transmission pricing provides a workable compromise. The economic analysis presented above weakens the argument. While LRIC transmission pricing provides directionally correct incentives, it is unlikely to be optimal. Furthermore, computing robust and correct deep-connection charges in a real power network is a complex process, both technically and as a regulatory undertaking. Producers and customers will attempt to influence the determination of the computation method in their favor. On balance, implementing LRIC transmission pricing is only slightly less challenging than implementing LMPs, and delivers only a share of the benefits.

## 6.5 Transmission Grid Expansion under Different Regimes

Optimal expansion of the transmission grid is a critical issue. In most countries, transmission is organized as a regulated monopoly, hence the issue is to provide incentives for optimal expansion in the regulatory contract. An alternative model is merchant transmission, under which independent transmission providers develop interconnections.

### 6.5.1 The Congestion Rent Does Not Induce Optimal Grid Expansion

The first observation is that leaving the congestion rent to a monopoly grid company does not provide optimal incentives. If it receives the entire congestion rent

$$R(\Phi^+) = \Lambda(\Phi^+) \times \Phi^+,$$

it finds itself in the situation of a classical monopoly. It maximizes its rent by setting the marginal revenue equal to the marginal cost:

$$\Lambda(\Phi^+) + \frac{d\Lambda(\Phi^+)}{d\Phi^+} \times \Phi^+ = \gamma.$$

As we have seen, $\frac{d\Lambda(\Phi^+)}{d\Phi^+} < 0$: increasing the capacity on a line reduces its value. Thus if the transmission company increases the interconnection, it captures the expected price difference on the marginal capacity but also reduces the expected price difference, hence the rent, on all inframarginal units.

The result extends to a Transco that owns and operates multiple lines. If the owner is granted the congestion rent, he will behave as a multiproduct monopolist and invest suboptimally.

Thus leaving the congestion rent to the monopolist was quickly abandoned as an approach to induce optimal investment. In the United States, transmission-asset owners receive the congestion rent through the auction of FTRs, but this rent is deducted from their required revenues so as not to provide incentives to increase congestion. A similar mechanism is in place in Europe, where transmission-asset owners receive the congestion rent associated with crossborder interconnections, which is then deducted from their required revenues.

### 6.5.2 Optimal Regulation of a Monopoly Transmission Company

The regulatory contract aims to induce (a) the optimal cost reduction or rent extraction trade-off and (b) the optimal grid expansion.

#### 6.5.2.1 Optimal cost reduction or rent extraction trade-off

This is the core of modern regulation theory, exposed for example by Laffont and Tirole (1993). The starting point is information asymmetry between the regulator (the principal) and the regulated firm (the agent). The latter has better information than the former concerning its cost and the intensity of its cost-reducing activities.

Historically, the regulator would cover the firm's reported costs, an approach known as cost-of-service regulation. In that case, the firm had no incentives to reduce costs. Since the regulator did not know the firm's cost-saving potential, there was widespread suspicion of inefficiency. This situation was captured by John Hicks's (Hicks 1935) famous observation: "The best of all monopoly profits is a quiet life." A completely opposite approach is to fix the regulated firm's revenue at a given level for the entire regulatory period, for example, five years. If the firm reduces its costs below the fixed level, it is allowed to keep the savings realized; hence it captures significant profit. If it fails to do so, it may incur a loss. Thus the firm faces strong incentives to reduce costs.

Laffont and Tirole (1993) formally study this problem. Building on previous academic analyses (e.g., Baron and Myerson 1982), they propose a very simple representation of the situation. The variable cost for a firm to produce output $q$ is $C(q)$ defined by

$$C(q) = (\beta - e)q,$$

where $\beta$ is the natural cost efficiency of the firm and $e$ a cost-reducing effort, both unobservable to the regulator. The regulator observes the cost $C(q)$ in the regulated firm's financial statements. However, if the regulator observes low cost $C(q)$, he does not know whether the firm is naturally efficient (low $\beta$) or exerts a very high cost-reducing effort (high $e$). Reducing cost generates private disutility for the managers of the firm, who no longer enjoy Hicks's "quiet life." Since effort is unobservable, the regulator cannot force managers to exert a certain level; rather she needs to induce them by leaving them an information rent.

Using this model, Laffont and Tirole (1993) find that the regulator should propose a menu of (linear) contracts and let the regulated firm choose its preferred contract within this menu. A naturally efficient firm (low $\beta$) optimally chooses a contract close to a fixed-price contract and exerts high cost-reducing effort $e$, while a naturally inefficient firm chooses a contract close to a cost-plus contract and exerts low cost-reducing effort.

This model is simple yet powerful. It has exerted a profound influence on academic economists and policy makers and figured prominently in the citation for Jean Tirole's Nobel prize.

#### 6.5.2.2 Optimal expansion

Laffont and Tirole (1993) do not discuss in great depth how to induce the regulated firm to produce the optimal output. They prove a dichotomy property, that is, they derive a sufficient condition on the cost function $C(q)$ for the output choice to be treated independently from a cost-reduction or rent-extraction problem.

Building on this analysis, Léautier (2000) proves that, if demand is completely price inelastic (which was a reasonable representation of power markets in the late 1990s), the redispatching cost is an efficient congestion metric, that is, having the regulated firm responsible for the redispatching cost at the margin gives it incentives to optimally expand the grid. This result also holds if demand is price responsive. The proof is slightly technical, but the intuition is simple: as shown in figures 6.8 and 6.9, the redispatching cost is a triangle, hence, increasing transmission capacity generates no inframarginal effect, contrary to the congestion rent. This argument is formalized below:

**Lemma 6.1.** The marginal redispatching cost is equal to the marginal value of transmission capacity:

$$\frac{d}{d\Phi^+}\mathbb{E}[R(\Phi^+,\theta)-C(\Phi^+,\theta)]=\Lambda(\Phi^+).$$

***Proof.*** From equation

$$\begin{aligned}\frac{d}{d\Phi^+}\mathbb{E}[R(\Phi^+,\theta)-C(\Phi^+,\theta)] &= -\mathbb{E}\left[\left(\begin{array}{c}(p^U-p_1)\frac{dQ_1^d}{d\Phi^+}-(p_2-p^U)\frac{dQ_2^d}{d\Phi^+}\\ -(p^U-p_1)\frac{dQ_1^s}{d\Phi^+}+(p_2-p^U)\frac{dQ_2^s}{d\Phi^+}\end{array}\right)\right]\\ &= -\mathbb{E}\left[(p^U-p_1)\frac{dQ_1}{d\Phi^+}+(p^U-p_2)\frac{dQ_2}{d\Phi^+}\right]\\ &= \mathbb{E}\left[-\sum_{m=1}^{2}p_m(\Phi^+,\theta)\frac{dQ_m}{d\Phi^+}+p^U\sum_{m=1}^{2}\frac{dQ_m}{d\Phi^+}\right]\\ &= -\mathbb{E}\left[\sum_{m=1}^{2}p_m(\Phi^+,\theta)\frac{dQ_m}{d\Phi^+}\right].\end{aligned}$$

When the interconnection is not congested, $\frac{dQ_1}{d\Phi^+}=\frac{dQ_2}{d\Phi^+}=0$. When the interconnection is congested, $Q_1=\Phi^+=-Q_2$, hence $\frac{dQ_1}{d\Phi^+}=1=-\frac{dQ_2}{d\Phi^+}$. Therefore,

$$\frac{d}{d\Phi^+}\mathbb{E}[R(\Phi^+,\theta)-C(\Phi^+,\theta)]=\mathbb{E}[(p_2(\Phi^+,\theta)-p_1(\Phi^+,\theta))^+]=\Lambda(\Phi^+),$$

which proves the lemma. □

Lemma 6.1 enables us to prove the main result of this section:

**Proposition 6.4.** If a for-profit transmission asset owner and operator is responsible for the redispatching cost, he selects the optimal interconnection capacity.

***Proof.*** The Transco profit is

$$\Pi(\Phi^+)=\mathbb{E}[R(\Phi^+,\theta)-C(\Phi^+,\theta)]-(F+\gamma\Phi^+).$$

The Transco chooses $\Phi^+$ to maximize its profit. The first-order condition is

$$\frac{d\Pi}{d\Phi^+}=0\Leftrightarrow\frac{d}{d\Phi^+}\mathbb{E}[R(\Phi^+,\theta)-C(\Phi^+,\theta)]=\Lambda(\Phi^+)=\gamma,$$

which characterizes the optimal interconnection capacity. The profit function is concave since $\frac{d^2\Pi}{(d\Phi^+)^2}=\frac{d\Lambda}{d\Phi^+}<0$; hence the first-order condition defines the unique maximum.

Section 6.7 below shows Proposition 6.4 holds for a general $N$-node network. This result is particularly powerful in the case of a Transco, which owns and operates the transmission grid. Increasing the capacity transfer on the grid can be achieved by physical investment but also by improving the operating procedures, for example, monitoring asset conditions in real time. Transcos may object to be exposed to the full redispatching cost; hence incentive regulation may expose them only to the deviation from a target. This approach was implemented in England and Wales in the 1990s and led to significant reduction in congestion cost. It stopped in 2006 when the National Grid Company, the Transco in England and Wales, could no longer beat the target and had to pay penalties.

The problem is more challenging if one firm owns the grid while a not-for-profit entity is responsible for its operation, as is the case in the United States. There, financial incentives cannot be included in the regulatory compact, and other approaches must be implemented.

However, even with a Transco, implementing Proposition 6.4 may not be straightforward: if a market is organized around nodal prices (as it should), a separate computation is required to compute the congestion cost. As we have seen before, both approaches to managing congestion require the same inputs (costs and demand curves at every node, available transmission capacities) and produce the same production and consumption plan. Still, the regulator should perform a separate analysis to compute the redispatching cost.

To address this issue, Bill Hogan, Juan Rosellon, and Ingo Vogelsang have attempted in a series of articles to design a mechanism that relies on FTRs, summarized in Hogan et al. (2010). A Transco is a multiproduct monopoly, which, if left on its own, would likely produce less output than is socially optimal to increase its profit. This problem was extensively studied in the late 1970s and 1980s (e.g., by Vogelsang and Finsinger [1979] and Sappington and Sibley [1988]). Therefore, Hogan and his colleagues hoped to be able to transpose these results to a Transco. This has proved to be a harder task than expected. One of the issues is that the cost function for the production of FTRs (which is derived from the cost function of the expansion of the interconnections on the network but is different from it) is not well behaved: for example, increasing the amount of FTRs available between two nodes may reduce the amount of available FTRs between two other nodes.

### 6.5.3 Coordination between Generation and Transmission Investment

How does coordination occur when generation and transmission are no longer integrated?

Investment is a dynamic process: the generation mix evolves for a variety of reasons, including the configuration of the grid (existing and anticipated additions), then the grid adapts to the generation mix (existing and anticipated changes), then the generation mix adapts to the network, and so on. If competition, regulation, and foresight were

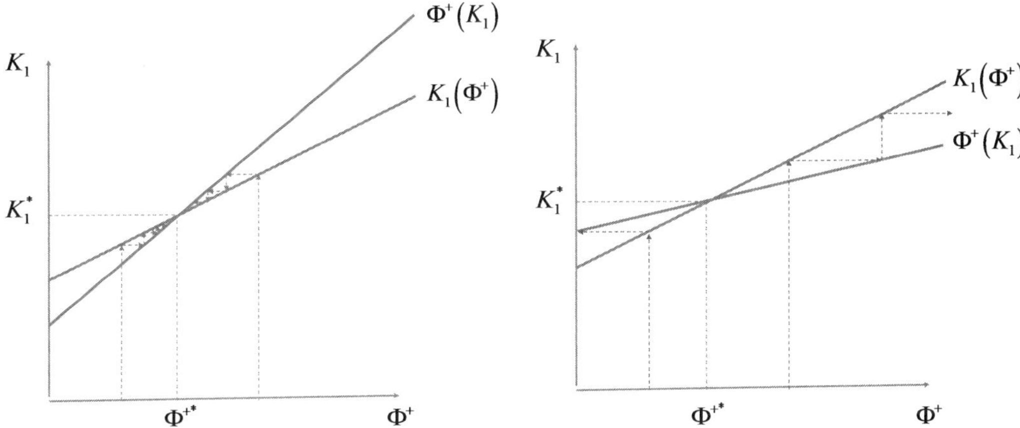


**Figure 6.14**
The function $\Phi^+(K_1)$ crosses $K_1(\Phi^+)$ from below (left panel) or from above (right panel). Convergence toward the fixed point $\Phi^{+*}$ occurs on the left panel only.

perfect, if capacity increments could be positive and negative and made arbitrarily small, this dynamic investment process would converge to the optimal long-term equilibrium defined by equations (6.1), (6.2), and (6.4).

The proof that this dynamic process converges is slightly technical but instructive.

**Convergence toward the optimum** The proof is illustrated in figure 6.14.

The optimal interconnection capacity $\Phi^{+*}$ is uniquely defined by $\Phi^+(K_1(\Phi^{+*})) = \Phi^{+*}$, that is, $\Phi^{+*}$ is the unique fixed point of the function $x \to \Phi^+(K_1(x))$. To examine convergence, we construct a sequence $\{u_n\}_{n\geq 0}$, defined by induction as $u_{n+1} = \Phi^+(K_1(u_n))$. Suppose the first term $u_0 > \Phi^{+*}$. The sequence $u_n$ converges toward $\Phi^{+*}$ on the left panel of figure 6.14 and diverges on the right panel of figure 6.14. The same occurs if $u_0 < \Phi^{+*}$. Thus convergence occurs if and only if $\Phi^+(K_1)$ crosses $K_1(\Phi^+)$ from below, that is, $\frac{d\Phi^+}{dK_1} < \frac{dK_1}{d\Phi^+}$.

This leads to comparing the slopes of functions $\Phi^+(K_1)$ and $K_1(\Phi^+)$. From equation (6.2), $K_1(\Phi^+)$ is a straight line, of slope 1. Therefore convergence occurs if and only if $\frac{d\Phi^+}{dK_1} < 1$.

$$\frac{d\Phi^+}{dK_1} = \frac{1}{1 + \frac{\alpha_1 J(X^+)}{\alpha_2 J(X_1)}}.$$

Therefore, $0 < \frac{d\Phi^+}{dK_1} < 1$, and the iterative process converges toward the optimal transmission capacity.

The conditions leading to convergence are not met in practice. Our two-market example illustrates the point. Suppose the interconnection is owned and operated by a regulated

Transco, and that the Transco increases the interconnection's capacity to $\Phi^+ > \Phi^{+*}$. This could arise since transmission capacities are built in discrete and large increments. For example, the Transco could build one new line, which would result in interconnection capacity lower than $\Phi^*$; hence it decides to build two new lines, which results in interconnection capacity higher than $\Phi^*$. Then, baseload capacity is built in market 1 up to $K_1(\Phi^+)$. Since the Transco will not reduce transmission capacity, the story ends here. For the simplest possible market configuration, with no uncertainty, overinvestment caused by the indivisibility of transmission capacity precludes us from reaching the optimum. Multiple other factors make coordination between transmission and generation expansion difficult.

## 6.6 Loop Flows and Nodal Pricing on a Three-Node Network

We now consider a three-market (or three-node) network, which enables us to examine a phenomenon called "loop flows." Power flows on a network do not follow contractual arrangements. Rather, they obey Kirchhoff's circuit laws, derived in 1845 by the German physicist Gustav Kirchhoff (box 6.1). A Megawatt-hour produced at point $A$ to be consumed at point $B$ does not solely travel on the line from $A$ to $B$. Rather, it travels on the entire transmission grid, following the paths of least resistance. The power flows thus trace loops on the grid. These loop flows produce surprising results and have profound implications for market design. Market designers have attempted to ignore them at their peril. However, including loop flows does not fundamentally alter the economic intuition.

### 6.6.1 Set-Up and Transmission Constraint

Consider a three-market network. Demand $D(p, \theta)$ is entirely located in market 3, and generation is located in markets 1 (constant variable production cost $c_1$) and 2 (constant variable production cost $c_2 > c_1$). A transmission line connects each node. We assume

**Box 6.1**
Kirchhoff's laws

Gustav Kirchhoff was an impressive scientist: not only did he derive in 1845 his circuit laws when he was still a student (these became his dissertation), but he also derived three laws in spectroscopy and one in thermochemistry. Kirchhoff's circuit laws were later proved to be a specific case of Maxwell equations, which were published in 1860–1861.

This historical digression proves that the laws that rule power flows on a grid have been established long before the laws of economics. A story goes around among specialists of the power industry. I do not know whether it is true, but I have heard it many times, so here it is: the chieftain of a power company testifies in front of Congress. When asked by a congressman why utilities do things a certain way, he responds, "There is no other way. The laws of Kirchhoff force us to do it this way." The then-congressman helpfully answers, "We can change this Kirchhoff's laws. This is precisely the place where we make laws."

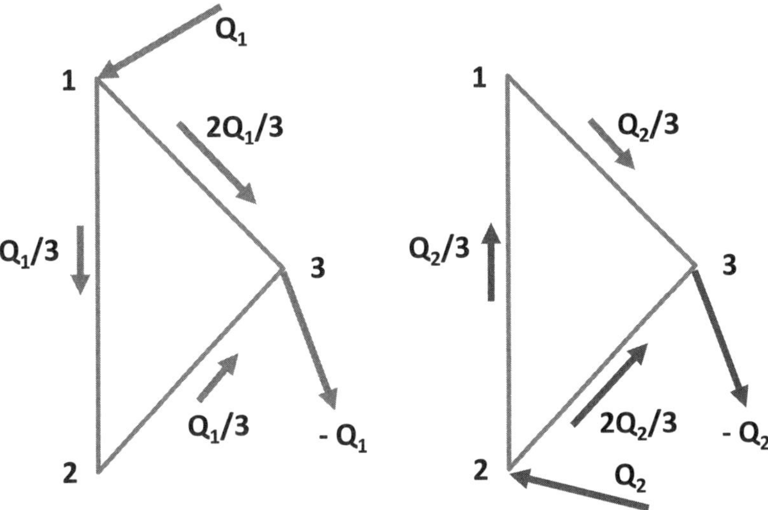


**Figure 6.15**
Power flows on a three-node network. Energy injected at one node and consumed at another uses all the lines, not only the direct path.

that the interconnection linking markets 1 and 2 can be congested. Since $c_1 < c_2$, power is flowing from market 1 to market 2. Denote $\Phi^+$ the transfer capacity from market 1 to market 2. Other interconnections are large enough never to be congested.

Kirchhoff's circuit laws require that, when $Q_1$ MWh is produced in market 1 and consumed in market 3, they travel directly on interconnection $(1, 3)$, but also through market 2, on interconnections $(1, 2)$ and $(2, 3)$. In other words, power flows cannot be directed on a grid. The presence of two paths between two markets is called a loop. Sometimes people use the term "meshed networks."

To simplify the analysis, suppose all three interconnections offer the same resistance to power flows. Thus the direct path from market 1 to market 3 is half as long as the indirect path through market 2. In this case, Kirchhoff's laws require that the volume traveling on each route is inversely proportional to the distance: $\frac{2}{3}Q_1$ go through the short direct path, and $\frac{1}{3}Q_1$ go through the twice-longer indirect path, as illustrated in the left panel of figure 6.15. This natural allocation minimizes the travel cost.

Similarly, if $Q_2$ MW are produced in market 2 and consumed in market 3, $\frac{2}{3}Q_2$ flow directly on interconnection $(2, 3)$, and $\frac{1}{3}Q_2$ through market 1, on interconnections $(2, 1)$ and $(1, 3)$, as illustrated in the right panel of figure 6.15.

Thus as illustrated in figure 6.16, the flow from market 1 to market 2 on interconnection $(1, 2)$ is

$$z = \frac{Q_1 - Q_2}{3}.$$

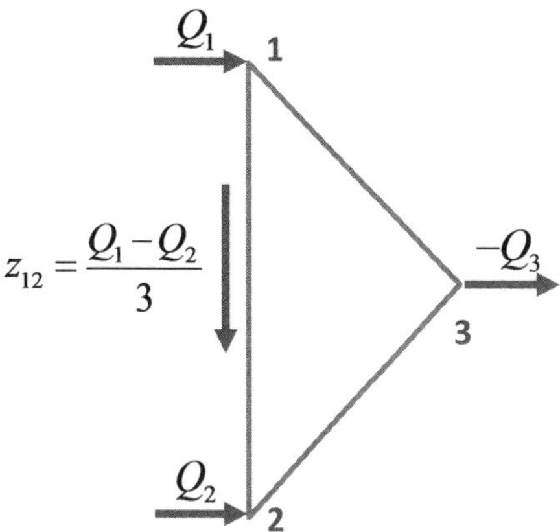


**Figure 6.16**
Symmetric three-market network. Producers are located in market 1 and 2. Demand is located in market 3. Interconnection (1, 2) is potentially congested.

The constraint is thus

$$z \le \Phi^+ \Leftrightarrow Q_1 - Q_2 \le 3\Phi^+.$$

To simplify the notation, we define and use $X^+ = 3\Phi^+$.

### 6.6.2 Production and Consumption

#### 6.6.2.1 Interconnection (1, 2) not congested

The cheapest technology located in market serves the entire demand. Prices are identical in all three markets, equal to the marginal production cost $c_1$. Production in market 1 is equal to demand in market 3:

$$Q_1 = D(c_1, \theta).$$

Since technology 2 does not produce, power flow on interconnection $(1, 2)$ is $z = \frac{D(c_1,\theta)}{3}$. Hence, the interconnection $(1, 2)$ is not congested as long as

$$D(c_1, \theta) \le X^+ \Leftrightarrow P(X^+, \theta) \le c_1 \Leftrightarrow \theta \le \hat{\theta}_0(X^+, c_1).$$

#### 6.6.2.2 Interconnection (1, 2) congested

**Technology 2 not producing** For $\theta > \hat{\theta}_0(X^+, c_1)$, the interconnection $(1, 2)$ is congested. Suppose first technology 2 is not turned on: production in market 1 is equal to demand in market 3, $Q_1 = X^+ = D(p_3(\theta), \theta)$. Price in market 1 is the marginal production cost $p_1(\theta) = c_1$, while the price in market 3 is the value of the marginal MWh consumed $p_3(\theta) = P(X^+, \theta) > c_1$. Prices are different in markets 1 and 3, even though the interconnection $(1, 3)$ is not congested. This surprising result arises because the loop flow creates an externality: power flowing from market 1 to market 3 also travels through the congested interconnection $(1, 2)$; hence a consumer in market 3 must pay for this externality.

To compute the price in market 2, suppose we inject $\delta Q > 0$ MWh in market 2 and consume them in market 3. Their value is the sum of two terms: first, $\delta Q$ are consumed in market 3, valued at $P(X^+, \theta)$. Second, $\frac{\delta Q}{3}$ flows on interconnection $(1, 2)$ from market 2 to market 1. This frees up $\frac{\delta Q}{3}$ on interconnection $(1, 2)$ in the direction from market 1 to market 2. Thus injecting $\delta Q > 0$ in market 2 enables us to produce an extra $\delta Q > 0$ in market 1 and consume it in market 3. Their net value is $(P(X^+, \theta) - c_1)$. Summing both effects, the price in market 2 is

$$p_2(\theta) = 2P(X^+, \theta) - c_1 > P(X^+, \theta).$$

Again, price in market 2 exceeds price in market 3, even though the interconnection $(2, 3)$ is not congested. This arises because of the positive congestion externality generated by an injection in market 2.

Technology 2 does not produce as long as its marginal cost of production exceeds the price in market 2, that is, when

$$c_2 > p_2(\theta) \Leftrightarrow P(X^+, \theta) < \frac{c_1 + c_2}{2} \Leftrightarrow \theta < \hat{\theta}_0 \left( X^+, \frac{c_1 + c_2}{2} \right).$$

We now compute the marginal value of capacity on interconnection $(1, 2)$. A $\delta\Phi^+ > 0$ capacity increase enables us to produce $3\delta\Phi^+$ more MWh in market 1 and consume them in market 3. The net value per unit of a capacity increase is

$$\eta(\theta) = 3(P(X^+, \theta) - c_1) = 3\left( \frac{p_2(\theta) + c_1}{2} - c_1 \right) = \frac{3}{2}(p_2(\theta) - c_1).$$

**Technology 2 producing** For $\theta > \hat{\theta}_0 \left( X^+, \frac{c_1+c_2}{2} \right)$, technology 2 also produces. Price in market 2 is equal to the marginal production cost: $p_2(\theta) = c_2$. Price in market 1 is unchanged: $p_1(\theta) = c_1$.

To compute price in market 3, suppose we consume an additional $\delta Q_3 > 0$ in market 3. To satisfy this extra demand while respecting the transmission constraint, we have to produce $\frac{\delta Q_3}{2}$ in each market. The marginal cost of serving this extra demand is therefore

$$p_3(\theta) = \frac{c_1 + c_2}{2}.$$

Finally, a $\delta\Phi^+ > 0$ increase in the capacity of the congested line enables us to produce $3\delta\Phi^+ > 0$ more MWh in market 1 and consume them in market 3. The net value per unit of a capacity increase is

$$\eta(\theta) = 3\left( \frac{c_1 + c_2}{2} - c_1 \right) = \frac{3}{2}(c_2 - c_1).$$

An alternative derivation of $\eta(\theta)$ comes from computing the production in both markets. We know that total production equals total demand:

$$Q_1(\theta)+Q_2(\theta)=D(p_3(\theta),\theta);$$

and interconnection $(1,2)$ is congested:

$$Q_1(\theta)-Q_2(\theta)=X^+.$$

This yields

$$\begin{cases} Q_1(\theta)=\frac{D(p_3(\theta),\theta)+X^+}{2} \\ Q_2(\theta)=\frac{D(p_3(\theta),\theta)-X^+}{2} \end{cases}. \tag{6.6}$$

A $\delta\Phi^+>0$ capacity increase on interconnection $(1,2)$ enables us to substitute $\frac{3}{2}\delta\Phi^+$ MWh produced at cost $c_1$ in market 1 for $\frac{3}{2}\delta\Phi^+$ MWh produced at cost $c_2$ in market 2, thereby saving $\frac{3}{2}(c_2-c_1)\delta\Phi^+$.

#### 6.6.2.3 No interconnection

Suppose no interconnection exists between markets 1 and 2, that is, $\Phi^+=0$. In that case, there is no loop flow, hence no congestion. We are back to the standard two-technology model described in chapter 2. Prices in all markets are equal. We thus obtain the surprising result that creating a transmission line with low capacity between markets 1 and 2 produces congestion.

### 6.6.3 Optimal Generation Mix

The above dispatch is valid as long as

$$Q_1(\theta)\le K_1 \Leftrightarrow \frac{D(p_3(\theta),\theta)+X^+}{2}\le K_1 \Leftrightarrow D\left(\frac{c_1+c_2}{2},\theta\right)$$
$$\le 2K_1-X^+ \Leftrightarrow \theta\le\hat{\theta}_0\left(2K_1-X^+,\frac{c_1+c_2}{2}\right).$$

To simplify the notation, define and use $X_1=2K_1-X^+$. For $\theta>\hat{\theta}_0\left(X_1,\frac{c_1+c_2}{2}\right)$, output in market 1 is $Q_1(\theta)=K_1$. Since interconnection $(1,2)$ is at capacity, output in market 2 is given by

$$K_1-Q_2(\theta)=X^+ \Leftrightarrow Q_2(\theta)=K_1-X^+,$$

and $p_2(\theta)=c_2$. Therefore, demand in market 3 is $Q_3^d(\theta)=Q_1(\theta)+Q_2(\theta)=X_1$, hence $p_3(\theta)=P(X_1,\theta)$.

To compute price in market 1, suppose we inject an additional $\delta Q>0$ in market 1. To respect the transmission constraint, we must also produce and inject an additional $\delta Q>0$ in market 2. The value generated is $(2p_3(\theta)-c_2)\times\delta Q$, hence $p_1(\theta)=2P(2X_1,\theta)-c_2$. An alternative derivation recognizes that to consume an additional $\delta Q>0$ in market 3,

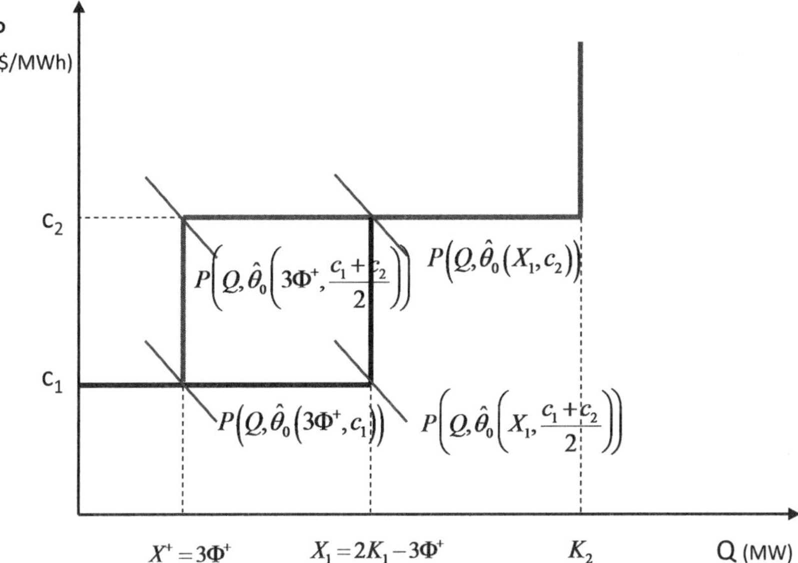


**Figure 6.17**
Nodal prices in markets 1 (bottom line) and 2 (top line).

$\frac{\delta Q}{2}$ must be produced in each market. Thus,

$$p_3(\theta)=\frac{p_1(\theta)+p_2(\theta)}{2}.$$

This lasts as long as

$$p_1(\theta)=2P(X_1,\theta)-c_2\leq c_2\Leftrightarrow P(X_1,\theta)\leq c_2\Leftrightarrow\theta\leq\hat{\theta}_0(X_1,c_2).$$

For $\theta>\hat{\theta}_0(X_1,c_2)$, interconnection (1, 2) is no longer congested, and we are back to the uncongested case described in chapter 2. The price paths in each market are illustrated in figure 6.17.

Since there is no congestion on-peak, the optimal cumulative capacity is equal to the optimal cumulative uncongested capacity: $K_2=K_2^U$.

Determination of the optimal baseload capacity involves a few algebraic manipulations. From the above discussion, the optimum baseload technology is uniquely defined by

$$\int_{\hat{\theta}_0\left(X_1,\frac{c_1+c_2}{2}\right)}^{\hat{\theta}_0(X_1,c_2)}(2P(X_1,\theta)-c_2-c_1)f(\theta)d\theta+\int_{\hat{\theta}_0(X_1,c_2)}^{\hat{\theta}_0(K_2,c_2)}(c_2-c_1)f(\theta)d\theta$$
$$+\int_{\hat{\theta}_0(K_2,c_2)}^{+\infty}(P(K_2,\theta)-c_2)f(\theta)d\theta=r_1.$$

As usual, this expression can be modified to make $\Psi_0(K_2, c_2)$ appear. Observing that, at the optimum, $\Psi_0(K_2, c_2) = r_2$ and rearranging yields

$$\int_{\hat{\theta}_0\left(X_1, \frac{c_1+c_2}{2}\right)}^{\hat{\theta}_0(X_1, c_2)} \left(P(X_1, \theta) - \frac{c_1+c_2}{2}\right) f(\theta)d\theta + \int_{\hat{\theta}_0(X_1, c_2)}^{+\infty} \frac{c_2 - c_1}{2} f(\theta)d\theta = \frac{r_1 - r_2}{2}.$$

A final modification to make $\Psi_0\left(X_1, \frac{c_1+c_2}{2}\right)$ appear yields

$$\Psi_0\left(X_1, \frac{c_1+c_2}{2}\right) - \Psi_0(X_1, c_2) = \frac{r_1 - r_2}{2}.$$

Baseload capacity on the three-node network is uniquely defined by an expression similar to the two-node case.

### 6.6.4 Optimal Interconnection Capacity

The marginal value of the capacity on interconnection (1, 2) is

$$\Lambda(\Phi^+) = \frac{3}{2}\mathbb{E}[p_2(\theta) - p_1(\theta)].$$

The marginal value is proportional but not equal to the difference in prices at the extremities of the interconnection. Increasing interconnection (1, 2) capacity by 1 MW generates 3/2 MW of FTRs between markets 2 and 1.

A closed form solution of $\Lambda(\Phi^+)$ is available. We have

$$\Lambda(\Phi^+) = \frac{3}{2}\left(\begin{array}{c} \int_{\hat{\theta}_0(X^+, c_1)}^{\hat{\theta}_0\left(X^+, \frac{c_1+c_2}{2}\right)} 2(P(X^+, \theta) - c_1) f(\theta)d\theta + \int_{\hat{\theta}_0\left(X^+, \frac{c_1+c_2}{2}\right)}^{\hat{\theta}_0\left(X_1, \frac{c_1+c_2}{2}\right)} (c_2 - c_1) f(\theta)d\theta \\ + \int_{\hat{\theta}_0\left(X_1, \frac{c_1+c_2}{2}\right)}^{\hat{\theta}_0(X_1, c_2)} 2\left(c_2 - P(X_1, \theta)\right) f(\theta)d\theta \end{array}\right).$$

The usual algebraic manipulations lead to

$$\Lambda(\Phi^+) = 3\left[\Psi_0(X^+, c_1) - \Psi_0\left(X^+, \frac{c_1+c_2}{2}\right) - \left(\Psi_0\left(X_1, \frac{c_1+c_2}{2}\right) - \Psi_0(X_1, c_2)\right)\right].$$

This expression has the same structure as the two-node case. If baseload capacity is optimal given the interconnection capacity, the optimal interconnection capacity is uniquely defined by

$$\Psi_0(X^+, c_1) - \Psi_0\left(X^+, \frac{c_1+c_2}{2}\right) = \frac{\gamma}{3} + \frac{r_1 - r_2}{2}.$$

## 6.7 Extension to a General Network

We now extend the derivation of nodal prices and the previous results to a general network, following Schweppe et al. (1988). This presentation should be sufficient for most economists. Readers interested in a discussion of the physics underlying this discussion are referred to the appendix D in Schweppe et al. (1988).

### 6.7.1 Set-Up

#### 6.7.1.1 Notation

Consider a power network consisting of $M$ markets, linked by $L$ interconnections. For every market $m = 1, \ldots, M$, $Q_m^s(\theta)$, $Q_m^d(\theta)$, and $Q_m(\theta) = Q_m^s(\theta) - Q_m^d(\theta)$ are the production, demand, and net injection. For every interconnection, $l = 1, \ldots, L$, $\varphi_l(\theta)$ is the power flowing on the interconnection in state $\theta$, $\Phi_l^+$ and $\Phi_l^-$ are the maximum power flow in each direction. Without loss of generality, we assume $\varphi_l(\theta)$ measures the flow and $\Phi$ the capacity from the lowest-number node to the highest-number node. The vector of oriented flows is $\boldsymbol{\varphi}(\boldsymbol{\theta}) \in \mathbb{R}^L$.

#### 6.7.1.2 Energy balance

In every state $\theta$, energy balance requires that total production is equal to total demand plus thermal losses $\tilde{LO}(\varphi(\theta))$:

$$\sum_{m=1}^{M} Q_m^s(\theta) = \sum_{m=1}^{M} Q_m^d(\theta) + \tilde{LO}(\boldsymbol{\varphi}(\boldsymbol{\theta})).$$

As shown below, the $M$ net injections are linked by the above equation, thus only $(M-1)$ net injections are independent. We thus define a reference node (e.g., node 1) and write the flow equations as a function of the vector $\underline{\mathbf{Q}}(\theta) \in \mathbb{R}^{M-1}$ of the $(M-1)$ remaining net injections.

#### 6.7.1.3 Direct current load-flow approximation

Electric power is, in fact, a wave. It is represented by two dimensions, an amplitude and a phase angle, or normal and reactive power. The (DC) load-flow approximation assumes that the phase angles at the extremities of all lines are very close. Neglecting second-order terms yields a linear relationship between the vector of flux and the vector of net injections

$$\boldsymbol{\varphi}(\boldsymbol{\theta}) = \mathbf{H} \cdot \underline{\mathbf{Q}}(\theta),$$

where $\mathbf{H} \in \mathbb{R}^L \times \mathbb{R}^{M-1}$ is the admittance-transfer matrix.

Under the DC load-flow approximation, line losses are a quadratic function of the flows, hence of net injections

$$\tilde{LO}(\boldsymbol{\varphi}(\boldsymbol{\theta})) = \underline{\mathbf{Q}}^T(\theta) \cdot \mathbf{B} \cdot \underline{\mathbf{Q}}(\theta) = LO(\underline{\mathbf{Q}}(\theta)),$$

where the matrix $\mathbf{B} \in \mathbb{R}^{M-1} \times \mathbb{R}^{M-1}$ is symmetric. Global energy balance is thus

$$\sum_{m=1}^{M} Q_m^s(\theta) = \sum_{m=1}^{M} Q_m^d(\theta) + LO(\underline{\mathbf{Q}}(\theta)).$$

Denote $\mu_e(\theta)$ the Lagrange multiplier associated with this constraint. $\mu_e(\theta)$ represents the value of a marginal MWh in state $\theta$.

#### 6.7.1.4 Maximum flow on each interconnection

Since every line has a maximum transfer capacity,

$$-\Phi_l^- \leq \varphi_l(\theta) \leq \Phi_l^+ \Leftrightarrow -\Phi_l^- \leq \sum_{m=2}^{M} H_{lm} Q_m(\theta) \leq \Phi_l^+.$$

At the optimal dispatch, only one constraint is binding. Without loss of generality, denote $\Phi_l$ the binding constraint in state $\theta$ and $\eta_l(\theta)$ the associated Lagrange multiplier.

As previously mentioned, capacity on interconnections is increasingly determined dynamically. The transmission-capacity constraint would then be written as $-\Phi_l^-(\theta) \leq \varphi_l(\theta) \leq \Phi_l^+(\theta)$. The economic intuition would be unchanged.

### 6.7.2 Characterization of the Optimal Dispatch

#### 6.7.2.1 Optimization problem

Introduce $C_m(Q)$ and $c_m(Q)$, respectively, the total cost and marginal cost of producing quantity $Q$.

Schweppe et al. (1988) state the optimization problem using $U(Q,\theta)$ as the utility consumers derive in state $\theta$ from quantity $Q$. The marginal utility is equal to the value of the marginal Megawatt-hour: $\frac{\partial U(Q,\theta)}{\partial Q} = P(Q,\theta)$.

The optimization problem is thus to chose $\{Q_m^s(\theta), Q_m^d(\theta)\}_{\{m,\theta\}}$, the production and consumption at each node $m$ and in each state $\theta$, to maximize the net surplus, subject to the constraints imposed by the laws of physics:

$$\begin{aligned} &max_{Q_m^s(\theta), Q_m^d(\theta)} \mathbb{E}\left[\sum_{m=1}^{M} \{U_m(Q_m^d(\theta),\theta) - C_m(Q_m^s(\theta))\}\right] \\ &st: \begin{cases} \sum_{m=1}^{M} Q_m^s(\theta) = \sum_{m=1}^{M} Q_m^d(\theta) + LO(\underline{\mathbf{Q}}(\theta)) \ \forall\theta & (\mu_e(\theta)) \\ \sum_{m=2}^{M} H_{lm} Q_m(\theta) \leq \Phi_l \ \forall(l,\theta) & (\eta_l(\theta)) \end{cases} . \end{aligned} \tag{6.7}$$

#### 6.7.2.2 Locational marginal prices

The Lagrangian of the optimization program is

$$\mathcal{L} = \mathbb{E}\left[\begin{array}{c} \sum_{m=1}^{M} \{U_m(Q_m^d(\theta),\theta) - C_m(Q_m^s(\theta))\} \\ +\mu_e(\theta)\left(\sum_{m=1}^{M} Q_m^s(\theta) - \sum_{m=1}^{M} Q_m^d(\theta) - LO(\underline{\mathbf{Q}}(\theta))\right) \\ +\sum_{l=1}^{L} \eta_l(\theta)\left(\Phi_l - \sum_{m=2}^{M} H_{lm} Q_m(\theta)\right) \end{array}\right].$$

The first-order derivatives are

$$\frac{\partial \mathcal{L}}{\partial Q^d_m(\theta)} = P_m(Q^d_m(\theta),\theta) - \mu_e(\theta)\left(1 - \frac{\partial LO}{\partial Q_m}\right) + \sum_{l=1}^{L} \eta_l(\theta) H_{lm}$$

for every $m$ such that $Q^d_m(\theta) > 0$, and

$$\frac{\partial \mathcal{L}}{\partial Q^s_m(\theta)} = -c_m(Q^s_m(\theta)) + \mu_e(\theta)\left(1 - \frac{\partial LO}{\partial Q_m}\right) - \sum_{l=1}^{L} \eta_l(\theta) H_{lm}$$

for every $m$ such that $Q^s_m(\theta) > 0$, with the convention that $\frac{\partial LO}{\partial Q_1} = H_{l1} = 0$.

The optimal prices $p_m(\theta)$ are defined for all $m \geq 1$ by the local-equilibrium conditions

$$p_m(\theta) = \begin{cases} P_m(Q^d_m(\theta),\theta) & \textit{if } Q^d_m(\theta) > 0 \\ c_m(Q^s_m(\theta)) & \textit{if } Q^s_m(\theta) > 0 \end{cases} \tag{6.8}$$

and the global-equilibrium conditions

$$\begin{cases} p_1(\theta) = \mu_e(\theta) \\ p_m(\theta) = \mu_e(\theta)\left(1 - \frac{\partial LO}{\partial Q_m}\right) - \sum_{l=1}^{L} H_{lm}\eta_l(\theta) \; for \; m > 1 \end{cases} . \tag{6.9}$$

The local-equilibrium conditions (6.8) imply that price at a node $m$ is equal to the marginal surplus from consumption or the marginal cost of production, or both.

The global-equilibrium conditions (6.9) take into account the network externalities: thermal losses and congestion. They are defined relative to the reference node, that is, no externality is included in the price at node 1, which is the global reference price $\mu_e(\theta)$.

Consider the consumer's perspective for a market $m > 1$ where $Q^d_m > 0$. She pays the global reference price $\mu_e(\theta)$, plus her marginal contribution to thermal losses $\frac{\partial LO}{\partial Q^d_m} = -\frac{\partial LO}{\partial Q_m}$, where losses are valued at the global reference price, plus for every line her contribution to congestion on this line $\frac{\partial \varphi_l}{\partial Q^d_m} = -\frac{\partial \varphi_l}{\partial Q_m} = -H_{lm}$, valued at the (virtual) value of the congestion $\eta_l(\theta)$. Symmetrically, if $Q^s_m > 0$, a producer sells her energy at the reference price $\mu_e(\theta)$, minus her marginal contribution to thermal losses $\frac{\partial LO}{\partial Q^s_m} = \frac{\partial LO}{\partial Q_m}$ valued at $\mu_e(\theta)$, minus her marginal contribution to congestion on every line $l$ $\frac{\partial \varphi_l}{\partial Q^s_m} = \frac{\partial \varphi_l}{\partial Q_m} = H_{lm}$ valued at $\eta_l(\theta)$.

Every market participant thus faces the marginal externalities created by her decisions. Under perfect competition, nodal pricing decentralizes the social optimum. Conversely, if market participants face prices other than nodal prices, net surplus is reduced.

#### 6.7.2.3 Equivalent formulations of the problem

**Prices as the optimization variables** The rest of this text uses a slightly different formulation. We use the surplus $S_m(p,\theta) = U_m(D_m(p,\theta),\theta)$ and choose the production $Q^s_m(\theta)$ and price $p_m(\theta)$ at each node to maximize the net surplus, recognizing that

$Q_m^d(\theta) = D_m(p_m(\theta), \theta)$. The resulting problem is

$$max_{Q_m^s(\theta), p_m(\theta)} \mathbb{E}\left[\sum_{m=1}^{M} \{S_m(p_m(\theta), \theta) - C_m(Q_m^s(\theta))\}\right]$$

$$st: \begin{cases} \sum_{m=1}^{M} Q_m^s(\theta) = \sum_{m=1}^{M} D_m(p_m(\theta), \theta) + LO(\underline{\mathbf{Q}}(\theta)) \ \forall\theta & (\mu_e(\theta)) \\ \sum_{m=2}^{M} H_{lm} Q_m(\theta) \leq \Phi_l \ \forall(l, \theta) & (\eta_l(\theta)) \end{cases}.$$

Both specifications yield the same outcome.

**Net injections as the optimization variables** Equation (6.8) states that, in every state $\theta \geq 0$ and at every node where both production and demand occur, the cost of the last Megawatt-hour produced is equal to the value of the last Megawatt-hour consumed: $p_m(\theta) = P_m(Q_m^d(\theta), \theta) = c_m(Q_m^s(\theta)) \Leftrightarrow Q_m^d(\theta) = D(c_m(Q_m^s(\theta)), \theta)$. Demand $Q_m^d(\theta)$ and supply $Q_m^s(\theta)$ at such a node are not independent. Only one quantity needs to be chosen at every node. For simplicity, we choose net injection $Q_m$. If only production occurs at node $m$, net injection is $Q_m(\theta) = Q_m^s(\theta)$; if only demand occurs, net injection is $Q_m(\theta) = -Q_m^d(\theta)$. Denote $w_m(Q_m, \theta)$ the net surplus at node $m$ for net injection $Q_m$ in state $\theta$, and $W(\mathbf{Q}(\theta), \theta) = \sum_{m=1}^{M} w_m(Q_m(\theta), \theta)$.

Since the short-term optimization proceeds state by state, we omit the state of the world $\theta$. For a given state $\theta \geq 0$, the problem (given by equation 6.7) becomes

$$max_{\mathbf{Q}} W(\mathbf{Q})$$

$$st: \begin{cases} \sum_{m=1}^{M} Q_m - LO(\underline{\mathbf{Q}}) = 0 \\ \sum_{m=2}^{M} H_{lm} Q_m \leq \Phi_l \ \forall l \end{cases}.$$

The associated Lagrangian is

$$\mathcal{L} = W(\mathbf{Q}) + \mu_e \left( \sum_{m=1}^{M} Q_m - LO(\underline{\mathbf{Q}}) \right) + \sum_{l=1}^{L} \eta_l (\Phi_l - \sum_{m=2}^{M} H_{lm} Q_m).$$

The first-order conditions are

$$\frac{\partial \mathcal{L}}{\partial Q_m} = \frac{\partial W}{\partial Q_m} + \mu_e \left(1 - \frac{\partial LO}{\partial Q_m}\right) - \sum_{l=1}^{L} \eta_l H_{lm} = 0 \Leftrightarrow p_m = -\frac{\partial W}{\partial Q_m}.$$

#### 6.7.2.4 Three-market example

As an illustration, consider the three-market example presented in section 6.6. The optimization program is

$$max_{Q_1^s(\theta), Q_2^s(\theta), p_3(\theta)} \mathbb{E}\left[S(p_3(\theta), \theta) - \sum_{m=1}^{M} c_m Q_m^s(\theta)\right]$$

$$st: \begin{cases} Q_1^s(\theta) + Q_2^s(\theta) = D_3(p_3(\theta), \theta) \ \forall\theta & (\mu_e(\theta)) \\ \frac{Q_1^s(\theta) - Q_2^s(\theta)}{3} \leq \Phi^+ \ \forall\theta & (\eta(\theta)) \end{cases}.$$

The associated Lagrangian is

$$\mathcal{L} = \mathbb{E}\begin{bmatrix} S(p_3(\theta),\theta) - \sum_{m=1}^{M} c_m Q_m^s(\theta) \\ +\mu_e(\theta)(Q_1^s(\theta) + Q_2^s(\theta) - D_3(p_3(\theta),\theta)) \\ +\eta(\theta)\left(\Phi^+ - \frac{Q_1^s(\theta)-Q_2^s(\theta)}{3}\right) \end{bmatrix}.$$

The first-order derivatives are

$$\begin{cases} \frac{\partial \mathcal{L}}{\partial Q_1^s} = -c_1 + \mu_e(\theta) - \frac{\eta(\theta)}{3} \\ \frac{\partial \mathcal{L}}{\partial Q_1^s} = -c_2 + \mu_e(\theta) + \frac{\eta(\theta)}{3} \\ \frac{\partial \mathcal{L}}{\partial p_3} = (p_3(\theta) - \mu_e(\theta))\frac{\partial D}{\partial p_3} \end{cases}.$$

The global-equilibrium conditions are

$$\begin{cases} p_1(\theta) = \mu_e(\theta) - \frac{\eta(\theta)}{3} \\ p_2(\theta) = \mu_e(\theta) + \frac{\eta(\theta)}{3} \\ p_3(\theta) = \mu_e(\theta) \end{cases}.$$

We solve for various configurations of the system. Suppose first only producers in market 1 produce, and the interconnection is uncongested. Since producers in market 1 produce, and $\eta(\theta) = 0$, the local-equilibrium condition in market 1 is $p_1(\theta) = c_1$. Thus $p_2(\theta) = p_3(\theta) = c_1$, and $Q_3^d(\theta) = D(c_1,\theta) = Q_1^s(\theta)$.

This lasts as long as the interconnection remains uncongested, i.e., $D(c_1,\theta) \le 3\Phi^+ \Leftrightarrow \theta \le \theta_0(\hat{X}^+, c_1)$. For $\theta \ge \theta_0(\hat{X}^+, c_1)$, the interconnection is congested, thus

$$Q_1^s(\theta) = Q_3^d(\theta) = 3\Phi^+ = D(p_3(\theta),\theta) \Leftrightarrow p_3(\theta) = P(X^+,\theta).$$

Thus

$$p_1(\theta) = c_1 = P(X^+,\theta) - \frac{\eta(\theta)}{3} \Leftrightarrow \eta(\theta) = 3(P(X^+,\theta) - c_1).$$

Finally,

$$p_2(\theta) = \mu_e(\theta) + \frac{\eta(\theta)}{3} = 2P(X^+,\theta) - c_1.$$

This lasts as long as technology 2 is not turned on, i.e., $2P(X^+,\theta) - c_1 \le c_2 \Leftrightarrow \theta \le \theta_0\left(\hat{X}^+, \frac{c_1+c_2}{2}\right)$. For $\theta \ge \theta_0(\hat{X}^+, c_1)$, technology 2 starts producing, thus

$$\begin{cases} p_2(\theta) = c_2 = \mu_e(\theta) + \frac{\eta(\theta)}{3} \\ p_1(\theta) = c_1 = \mu_e(\theta) - \frac{\eta(\theta)}{3} \end{cases}.$$

This system of two equations for two unknowns yields

$$\begin{cases} \eta(\theta) = \frac{3}{2}(c_2 - c_1) \\ \mu_e(\theta) = \frac{1}{2}(c_2 + c_1) \end{cases}.$$

### 6.7.3 Merchandizing Surplus

The merchandizing surplus in state $\theta$ is

$$MS(\theta) = \sum_{m=1}^{M} (Q_m^d(\theta) - Q_m^s(\theta)) p_m(\theta) = -\sum_{m=1}^{M} Q_m(\theta) p_m(\theta).$$

Schweppe et al. (1988) show that

$$MS(\theta) = \mu_e(\theta) LO(\underline{\mathbf{Q}}(\theta)) + \sum_{l=1}^{L} \eta_l(\theta)\Phi_l.$$

The proof proceeds as follows. Substituting the LMPs given by equation (6.9) in the expression of the merchandizing surplus yields

$$\begin{aligned} MS(\theta) &= -\sum_{m=1}^{M} Q_m(\theta) \left[ \mu_e(\theta) \left(1 - \frac{\partial LO}{\partial Q_m}\right) - \sum_{l=1}^{L} H_{lm}\eta_l(\theta) \right] \\ &= -\mu_e(\theta) \sum_{m=1}^{M} Q_m(\theta) \left(1 - \frac{\partial LO}{\partial Q_m}\right) + \sum_{m=1}^{M} Q_m(\theta) \sum_{l=1}^{L} H_{lm}\eta_l(\theta). \end{aligned}$$

Consider the first term on the right-hand side. Observe that (a) the energy balance yields $\sum_{m=1}^{M} Q_m(\theta) = LO(\underline{\mathbf{Q}}(\theta))$, and (b) since $LO(\underline{\mathbf{Q}}(\theta))$ is a quadratic function of net injections, $LO(\underline{\mathbf{Q}}(\theta)) = \frac{1}{2}\sum_{m=1}^{M} Q_m(\theta)\frac{\partial LO}{\partial Q_m}$. Therefore,

$$\sum_{m=1}^{M} Q_m(\theta) \left(1 - \frac{\partial LO}{\partial Q_m}\right) = LO(\underline{\mathbf{Q}}(\theta)) - 2LO(\underline{\mathbf{Q}}(\theta)) = -LO(\underline{\mathbf{Q}}(\theta)).$$

Consider now the second term on the right-hand side. Inverting the order of summations yields

$$\sum_{m=1}^{M} Q_m(\theta) \sum_{l=1}^{L} H_{lm}\eta_l(\theta) = \sum_{l=1}^{L} \eta_l(\theta) \sum_{m=1}^{M} H_{lm} Q_m(\theta) = \sum_{l=1}^{L} \eta_l(\theta)\varphi_l(\theta) = \sum_{l=1}^{L} \eta_l(\theta)\Phi_l,$$

since $\eta_l(\theta) > 0 \Leftrightarrow \varphi_l(\theta) = \Phi_l$.

Putting the two pieces together proves the result.

Nodal pricing thus generates a positive surplus for the market maker, equal to the value of the losses (valued at $\mu_e(\theta)$), plus value of the congestion on the network.

### 6.7.4 Revenue Adequacy of Financial Transmission Rights

In the two-node and lossless network examined in section 6.2, the merchandizing surplus is exactly equal to the revenues to FTR holders, thus the ISO could always cover her financial obligations. This property extends a general network—although with some caveats, which the proof clarifies.

The initial proof, proposed by Hogan (1992), starts from the fact that, since an FTR (in these days called a transmission congestion contract, TCC) is a pair of net injections, one positive and one negative, a set of FTR is a dispatch. An FTR of volume $t$ between nodes $i$ and $j$, which pays its holder $t(p_j - p_i)$, is represented by a vector $\mathbf{t} \in \mathbb{R}^M$ composed of 0 on all positions, except $-t$ in position $i$ and $+t$ in position $j$. Payment to the FTR holder is thus $\mathbf{p}^T \cdot \mathbf{t}$ , where $\mathbf{p} \in \mathbb{R}^M$ is the vector of nodal prices, and $x^T$ is the transpose of vector $x \in \mathbb{R}^M$. The dispatch associated with this FTR is $d = -t$, since net injections are $+t$ MWh at node $i$ and $-t$ MWh at node $j$. The set of all FTR granted by an ISO is a vector $\mathbf{T} \in \mathbb{R}^M$. Its associated dispatch is $\mathbf{D} = -\mathbf{T} \in \mathbb{R}^M$. Payment by the ISO to FTR holders is $\mathbf{p}^T \cdot \mathbf{T} = -\mathbf{p}^T \cdot \mathbf{D}$.

A set of FTR $\mathbf{T}$ is feasible if and only if the associated dispatch $\mathbf{D}$ is within the constraint set, that is, the set $\mathcal{C}$ defined by the constraints of optimization problem (6.7).

The next step is to prove the convexity of the constraint set $\mathcal{C}$. A set is convex if and only if, for all $x_1$and $x_2$ in the set, for all $t \in [0, 1]$, $t \cdot x_1 + (1-t) \cdot x_2$ is also in the set. Hogan (1992) observes that the constraint set is convex for a lossless network. Bushnell and Stoft (1996) extend the proof to include losses.

To see that, it is helpful to express the constraints in matrix form. The energy-balance constraint is

$$\mathbf{e}^T \cdot \mathbf{Q}(\theta) - \underline{\mathbf{Q}}(\theta)^T \cdot \mathbf{B} \cdot \underline{\mathbf{Q}}(\theta)^T = 0,$$

where $\mathbf{e} \in \mathbb{R}^{\mathbf{M}}$ is the vector of 1, "$\cdot$" is the matrix multiplication, and the matrix $\mathbf{B}$ is semi-definite positive. The flows constraint is

$$\mathbf{H} \cdot \underline{\mathbf{Q}}(\theta) \leqq \mathbf{\Phi},$$

where the symbol "$\leqq$" in the flow constraint means that every element of the vector on the left of "$\leqq$" is lower than the corresponding element of the vector on the right.

Bushnell and Stoft (1996) observe that the energy-balance constraint can be rewritten as an inequality: $\mathbf{e}^T \cdot \mathbf{Q} - \underline{\mathbf{Q}}^T \cdot \mathbf{B} \cdot \underline{\mathbf{Q}}^T \geq 0$. Since energy is costly to produce, the constraint will be binding at the optimum. This apparently innocuous change proves extremely important. The constraint set $\mathcal{C}$ is the subset of $\mathbb{R}^M$ that satisfies both inequalities.

Since $\mathbf{B}$ is semidefinite positive, the energy-balance constraint defines a convex set. An example illustrates the intuition. In one dimension, the constraint is $x - bx^2 \geq 0$, for

$b \geq 0$. We immediately have

$$x - bx^2 \geq 0 \Leftrightarrow x(1 - bx) \geq 0 \Leftrightarrow x \in \left[0, \frac{1}{b}\right].$$

The constraint set is a segment, hence it is convex. The same intuition applies in $\mathbb{R}^M$.

Since they are linear, the flow constraints also define a convex set. If $\mathbf{H} \cdot \underline{\mathbf{Q}}_1 \leq \mathbf{\Phi}$ and $\mathbf{H} \cdot \underline{\mathbf{Q}}_2 \leq \mathbf{\Phi}$, then

$$\mathbf{H} \cdot (t\underline{\mathbf{Q}}_1 + (1-t)\underline{\mathbf{Q}}_2) = t\mathbf{H} \cdot \underline{\mathbf{Q}}_1 + (1-t)\mathbf{H} \cdot \underline{\mathbf{Q}}_2 \leq t\mathbf{\Phi} + (1-t)\mathbf{\Phi} = \mathbf{\Phi}.$$

Bushnell and Stoft (1996) thus conclude that the constraint set $\mathcal{C}$ is convex even when losses are included.

The final step proceeds by contradiction. Suppose there exists a feasible set of FTRs $\mathbf{T}$ such that payments to FTR holders exceed the merchandizing surplus, that is, $-\mathbf{p}^{*T} \cdot \mathbf{T} - \mathbf{p}^{*T} \cdot \mathbf{Q}^* < 0$. Thus there exists a dispatch $\mathbf{D} = -\mathbf{T} \in \mathcal{C}$ such that $\mathbf{p}^{*T} \cdot (\mathbf{D} - \mathbf{Q}^*) < 0$. Since $p_m^* = -\frac{\partial W}{\partial Q_m}$, we have $\mathbf{p}^* = \nabla W(\mathbf{Q}^*)$, where $\nabla W(\mathbf{Q}) \in \mathbb{R}^N$ is the gradient of net surplus $W(\mathbf{Q})$. Thus we have $\mathbf{D} \in \mathcal{C}$ such that

$$\nabla W(\mathbf{Q}^*)^T \cdot (\mathbf{D} - \mathbf{Q}^*) > 0.$$

Since $\mathbf{D} \in \mathcal{C}$ and $\mathbf{Q}^* \in \mathcal{C}$, the direction $(\mathbf{D} - \mathbf{Q}^*)$ is feasible. However, since $\mathbf{Q}^*$maximizes $W(\mathbf{Q}^*)$, $\nabla W(\mathbf{Q}^*)^T \cdot \mathbf{V} \leq 0$ for all feasible directions. This constitutes a contradiction. Thus if a set of FTRs is feasible, the merchandizing surplus exceeds payments to FTR holders.

As suggested by the proof, revenue adequacy holds as long as the optimization program is convex and may no longer hold if the program is not convex (see, e.g., Philpott and Pritchard 2004).

### 6.7.5 Generation Investment

We have seen in chapter 2 that if competition is perfect, peak-load pricing leads to optimal investment in generation assets in a single market. How does that result extend to multiple markets separated by congested transmission lines?

**Optimization program** Peak-load pricing extends naturally to multiple markets. To simplify, we assume that only one technology is available at each node $m$, characterized by marginal production cost $c_m$ and fixed cost per unit $r_m$.

The social optimum is characterized by

$$max_{Q^s_m(\theta), Q^d_m(\theta), k_m} \mathbb{E}\left[\sum_{m=1}^M \left\{U_m(Q^d_m(\theta), \theta) - c_m Q^s_m(\theta)\right\}\right] - \sum_{m=1}^M r_m k_m$$

$$st: \begin{cases} Q^s_m(\theta) \leq k_m \ \forall (m, \theta) & (\lambda_m(\theta)) \\ \sum_{m=1}^M Q^s_m(\theta) = \sum_{m=1}^M Q^d_m(\theta) + LO(\underline{\mathbf{Q}}(\theta)) \ \forall \theta & (\mu_e(\theta)). \\ -\Phi_l^- \leq \sum_{m=2}^M H_{lm} Q_m(\theta) \leq \Phi_l^+ \ \forall (l, \theta) & (\eta_l(\theta)) \end{cases} \quad (6.10)$$

This optimization program is the combination of program (6.7) leading to nodal prices and a peak-load pricing program for multiple generation technologies (2.7). The costs and constraints present in program (2.7) have been added to the program (6.7). This leaves the structure of the solution unchanged and specifies the form of the long-term marginal cost $c_m(Q_m^S(\theta))$.

**Locational marginal prices** Nodal prices are defined by the local equilibrium conditions (6.8). Marginal costs being constant, these are

$$p_m(\theta) = \begin{cases} c_m & \text{if} \quad 0 < Q_m^S(\theta) < k_m \\ c_m + \lambda_m(\theta) & \text{if} \quad Q_m^S(\theta) = k_m \end{cases},$$

where $\lambda_m(\theta)$ is the multiplier associated with constraint $Q_m^S(\theta) \leq k_m$.

**Optimal production capacity** At node $m$, equilibrium production capacity satisfies

$\mathbb{E}[(p_m(\theta) - c_m)^+] = r_m$.

In general, the price $p_m(\theta)$ depends on the vector of capacities $\mathbf{K} \in \mathbb{R}^M$. The optimal capacities vector $\mathbf{K}^* \in \mathbb{R}^M$ is thus uniquely defined by the first-order conditions

$$\mathbb{E}[(p_m(\mathbf{K}^*, \theta) - c_m)^+] = r_m \ \forall m. \tag{6.11}$$

As before, these relations characterize the social optimum but also free entry. Nodal pricing thus leads to optimal generation mix in each market. This is a specific case of more general result: short-term optimality leads to long-term optimality, that is, if prices correctly reflect short-term market conditions, they generate the correct investment signals.

### 6.7.6 Transmission Grid Investment

#### 6.7.6.1 Long-term optimum

We first suppose that increasing capacity on one line has no impact on other lines' capacities, that is, $\frac{\partial \Phi_j}{\partial \Phi_l} = 0$ for $j \neq l$. The optimum is defined by

$$max_{Q_m^s(\theta), Q_m^d(\theta), \Phi_l} \mathbb{E}\left[\sum_{m=1}^M \{U_m(Q_m^d(\theta), \theta) - c_m Q_m^s(\theta)\}\right] - \sum_{m=1}^M r_m k_m - \Gamma(\mathbf{\Phi})$$

$$st: \begin{cases} Q_m^s(\theta) \leq k_m \ \forall m & (\lambda_m(\theta)) \\ \sum_{m=1}^M Q_m^s(\theta) = \sum_{m=1}^M Q_m^d(\theta) + LO(\underline{\mathbf{Q}}(\theta)) & (\mu_e(\theta)) \\ \sum_{l=1}^m H_{lm} Q_m(\theta) \leq \Phi_l \ \forall l & (\eta_l(\theta)) \end{cases}. \tag{6.12}$$

Problem (6.12) is problem (6.10) to which has been added the determination of transmission capacities $\Phi_l$. Differentiating the Lagrangian with respect to $\Phi_l$, necessary conditions satisfied by the optimal grid are

$$\Lambda_l(\mathbf{\Phi}) = \mathbb{E}[\eta_l(\theta)] = \frac{\partial \Gamma(\mathbf{\Phi}^*)}{\partial \Phi_l} \ \forall l. \tag{6.13}$$

As mentioned previously, the cost function $\Gamma(\mathbf{\Phi})$ may not be well behaved. The system of equations (6.13) may not admit a solution, or might admit multiple solutions.

#### 6.7.6.2 Optimal regulatory contract

For the two-node network, the regulatory contract presented in Proposition 6.4 leads to optimal investment since the marginal redispatching lost is equal to minus the marginal value of transmission capacity (Lemma 6.1). This result extends to a general network. Assuming the system of equations (6.13) admits a unique solution.

**Redispatch cost** The logic of the two-market network applies. If the power flows exceed the interconnection capacity, the market operator must modify the transactions to adjust the flows. At import-constrained nodes, she purchases power from the producers (who are constrained on) and consumers (who are constrained off) in order to increase production and hence reduce imports. Simultaneously, at export-constrained nodes, she reduces production and increases demand. The market operator adjusts until the power flows on congested interconnections are exactly equal to the capacity of these lines. The dispatch is thus exactly identical to the nodal pricing one.

We now compute the redispatching cost. Consider an export-constrained node, characterized by $Q_m^s(\mathbf{\Phi}) < Q_m^{sU}$ and $Q_m^d(\mathbf{\Phi}) > Q_m^{dU}$. For $x \in [Q_m^{dU}, Q_m^d(\mathbf{\Phi})]$, constrained-on consumers are compensated for their surplus loss from consuming; hence they are paid $(p^U - P_m(x, \theta))$. For $x \in [Q_m^s(\mathbf{\Phi}), Q_m^{sU}]$, constrained-off producers are paid the net operating profit they would have received: $(p^U - c_m(x))$. The total redispatching cost at export-constrained nodes in state $\theta$ is

$$\int_{Q_m^{dU}}^{Q_m^d(\mathbf{\Phi})} (p^U - P_m(x, \theta))dx + \int_{Q_m^s(\mathbf{\Phi})}^{Q_m^{sU}} (p^U - c_m(x))dx.$$

Consider an import-constrained node, characterized by $Q_m^s(\mathbf{\Phi}) > Q_m^{sU}$ and $Q_m^d(\mathbf{\Phi}) < Q_m^{sU}$. For $x \in [Q_m^{sU}, Q_m^s(\mathbf{\Phi})]$, constrained-on producers are paid their cost $c_m(x)$; hence their constrained-on payment is $(c_m(x) - p^U)$. For $x \in [Q_m^d(\mathbf{\Phi}), Q_m^{dU}]$, constrained-off consumers receive the net surplus they would have derived from consuming: $(P_m(x, \theta) - p^U)$. The total redispatching cost at import-constrained nodes in state $\theta$ is

$$\int_{Q_m^{sU}}^{Q_m^s(\Phi)} (c_m(x) - p^U(\theta))dx + \int_{Q_m^d(\Phi)}^{Q_m^{dU}} (P_m(x, \theta) - p^U(\theta))dx.$$

Comparing the expressions above, we observe redispatching costs are formally identical at export- and import-constrained nodes.

In state $\theta$, the market operator purchases and sells the constrained quantities $Q_m^s(\mathbf{\Phi}, \theta)$ and $Q_m^d(\mathbf{\Phi}, \theta)$ at the unconstrained price $p^U(\theta)$ and pays the redispatching cost. Observing that $\sum_{m=1}^{M} Q_m(\theta) = LO(\underline{\mathbf{Q}}(\theta))$, the operator's net profit is

$$\Pi^{CT}(\boldsymbol{\Phi}) = -\mathbb{E}\left[\begin{array}{c} p^U LO(\underline{\mathbf{Q}}(\theta)) + \\ \sum_{m=1}^{M}\left(\int_{Q_m^{sU}}^{Q_m^s(\Phi)}(c_m(x) - p^U(\theta))dx + \int_{Q_m^d(\Phi)}^{Q_m^{dU}}(P_m(x,\theta) - p^U(\theta))dx\right) \end{array}\right].$$

The market operator pays the redispatch cost and the transmission losses, valued at the unconstrained price.

**Marginal redispatching cost**

**Lemma 6.2.** Suppose $\frac{\partial H_{jm}}{\partial \Phi_l} = 0$ and $\frac{\partial \Phi_j}{\partial \Phi_l} = 0$ for $j \neq l$. The marginal redispatching cost with respect to capacity on line $l$ is equal to the marginal value of line $l$:

$$\frac{\partial \Pi^{CT}}{\partial \Phi_l} = \Lambda_l(\boldsymbol{\Phi}).$$

***Proof.*** Since $LO(\underline{\mathbf{Q}}(\theta)) = \sum_{m=1}^{M} Q_m(\theta)$, differentiation with respect to $\Phi_l$ yields

$$\begin{aligned} \frac{\partial \Pi^{CT}}{\partial \Phi_l} &= -\mathbb{E}\left[p^U \sum_{m=1}^{M} \frac{\partial Q_m}{\partial \Phi_l} + \sum_{m=1}^{M}\left((p_m - p^U)\frac{\partial Q_m^s}{\partial \Phi_l} - (p_m - p^U)\frac{\partial Q_m^d}{\partial \Phi_l}\right)\right] \\ &= -\mathbb{E}\left[p^U \sum_{m=1}^{M} \frac{\partial Q_m}{\partial \Phi_l} + \sum_{m=1}^{M}(p_m - p^U)\frac{\partial Q_m}{\partial \Phi_l}\right] \\ &= -\mathbb{E}\left[\sum_{m=1}^{M} p_m \frac{\partial Q_m}{\partial \Phi_l}\right]. \end{aligned}$$

The next step follows the proof of the expression of the merchandizing surplus. Substituting the LMPs given by equations (6.9) in the expression of the marginal redispatching cost yields

$$\begin{aligned} \frac{\partial \Pi^{CT}}{\partial \Phi_l} &= -\mathbb{E}\left[\sum_{m=1}^{M}\left[\mu_e(\theta)\left(1 - \frac{\partial LO}{\partial Q_m}\right) - \sum_{j=1}^{L} H_{jm}\eta_j(\theta)\right]\frac{\partial Q_m}{\partial \Phi_l}\right] \\ &= -\mathbb{E}\left[\mu_e(\theta)\sum_{m=1}^{M}\left(1 - \frac{\partial LO}{\partial Q_m}\right)\frac{\partial Q_m}{\partial \Phi_l} - \sum_{m=1}^{M}\sum_{j=1}^{L} H_{jm}\eta_j(\theta)\frac{\partial Q_m}{\partial \Phi_l}\right]. \end{aligned}$$

Consider the first term on the right-hand side. Since $\sum_{m=1}^{M} Q_m(\theta) = LO(\underline{\mathbf{Q}}(\theta))$, $\sum_{m=1}^{M} \frac{\partial Q_m}{\partial \Phi_l} = \frac{\partial LO(\underline{\mathbf{Q}})}{\partial \Phi_l}$, and by construction $\frac{\partial LO(\underline{\mathbf{Q}})}{\partial \Phi_l} = \sum_{m=1}^{M} \frac{\partial LO}{\partial Q_m}\frac{\partial Q_m}{\partial \Phi_l}$. Therefore,

$$\sum_{m=1}^{M}\left(1 - \frac{\partial LO}{\partial Q_m}\right)\frac{\partial Q_m}{\partial \Phi_l} = \sum_{m=1}^{M}\frac{\partial Q_m}{\partial \Phi_l} - \sum_{m=1}^{M}\frac{\partial LO}{\partial Q_m}\frac{\partial Q_m}{\partial \Phi_l} = \frac{\partial LO(\underline{\mathbf{Q}})}{\partial \Phi_l} - \frac{\partial LO(\underline{\mathbf{Q}})}{\partial \Phi_l} = 0.$$

Consider now the second term on the right-hand side. Inverting the order of summations yields

$$\sum_{m=1}^{M}\sum_{j=1}^{L} H_{jm}\eta_j(\theta)\frac{\partial Q_m}{\partial \Phi_l} = \sum_{j=1}^{L}\eta_j(\theta)\sum_{m=1}^{M} H_{jm}\frac{\partial Q_m}{\partial \Phi_l}.$$

Since $\eta_j(\theta) > 0 \Leftrightarrow \varphi_j(\theta) = \Phi_j$, we limit the first sum to lines such that $\varphi_j(\theta) = \sum_{m=1}^{M} H_{jm}Q_m = \Phi_j$. Assuming $\frac{\partial H_{jm}}{\partial \Phi_l} = 0$,

$$\sum_{m=1}^{M} H_{jm}\frac{\partial Q_m}{\partial \Phi_l} = \frac{\partial\left(\sum_{m=1}^{M} H_{jm}Q_m\right)}{\partial \Phi_l} = \frac{\partial \Phi_j}{\partial \Phi_l}.$$

Thus if $\frac{\partial \Phi_j}{\partial \Phi_l} = 0$ for $j \neq l$,

$$\sum_{j=1}^{L}\eta_j(\theta)\frac{\partial \Phi_j}{\partial \Phi_l} = \eta_l(\theta) \Rightarrow \mathbb{E}\left[\sum_{j=1}^{L}\eta_j(\theta)\frac{\partial \Phi_j}{\partial \Phi_l}\right] = \Lambda_l(\mathbf{\Phi}).$$

Putting the two pieces together and taking the expectation proves the result. □

Lemma 6.2 is sufficient to extend Proposition 6.4 to a general $N$-node network.

Léautier (2000) shows that Lemma 6.2 continues to hold if we assume $\frac{\partial H_{jm}}{\partial \Phi_l} \neq 0$, which is realistic. Physically increasing the capacity on an interconnection is likely to modify its admittance, which then modifies the admittance matrix. If this occurs, new terms are added, but the economic intuition is unchanged.

**Related capacity on lines** As mentioned earlier, increasing the transfer capacity on line $l$ may modify the $(N-1)$ contingency on line $j$, hence impact the transfer capacity on line $j$. Full treatment of this issue is beyond the scope of this textbook. The main issue is likely to be that the cost function may no longer be well behaved. Assuming the cost function remains well behaved, the economic intuition and the main results hold.

Suppose for example transfer capacity on line 2 is impacted by transfer capacity on line 1 and another decision $X$: $\Phi_2(\Phi_1, X)$. Transfer capacities on all other lines are independent.

Differentiating the Lagrangian 6.12 of the optimal-grid problem with respect to $\Phi_1$ and $X$ yields the two first-order conditions:

$$\mathbb{E}\left[\eta_1(\mathbf{\Phi},\theta) + \eta_2(\mathbf{\Phi},\theta)\frac{\partial \Phi_2}{\partial \Phi_1}\right] = \frac{\partial \Gamma}{\partial \Phi_1} + \frac{\partial \Gamma}{\partial \Phi_2}\frac{\partial \Phi_2}{\partial \Phi_1}$$

and

$$\mathbb{E}\left[\eta_2(\mathbf{\Phi},\theta)\frac{\partial \Phi_2}{\partial X}\right] = \frac{\partial \Gamma}{\partial \Phi_2}\frac{\partial \Phi_2}{\partial X}.$$

Suppose this system admits a unique solution, which is a maximum.

Lemma 6.2's derivations yield

$$\frac{\partial \Pi^{CT}}{\partial \Phi_1} = \mathbb{E}\left[\sum_{j=1}^{L} \eta_j(\theta)\frac{\partial \Phi_j}{\partial \Phi_1}\right] = \mathbb{E}\left[\eta_1(\theta) + \eta_2(\mathbf{\Phi}, \theta)\frac{\partial \Phi_2}{\partial \Phi_1}\right]$$

and

$$\frac{\partial \Pi^{CT}}{\partial X} = \mathbb{E}\left[\sum_{j=1}^{L} \eta_j(\theta)\frac{\partial \Phi_j}{\partial X}\right] = \mathbb{E}\left[\eta_2(\mathbf{\Phi}, \theta)\frac{\partial \Phi_2}{\partial X}\right].$$

Proposition 6.4 thus continues to hold.

**Transfer capacity varying across states of the world** Suppose, for example, $\Phi_l = \Phi_l(X_l, \theta)$, where $X_l$ is an observable measure (e.g., voltage of the line) or possibly a vector of observable measures. The cost function is then $\Gamma(\mathbf{X})$, and equations (6.13) become

$$\mathbb{E}\left[\eta_l(\theta)\frac{\partial \Phi_l}{\partial X_l}\right] = \frac{\partial \Gamma(\mathbf{X})}{\partial X_l}.$$

Proposition 6.4 thus continues to hold.

# 7 Serious Games: Market Power and Transmission Constraints

## 7.1 Introduction

As discussed in chapter 6, constraints on the transmission grid separate a market into submarkets. This creates opportunities for producers to exercise local market power: a producer may be small in the global market but may find himself a monopoly on its local market for a few hours per year. The analysis of the interaction between market power and transmission constraints leads to three main results. First, transmission constraints increase the possibilities for producers to exercise market power. Second, financial transmission rights (FTRs) used to manage congestion impact behavior in the spot market: since they are forward contracts, the economic arguments presented in section 4.3 apply. Third, the microstructure of FTR markets also matters, for example, whether FTRs are owned by a single or by multiple owners or which kind of auction is used to allocate them.

This chapter's main lesson extends beyond these results. Despite my best efforts at simplifying it, the analysis presented here is quite technical. For example, even in the simple case of two markets linked by a single interconnection and identical monopolies in each market, only mixed-strategy equilibria exist for some values of the parameters. In the more complex—but more realistic—case of supply-function equilibria (SFE) for a general network, the bidding functions are given by complex differential equations that can only be solved numerically. Congestion is sometimes endogenous: Market power produces congestion that would not exist if markets were perfectly competitive, as producers may find it optimal to constrain the network, hence enjoy the protection of transmission constraints. Strategic games played by producers when transmission constraints are possible are very serious, indeed.

This observation is essential for corporate and policy decision makers. For the former, competing effectively in power markets requires sophisticated numerical models, hence significant investment in the quantitative talent required to develop and implement these models. Like investment banks and financial trading houses in the 1980s, electric power companies need to "quant-up," and most have done so. Policy makers need to be aware of this reality: firms exploit every opportunity to increase their profits—as their fiduciary responsibility dictates, hence bidding strategies in wholesale markets are highly sophisticated. Market monitoring and regulatory agencies also need to quant-up, and most have done so. As indicated earlier in this book, I believe effective market monitoring is possible. The complexity of the analyses presented in this chapter shows clearly the amount of highly quantitative resources it requires.

This chapter presents simple models developed by academics to illustrate the richness and the complexity of the games played. It is structured around two issues. First, how do transmission constraints interact with the exercise of market power? Presumably, when they are protected by a transmission constraint, producers are able to capture higher profit.

Answering this question intuitively is easy: congestion on the transmission grid splits the market into submarkets, sometimes called transmission islands by practitioners. Producers located in these isolated islands can then exercise their market power, even though they do not have market power in the global market. For example, a producer protected by an import constraint will find himself a monopoly on the residual demand curve, hence will be able to behave as one.

Providing analytical representation of strategic behavior in the presence of transmission constraints is technically challenging. The analytics depend on the microstructure of the power market assumed by the analyst: first- or second-price auction or supply function equilibrium. Section 7.2 presents a selection of models developed by academics.

These models share two common features. Transmission constraints have the impact of generation-capacity constraints in the Bertrand-Edgeworth games presented in section 3.5: in the presence of transmission constraints, even the highest bidder produces a residual quantity. This naturally leads to mixed-strategy equilibria. A second feature of these models is that producers strategically create congestion to increase their profits.

A second question is How do transmission rights, in particular financial transmission rights, which are the primary congestion management approach, interact with the exercise of market power? Does granting FTRs exacerbate or mitigate incentives to reduce output (or increase price)? A related question is Under which conditions do imperfectly competitive producers purchase FTRs?

Here again, the main intuition is straightforward: since FTRs are forward contracts, the insights on forward commitments presented in section 4.3 apply. For example, oligopoly producers protected by an import constraint benefit from an increase in their nodal price. If they purchase an FTR terminating in their own market, they gain additional exposure to their nodal price, hence faces stronger incentives to increase it.

However, depending on the microstructure of the FTR market, they may or may not purchase FTRs. In particular, if the SO sells FTRs using a uniform price auction, the producers face a prisoner's dilemma and may end up selling FTRs in the forward market, while they would do better in the spot market if they had bought them. On the other hand, if the SO uses a pay-your-bid auction, producers end up buying FTRs, which enhances their market power in the spot market.

Since SOs use uniform-price auctions to sell FTRs, these academic results suggest that auctioning FTRs enhances competition. However, one should be cautious about this result, which relies on many simplifying assumptions. The impact of FTRs on market power in a complex network is a fertile avenue for further research.

## 7.2 Models of Local Market Power

Multiple authors have developed models that illustrate how transmission constraints facilitate producers' exercise of market power. Three families of models are presented here, in order of increasing complexity. First, the simplest models examine price-inelastic demand and a symmetric three-market network. We examine the equilibrium under symmetric and asymmetric information, building on Léautier (2001). Second, we examine the asymmetric two-market network developed by Borenstein, Bushnell, and Stoft (2000). Somehow surprisingly, the analytics are more complex on the two-market than on the three-market network. In both cases, congestion arises endogenously, that is, producers create congestion to increase their profit. Finally, we briefly discuss how to define supply-function equilibria when transmission constraints are present.

### 7.2.1 Symmetric Constraint: Three-Market Network

Models presented in this and the next section focus on the equilibrium in the spot market; hence only one realization of demand is considered. To simplify the notation, the state of the world $\theta$ is omitted from the expressions. All models presented in this chapter assume congestion occurs on a single interconnection, the capacity of which is identical in both directions. To simplify the notation, denote $\Phi = \Phi^+ = \Phi^-$ the capacity on the (potentially) congested interconnection.

Consider the three-market network presented in figure 7.1. As in section 6.6, one producer is located in each market 1 and 2, and demand is located at node 3. Both producers are identical. Their variable production cost per unit is $c < 1$. Demand is $l > 0$ and does not vary with price.

Assume the SO runs a pay-as-bid auction, that is, each producer is paid her bid if she is dispatched. This setting is the only one compatible with nodal pricing, since producers are located at different nodes. Since demand is price inelastic, the SO imposes a cap $\bar{p}$ on bids; otherwise producers would submit arbitrarily large bids.

The other interconnections are never congested. If $\Phi \geq \frac{l}{3}$, a single producer can serve the entire demand, and the interconnection is not congested.

Consider now $\Phi < \frac{l}{3}$. The SO needs to dispatch both producers to serve demand. For $m = 1, 2$, denote $b_m$ generator $m$'s bid. Following the analysis presented in section 6.6, in

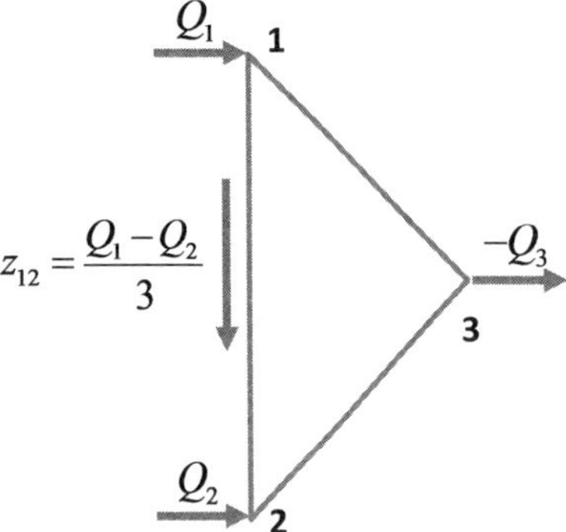


**Figure 7.1**
Symmetric three-market network. Producers are located in markets 1 and 2. Demand is located in market 3. Interconnection (1, 2) is potentially congested.

particular equation (6.6), the dispatch rule is

$$q_n(b_n, b_m) = \begin{cases} \frac{l+3\Phi}{2} \ if \ b_n < b_m \\ \frac{l-3\Phi}{2} \ if \ b_n > b_m \\ \frac{l}{2} \ if \ b_n = b_m \end{cases} :$$

the cheapest generator produces the competitive quantity $\left(\frac{l+3\Phi}{2}\right)$, while the most expensive generator is called to produce the residual "constrained" quantity $\left(\frac{l-3\Phi}{2}\right)$. When both generators bid the same price, they share production. The situation is similar to that of Fabra, von der Fehr, and Harbord (2006), presented in section 3.5: owing to the transmission constraint, the most expensive generator produces the "constrained" quantity. This modifies the equilibrium of the game.

#### 7.2.1.1 Perfect information

Suppose first the marginal cost $c$ is known to both firms. As in Fabra, von der Fehr, and Harbord (2006), no pure-strategy equilibrium exists, and the unique mixed-strategy equilibrium is distributed on the interval $\left[\bar{p} - \frac{6\Phi}{l+3\Phi}(\bar{p} - c), \bar{p}\right]$ according to distribution

$$\xi(b) = \frac{l+3\Phi}{6\Phi} \frac{b - \underline{b}}{b - c}.$$

Since producers play a mixed-strategy equilibrium, their expected profit does not depend on their bid. It is

$$\mathbb{E}[\pi(b)] = \frac{l-3\Phi}{2}(\bar{p} - c).$$

Each producer captures, on average, the maximum margin $(\bar{p} - c)$ on the constrained quantity $\left(\frac{1-3\Phi}{2}\right)$ and captures no margin on the competitive quantity.

Imperfect competition creates artificial congestion on the interconnection: since generators have identical marginal cost, the line is not congested if competition is perfect. However, generators (almost) always bid different prices, hence the line is (almost) always congested, in one direction or the other.

#### 7.2.1.2 Imperfect information

Léautier (2001) shows that a similar intuition holds if information on costs is imperfect. Léautier (2001) uses the standard auction information structure: Ex ante, firms do not know their and their rival's costs, but they know the distribution of these costs. In the interim stage, each firm learns its cost and submits its bid. Ex post, rival's cost is revealed. This constitutes a realistic description of reality. Firms have only imperfect information on their rivals' costs: for example, details of fuel supply contracts may not be publicly disclosed. They therefore devise bidding strategies to account for this uncertainty.

Without loss of generality, costs are assumed to be distributed on [0, 1]. Since the SO knows the distribution of costs, she sets the price cap at $\bar{p} = 1$.

Léautier (2001) proves that the optimal auction, that is, the auction that minimizes the expected cost for the SO can be implemented as follows: (a) the SO purchases the constrained quantity $\left(\frac{l-3\Phi}{2}\right)$ at the maximum possible cost $\bar{p}$, and (b) runs a pay-as-bid auction for the large quantity $\left(\frac{l+3\Phi}{2}\right)$. Again, the producers are able to capture the maximum margin on the constrained quantity. They capture their standard information rent on the competitive quantity.

### 7.2.2 Asymmetric Constraint: Two-Market Network

Consider now two markets linked by an interconnection, as presented in section 6.2. Congestion on the interconnection has an asymmetric impact on producers: one market is import constrained, hence producers located in this market are protected from competitive imports, while the other is export constrained, hence producers in that market are limited in their ability to export.

Borenstein, Bushnell, and Stoft (2000) characterize the equilibrium in this situation. They find results similar to the symmetric three-network case: congestion arises endogenously from the strategic behavior in each market, and only mixed equilibria exist—at least for some values of the parameters.

Before characterizing this equilibrium, it is helpful to derive properties of the import- and export-constrained inverse demands.

#### 7.2.2.1 Constrained equilibrium quantities

Consider the two-node market presented in chapter 6. At node $m = 1, 2$, production is $Q_m^s$ and demand is $\alpha_m D(p)$. We no longer require $\alpha_1 + \alpha_2 = 1$. The energy balance then becomes $Q_1^s + Q_2^s = (\alpha_1 + \alpha_2) D(p)$. To simplify the exposition we continue to assume that capacity on the interconnection is identical in both directions $\Phi^+ = \Phi^- = \Phi$ and that transmission losses are negligible.

The first step is to express the flow on the interconnection as a function of production in each market. Suppose the line is not congested, as is represented in figure 7.2.

The price is identical in both markets, denoted $p$. Equilibrium conditions are $Q_1^s = \alpha_1 D(p) + \varphi$ in market 1 and $\varphi + Q_2^s = \alpha_2 D(p)$ in market 2. Combining these conditions yields

$$(\alpha_1 + \alpha_2)\varphi = \alpha_2 Q_1^s - \alpha_1 Q_2^s \Leftrightarrow \varphi = \frac{\alpha_2 Q_1^s - \alpha_1 Q_2^s}{\alpha_1 + \alpha_2}.$$

The line is not congested as long as outputs at each node are close, that is,

$$-\Phi < \varphi < \Phi \Leftrightarrow -(\alpha_1 + \alpha_2)\Phi < \alpha_2 Q_1^s - \alpha_1 Q_2^s < (\alpha_1 + \alpha_2)\Phi.$$

Suppose now the interconnection is congested from market 1 to market 2, as represented in figure 7.3.

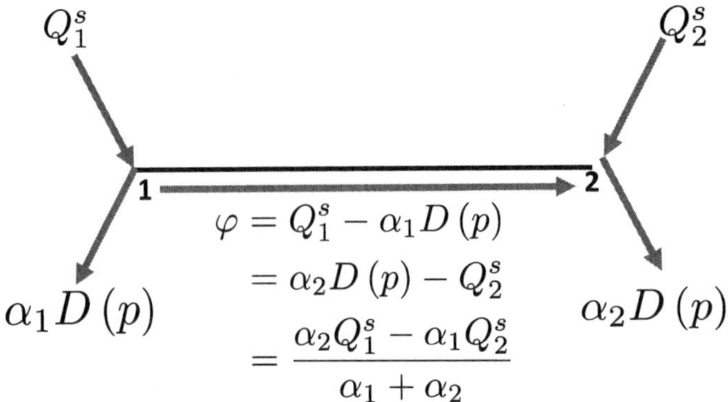


**Figure 7.2**
Production, consumption, and transmission flow on a unconstrained two-node market.

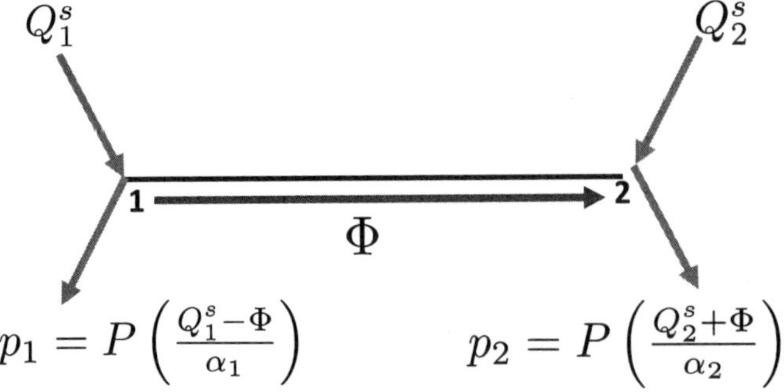


**Figure 7.3**
Production, consumption, and transmission flow on a constrained two-node market.

As discussed in chapter 6, $p_1 < p_2$, $Q_1^s = Q_1^d + \Phi = \alpha_1 D(p_1) + \Phi$, and $Q_2^s = Q_2^d - \Phi = \alpha_2 D(p_2) - \Phi$. Price is $P\left(\frac{Q_1^s - \Phi}{\alpha_1}\right)$ in the export-constrained market 1 and $P\left(\frac{Q_2^s + \Phi}{\alpha_2}\right)$ in the import-constrained market 2. Finally, we verify directly that, since $p_1 < p_2 \Leftrightarrow D(p_1) > D(p_2)$, we have

$$\alpha_2 Q_1^s - \alpha_1 Q_2^s = (\alpha_1 + \alpha_2)\Phi + \alpha_1\alpha_2(D(p_1) - D(p_2)) > (\alpha_1 + \alpha_2)\Phi.$$

When $\alpha_2 Q_1^s - \alpha_1 Q_2^s < -(\alpha_1 + \alpha_2)\Phi \Leftrightarrow \alpha_2 Q_1^s < \alpha_1 Q_2^s - (\alpha_1 + \alpha_2)\Phi$, the interconnection is congested from market 2 into market 1. Demand is $Q_1^d = (Q_1^s + \Phi) = \alpha_1 D(p_1)$ in market 1, and $Q_2^d = (Q_2^s - \Phi) = \alpha_2 D(p_2)$ in market 2. Inverse demand is $P\left(\frac{Q_1^s + \Phi}{\alpha_1}\right)$ in the import-constrained market 1 and $P\left(\frac{Q_2^s - \Phi}{\alpha_2}\right)$ in the export-constrained market 2.

From the previous discussion, import-protected producers face inverse demand $P\left(\frac{Q+\Phi}{\alpha_m}\right)$ for $m = 1, 2$, while export-constrained producers face inverse demand $P\left(\frac{Q-\Phi}{\alpha_m}\right)$. The constraint separates both markets: competitive dynamics are independent in each market.

We examine the properties of the Cournot equilibrium when $N$ identical producers with constant operating cost $c$ face inverse demand $P(Q + \Phi)$.

Producer $n$ profit is

$$\pi^n = q^n(P(Q + \Phi) - c).$$

Aggregate output $Q_I(\Phi)$ is uniquely defined by

$$P(Q_I + \Phi) + \frac{Q_I}{N}\partial_Q P(Q_I + \Phi) = c. \tag{7.1}$$

The industry profit at the equilibrium is

$$\Pi_I = Q_I(P(Q_I + \Phi) - c) = -\frac{(Q_I)^2}{N}\partial_Q P(Q_I + \Phi) > 0.$$

Applying the implicit function theorem to equation (7.1) yields

$$\frac{\partial Q_I}{\partial \Phi} = -\frac{\partial_Q P + \frac{Q_I}{N}\partial_{QQ} P}{\left(1 + \frac{1}{N}\right)\partial_Q P + \frac{Q_I}{N}\partial_{QQ} P} < 0.$$

Then,

$$\begin{aligned}
\frac{\partial \Pi_I}{\partial \Phi} &= -\frac{Q_I}{N}\left[2\frac{dQ_I}{d\Phi}\partial_Q P + \frac{Q_I}{\alpha}\left(\frac{dQ_I}{d\Phi} + 1\right)\partial_{QQ} P\right] \\
&= \frac{Q_I}{N}\frac{\partial_Q P\left(2\partial_Q P + \frac{Q_I}{\alpha}\partial_{QQ} P\right)}{\partial_Q P\left(1 + \frac{1}{N}\right) + \frac{Q_I}{N\alpha}\partial_{QQ} P} < 0.
\end{aligned}$$

Therefore, the equilibrium aggregate output $Q_I$ and industry profit $\Pi_I$ are decreasing as $\Phi$ increases.

Suppose now $\Phi > 0$. This corresponds to an import-constrained market, in which producers are protected by the maximum import capacity $\Phi$. Since quantities are strategic substitutes in Cournot games, when $\Phi$ increases, $Q_I$ decreases, although less than 1 for 1, and $\Pi_I$ decreases also.

Suppose now $\Phi < 0$. This corresponds to an export-constrained market, in which producers are limited by the maximum export capacity $\Phi$. As $\Phi$ increases, producers export and generate more, and profits increase.

#### 7.2.2.2 Characterization of the equilibrium

**Profit functions** Borenstein, Bushnell, and Stoft (2000) first derive the equilibrium in the simplest case: one (local) monopolist is present in each market $m = 1, 2$. To further simplify the analysis, Borenstein, Bushnell, and Stoft (2000) assume that markets have equal size, normalized to $\alpha_1 = \alpha_2 = 1$, and technologies are identical, characterized by constant per unit variable cost of production $c$.

Suppose first the interconnection is unconstrained. The two firms compete à la Cournot. Consider a firm that produces $X$ when its competitor produces $Y$. The energy balance is $X + Y = 2D(p) \Leftrightarrow p = P\left(\frac{X+Y}{2}\right)$. The firm's profit is thus

$$\pi^C(X, Y) = X\left(P\left(\frac{X+Y}{2}\right) - c\right).$$

Suppose now the interconnection is congested. The profit of an export-constrained monopoly that produces output $X$ is

$$\pi^E(X, \Phi) = X(P(X - \Phi) - c).$$

The profit of an import-constrained monopoly that produces output $X$ is

$$\pi^I(X, \Phi) = X(P(X + \Phi) - c).$$

Consider the monopoly in market 1. If $Q_2^s - Q_1^s > 2\Phi \Leftrightarrow Q_1^s < Q_2^s - 2\Phi$, market 1 is import constrained, and the monopoly's profit is $\pi^I(Q_1^s, \Phi)$. Symmetrically, if $Q_1^s - Q_2^s > 2\Phi \Leftrightarrow Q_1^s > Q_2^s + 2\Phi$, market 1 is export constrained, and the monopoly's profit is $\pi^E(Q_1^s, \Phi)$. In both cases, the monopoly profit does not depend on the other firm's output.

Finally, if $-2\Phi < Q_1^s - Q_2^s < 2\Phi \Leftrightarrow Q_2^s - 2\Phi < Q_1^s < Q_2^s + 2\Phi$, the interconnection is not congested. Both firms compete à la Cournot, firm 1's profit is $\pi^C(Q_1^s, Q_2^s)$. The profit function of the monopoly in market 1 is thus defined piecewise by

$$\pi_1(Q_1^s, Q_2^s, \Phi) = \begin{cases} \pi^I(Q_1^s, \Phi) & \textit{if } Q_1^s < Q_2^s - 2\Phi \\ \pi^C(Q_1^s, Q_2^s) & \textit{if } Q_2^s - 2\Phi < Q_1^s < Q_2^s + 2\Phi. \\ \pi^E(Q_1^s, \Phi) & \textit{if } Q_1^s > Q_2^s + 2\Phi \end{cases}$$

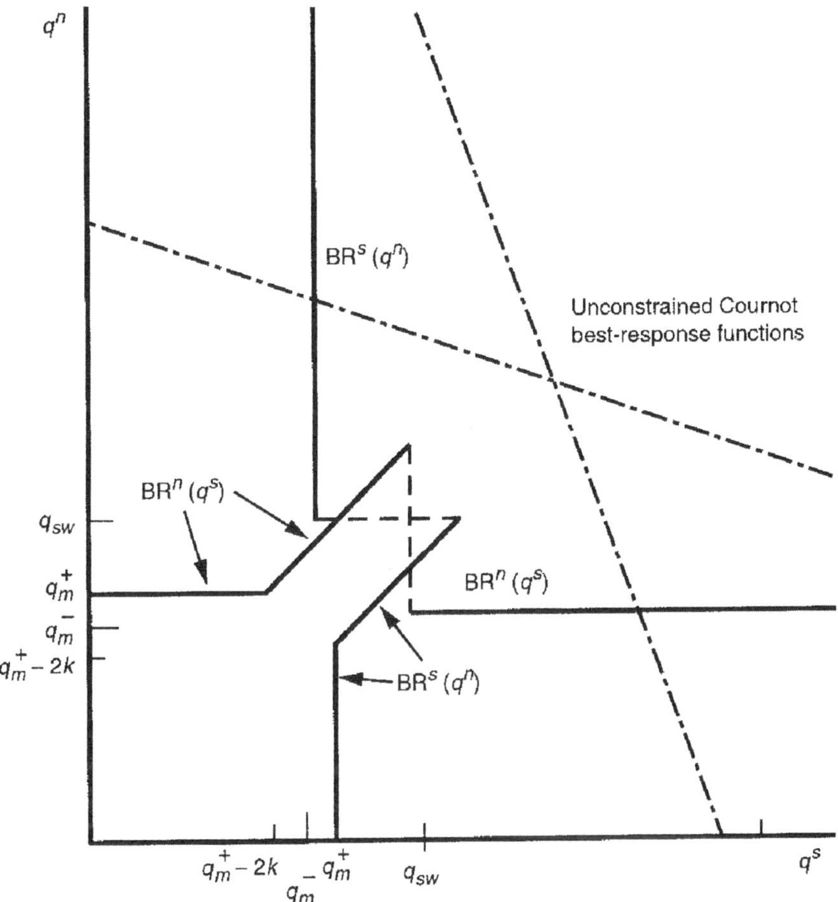


**Figure 7.4**
Best-response functions for a "thin" line.
*Source*: Borenstein, Bushnell, and Stoft (2000), figure 2.

**Best responses** Best responses are also defined piecewise and are presented in figures 7.4 and 7.5, which use different notation. Firm 1's best response is presented on the vertical axis and denoted $q^n$, while firm 2's best response is presented on the horizontal axis and denoted $q^s$, and transmission capacity $\Phi$ is denoted $k$.

Define $Q^E(\Phi) = argmax_Q \pi^E(Q_1^s, \Phi)$, the export-constrained optimal output $Q_E(\Phi)$ (denoted $q_m^+$) and $\pi^E(\Phi) = \pi^E(Q^E(\Phi), \Phi)$. Similarly, define $Q^I(\Phi) = argmax_Q \pi^I(Q_1^s, \Phi)$, the import-constrained optimal output $Q^I(\Phi)$ (denoted $q_m^-$ ) and $\pi^I(\Phi) = \pi^I(Q^I(\Phi), \Phi)$. We assume $Q^E(\Phi) > 2\Phi$.

Consider first low values of $Q_2^s$. As long as $Q_2^s < Q_E(\Phi) - 2\Phi$, market 1 is export constrained, and producer 1 produces $Q_E(\Phi)$ and captures profit $\pi^E(\Phi)$. This strategy

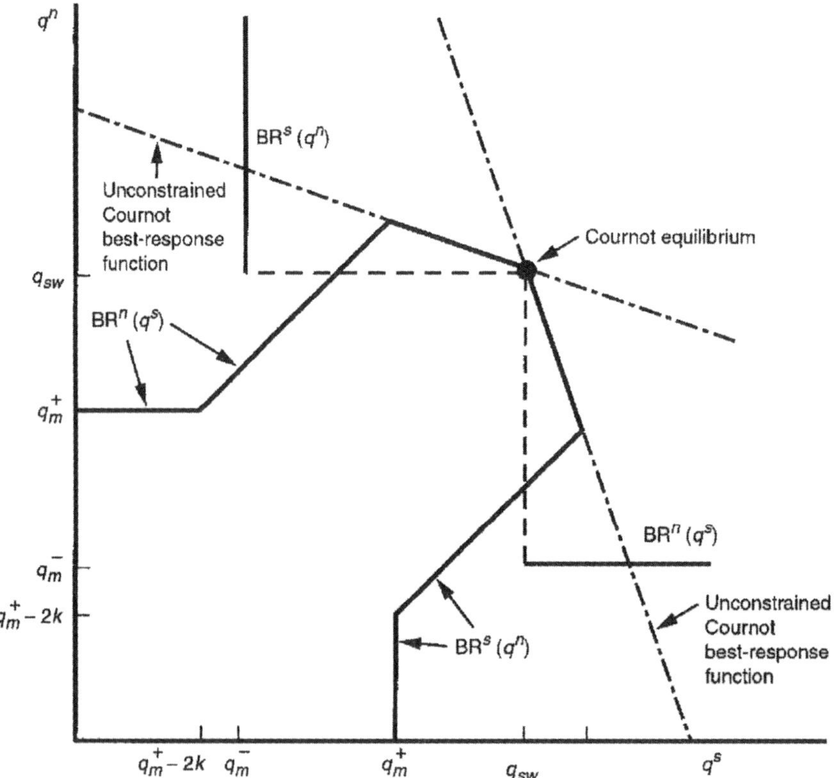


**Figure 7.5**
Best-response functions for a "thick" line.
*Source*: Borenstein, Bushnell, and Stoft (2000), figure 4.

is represented as the horizontal line on the left of figures 7.4 and 7.5. When $Q_2^s > Q_E(\Phi) - 2\Phi$, producer 1 produces $Q_1^s = Q_2^s + 2\Phi$ and maintains the interconnection export constrained. This strategy is represented as the upward sloping line on the left of figures 7.4 and 7.5.

After a certain threshold, the producer finds it more profitable to reduce output to relieve her export constraint. In that case, she either (a) produces her unconstrained Cournot best response $Q^C(Q_2^s) = argmax_Q \pi^C(Q, Q_2^s)$, or, when $Q_2^s$ exceeds the threshold $q_{sw}(\Phi)$, (b) discontinuously reduces output $Q^I(\Phi)$ so that market 1 becomes import constrained and captures $\pi^I(\Phi)$. This latter strategy is represented as the horizontal line on the right of figures 7.4 and 7.5.

For values of $\Phi$ lower than a threshold, $q_{sw}(\Phi)$ is small: the discontinuity occurs before firm 1's best response $Q_1^s = Q_2^s + 2\Phi$ crosses her unconstrained Cournot best response, and no pure-strategy equilibrium exists. This situation is presented in figure 7.4.

For values of $\Phi$ higher than a threshold, $q_{sw}(\Phi)$ is high enough that firm 1's best response $Q_1^s = Q_2^s + 2\Phi$ crosses her unconstrained Cournot best response, which crosses firm 2's unconstrained Cournot best response. The Cournot equilibrium constitutes the unique pure-strategy equilibrium. This is illustrated in figure 7.5.

For intermediate values of $\Phi$, firm 1's best response $Q_1^s = Q_2^s + 2\Phi$ crosses her unconstrained Cournot best responses. However, the latter does not cross firm 2's unconstrained Cournot best response before $q_{sw}(\Phi)$. No pure strategy equilibrium exists.

The proofs can be found in Borenstein, Bushnell, and Stoft (2000), which also proves the results are robust to various extensions, in particular asymmetric firms and markets.

This analysis illustrates that imperfect competition generates artificial congestion on transmission lines. Second, it provides another example of mixed-strategy equilibria arising from limited transmission capacities. Third, it can be reinterpreted using a constant interconnection capacity $\Phi$ and time-dependent demand $D(p, \theta)$. In this case, pure-strategy equilibrium exists for low demand, while only mixed-strategy equilibria exist for high demand.

Joung, Baldick, and Son (2008) extend Borenstein, Bushnell, and Stoft (2000) by allowing generators to own FTRs. They find that some allocation of FTRs is procompetitive, that is, it reduces the transmission capacity required for firms to play the unconstrained Cournot equilibrium. This finding is aligned with the analysis of the competitive impact of FTRs on the spot energy market presented in sections 7.3 and 7.4.

### 7.2.3 Supply-Function Equilibria with Transmission Constraints

As discussed in chapter 3, SFE constitute a closer description of centralized markets. However, they are more difficult to characterize than, for example, Cournot equilibria. Borenstein, Bushnell, and Stoft (2000) show that the presence of transmission constraints often leads to mixed-strategy equilibria, even in the simplest transmission network. Existence of SFE in the presence of transmission constraint is therefore not a priori guaranteed. A rich and growing academic literature examines that issue.

Holmberg and Philpott (2015) is a recent contribution that (a) derives conditions for computing the best response of a producer at a node of a general transmission network; (b) demonstrates the existence and uniqueness of SFE in a two-node network; (c) computes symmetric SFE in two-node and star networks and shows how these relate to a market integration function that can be computed from a model with price-taking agents; and (d) provides examples where SFE fail to exist.

## 7.3 FTRs and Local Market Power in the Simplest Setting

Joskow and Tirole (2000) propose an extremely simple model that captures the impact of FTR on market power. Their analysis yields two main results, derived below. First, following the analysis presented in section 4.3, since FTRs are forward contract, FTR ownership modifies strategic behavior in the spot market. Second, the amount of FTRs purchased by producers depends on the microstructure of the FTR market.

### 7.3.1 Imperfect Competition Given an FTR Position

Consider the two-market network presented in section 6.2. For $m = 1, 2$, producers with constant marginal cost $c_m$ are located in market $m$, with $c_2 > c_1$. Joskow and Tirole (2000) derive the results assuming increasing marginal cost. I assume constant marginal cost to maintain consistency with section 6.2, and to simplify the argument. Suppose demand is such that the interconnection is congested, that is, $\alpha_2 D(c_1) > \Phi$. To simplify the notation, Joskow and Tirole (2000) introduce $D_m(p, \theta) = \alpha_m D(p, \theta)$ for $m = 1, 2$.

Suppose market 1 is perfectly competitive, while a monopoly operates in market 2. Perfect competition in market 1 enables us to rule out equilibrium strategies leading to endogenous congestion as discussed in section 7.2, hence to focus on strategic behavior in the import-constrained market 2. Joskow and Tirole (2000) use price as a strategic variable, which is equivalent to using quantity for a monopoly. The monopoly in market 2 is protected by the constraint on imports. The resulting power flows are presented in figure 7.6.

Since the interconnection is congested from market 1 to market 2, the system operator issues $\Phi$ FTR paying $(p_2 - p_1)$. Suppose the monopoly owns a fraction $m_2 \in [0, 1]$ of the $\Phi$ FTRs available on the interconnection. Its profit is

$$\begin{aligned}\pi_2(p_2, m_2) &= (p_2 - c_2)(D_2(p_2) - \Phi) + m_2\Phi(p_2 - c_1)\\ &= (p_2 - c_2)(D_2(p_2) + (m_2 - 1)\Phi) + m_2\Phi(c_2 - c_1).\end{aligned}$$

In particular, since

$$\pi_2(p_2, 1) = (p_2 - c_2)D_2(p_2) + \Phi(c_2 - c_1),$$

if the monopoly buys all available transmission rights, it faces the entire demand $D_2(p, \theta)$.

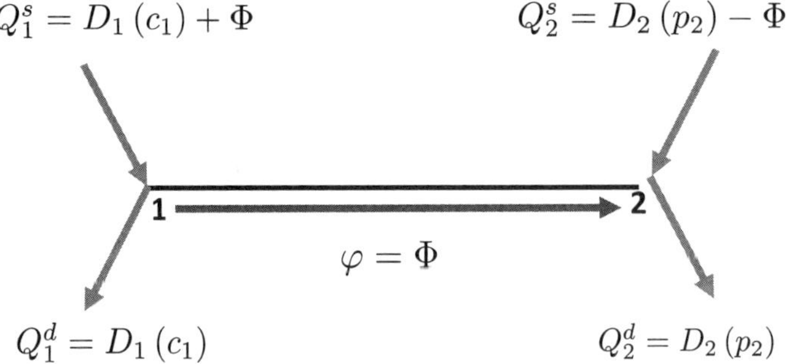


**Figure 7.6**
Power flows on the two-market network. Price in market 1 is competitively set at the variable per unit cost $c_1$, hence demand is $Q_1^d = D_1(c_1)$. The interconnection is congested: exports from market 1 into market 2 are equal to the interconnection capacity $\varphi = \Phi$. Production in market 1 is $Q_1^s = D_1(c_1) + \Phi$. Price in market 2 is set at $p_2$, hence demand is $Q_2^d = D_2(p_2)$. Producer(s) faces the residual demand $Q_2^s = D_2(p_2) - \Phi$.

Under our assumption on inverse demand, for any $m_2 \in [0, 1]$, there exists a unique price $p_2^M(m_2)$ that maximizes the monopoly's profit, that is, $p_2^M(m_2)$ is uniquely defined by $\frac{\partial \pi_2}{\partial p_2}(p_2^M, m_2) = 0$.

Define $\mathcal{G}(m_2) = (p_2^M(m_2) - c_2)(D_2(p_2^M(m_2)) - \Phi), \eta(m_2) = (p_2^M(m_2) - c_1)$, and $\mathcal{F}(m_2) = \Phi \times \eta(m_2)$, respectively the generating profit, the unit price of FTRs, and the congestion rent when the monopoly holds the fraction $m_2$ of available FTRs and selects $p_2 = p_2^M(m_2)$.

The monopoly's total profit is then

$$\pi_2(p_2^M(m_2), m_2) = \mathcal{G}(m_2) + m_2\mathcal{F}(m_2, \theta).$$

As $m_2$ increases, the monopoly's benefit from a high $p_2$ increases since $\frac{\partial^2 \pi_2}{\partial p_2 \partial m_2} = \Phi^+ > 0$; hence the monopoly price $p_2^M(m_2)$ also increases. This point can be proved formally by applying the implicit function theorem $\frac{dp_2^M}{dm_2} = \frac{\frac{\partial^2 \pi_2}{\partial p_2 \partial m_2}}{-\frac{\partial^2 \pi_2}{\partial p_2^2}}(p_2^M, m_2)$ and observing that $\frac{\partial^2 \pi_2}{\partial p_2^2}(p_2^M, m_2) < 0$ since $p_2^M(m_2)$ maximizes $\pi_2(p_2, m_2)$.

Thus $p_2^M(m_2)$ increases from $p_2^M(0)$ to $p_2^M(1)$. Since (a) $p_2^M(m_2) > p_2^M(0)$ if $m_2 > 0$, and (b) $p_2^M(0)$ maximizes operating profit by construction, to maximize total profit the monopoly reduces its operating profit compared with its maximum value and increases the congestion rent.

Since $p_2^M(m_2)$ increases as $m_2$ increases, so do $\eta(m_2)$ and $\mathcal{F}(m_2)$: the higher the share of FTRs held by the monopoly, the higher the unit price of FTRs and the value of the congestion.

### 7.3.2 Purchase of FTR Position

Joskow and Tirole (2000) then prove that the monopoly's FTR position depends on the microstructure of the FTR market.

This result is a priori surprising, since the profit of the import-constrained monopoly increases with her FTR holding. Applying the envelope theorem,

$$\begin{aligned}\frac{d\pi_2}{dm_2}(p_2^M(m_2), m_2) &= \frac{\partial \pi_2}{\partial m_2}(p_2^M(m_2), m_2)\\ &= \Phi(p_2^M(m_2) - c_1) > \Phi(p_2^M(0) - c_1) > \Phi(c_2 - c_1) > 0.\end{aligned}$$

Thus one would expect that the monopoly would buy all available transmission rights to maximize its profit.

While this intuition is correct, it ignores the incentives of initial FTR holders, who perfectly understand that the value of their rights increases with the share the monopoly ends up holding. Leveraging insights from the corporate finance literature, Joskow and Tirole (2000) observe that "the more the initial holders of rights can free ride on the ability of the monopoly to increase the value of these rights by exercising market power, the fewer rights the monopoly is likely to acquire through the rights market." Joskow and Tirole (2000) consider three different cases, which are presented below.

#### 7.3.2.1 FTRs initially held by a single nonstakeholder owner

If he does not sell any FTRs to the monopoly producer, the single initial owner receives $\mathcal{F}(0)$, and the monopoly receives $\mathcal{G}(0)$. If he sells all his FTRs to the monopoly producer for a total price $R$, the single initial owner receives $R$, and the monopoly producer's profit is $\mathcal{G}(1)+\mathcal{F}(1)-R$. A mutually beneficial trade is possible if and only if (a) the initial owner is willing to sell, that is, $R>\mathcal{F}(0)$ and (b) the monopoly producer is willing to buy, that is,

$$\mathcal{G}(1)+\mathcal{F}(1)-R>\mathcal{G}(0)\Leftrightarrow R<\mathcal{G}(1)+\mathcal{F}(1)-\mathcal{G}(0).$$

Thus a trade is possible if and only if

$$\mathcal{F}(0)+\mathcal{G}(0)<\mathcal{G}(1)+\mathcal{F}(1).$$

To prove this property holds, define $X(m)=\mathcal{G}(m)+\mathcal{F}(m)$. We have

$$\frac{dX}{dm}=\frac{d\mathcal{G}}{dm}+\frac{d\mathcal{F}}{dm}=\frac{d\mathcal{G}}{dm}+m\frac{d\mathcal{F}}{dm}+(1-m)\frac{d\mathcal{F}}{dm}.$$

As we have seen, $\frac{d\mathcal{F}}{dm}>0$, and

$$\frac{d\mathcal{G}}{dm}+m\frac{d\mathcal{F}}{dm}=\frac{\partial\pi_2}{\partial p_2}(p_2^M(m),m)\frac{dp_2^M}{dm}=0.$$

Thus

$$\frac{dX}{dm}=(1-m)\frac{d\mathcal{F}}{dm}>0\Rightarrow\mathcal{F}(0,\theta)+\mathcal{G}(0,\theta)<\mathcal{G}(1,\theta)+\mathcal{F}(1,\theta).$$

The monopoly producer purchases all the FTRs from their initial owner. The purchase price depends on the bargaining power of both parties.

#### 7.3.2.2 Unconditional tender offer to dispersed FTR owners

Suppose now the monopoly producer offers a price $p$ to dispersed owners, that is, commits to purchase all FTRs sold by their initial owners at price $p$. Joskow and Tirole (2000) prove that, in that case, no rights are purchased.

The proof proceeds by contradiction. Observe first that the monopoly producer never offers $p<\eta(0)$: initial owners would be better-off holding on to the certificates. Similarly, the monopoly producer never offers $p>\eta(1)$, since the FTRs cannot be worth more than $\eta(1)$.

Suppose the monopoly producer offers a price $p\in[\eta(0),\eta(1)]$. A no-arbitrage argument shows that the fraction of rights sold is such that their ex post value is equal to the price $p$, that is, $m_2$ is such that

$$\eta(m_2)=p.$$

Observing that $\mathcal{F}(m_2) = \eta(m_2)\Phi = p\Phi$, the monopoly producer's profit is

$$\mathcal{G}(m_2) + m_2\mathcal{F}(m_2) - pm_2\Phi = \mathcal{G}(m_2) < \mathcal{G}(0):$$

the monopoly producer finds it unprofitable to purchase FTRs.

The problem (for the monopoly producer) is that initial owners are able to capture ex ante through their sale price the entire increase in the value of the FTRs she will create ex post. Thus the monopoly realizes no net profits on the FTRs, once the purchase price is included. Then, since the monopoly producer reduces her operating profit to increase her overall profit, she does not enter in the trade.

#### 7.3.2.3 Pay-as-bid auction by the SO

Suppose now that the SO auctions off all rights, according the following procedure:

1. All bidders announce a price and a maximum quantity they are willing to buy at that price.
2. Rights are allocated to highest bidders.
3. Bidders pay their bids.

Suppose the market is deep, that is, that risk-neutral traders capture any positive arbitrage opportunity. Joskow and Tirole (2000) prove that the monopoly producer has no pure bidding strategy and derive the mixed-bidding strategy. The proof is quite technical, hence is not reproduced here.

### 7.3.3 Physical Transmission Rights

In some markets in the late 1990s, producers could purchase physical transmission rights, that is, the right to "physically" use an interconnection. A major difference with financial rights is that the holder of these physical rights is not obliged to use them entirely, for example, a producer may purchase 1,000 MW of interconnection capacity and schedule only 600 MW for a given hour.

Joskow and Tirole (2000) prove that a producer who holds physical transmission rights may exert market power by withholding these rights, that is, by reducing capacity on the interconnection actually used. Since this point is intuitive and physical rights are rarely used, the proof is not reproduced.

## 7.4 FTRs and Local Market Power in a Richer Setting

This section presents three extensions to the simplest model presented in section 7.3. First, Léautier (2014b) examines how transmission constraints affect oligopoly investment in generation, using the simple two-node market presented in section 6.4.

Second, Gilbert, Neuhoff, and Newbery (2004) extend Joskow and Tirole's (2000) analysis and consider an $N$-firm oligopoly that can be export- as well as import-constrained.

Their main insights are similar. First, a producer's ownership of FTRs changes her exposure to spot prices, which then changes her incentives to raise these prices. Second, a producer's purchase of FTRs depends on the microstructure of the FTR market.

Finally, Joskow and Tirole (2000) extend the analysis to three-market network. They illustrate how loop flows create slightly counterintuitive results, without altering the economic intuition.

In their conclusion, Gilbert, Neuhoff, and Newbery (2004) point out the limits of their analysis, which rely on symmetric information between generators and traders and the existence of a unique equilibrium in the spot market. These assumptions are unlikely to hold in reality.

### 7.4.1 Imperfectly Competitive Investment in the Presence of Transmission Constraints

The market is as described in sections 6.2 and 6.3: baseload technology is located at node 1, and peaking technology at node 2. $N$ symmetric producers, each having access to both technologies, compete à la Cournot in both markets, that is, each producer $n$ owns $\frac{1}{N}$ of aggregate baseload capacity $k_1$ installed in market 1, and $\frac{1}{N}$ of aggregate peaking capacity $k_2$ installed in market 2.

The interconnection has limited capacity. The power flowing from market 1 to market 2 in state $\theta$, denoted $\varphi(\theta)$, cannot exceed the transmission capacity:

$$-\Phi \leq \varphi(t) \leq \Phi.$$

Congestion on the interconnection is managed using financial transmission rights described in chapter 6. Léautier (2014b) assumes each firm owns (or has rights to) $\frac{1}{N}th$ of the available FTRs and that producers do not include the acquisition cost of FTRs in their analysis. For example, they are granted FTRs, as was the case in the mid-Atlantic market in the United States. This assumption leads to the very simple solution presented below. Strategic behavior in the FTR market discussed in section 7.3 is ignored.

If the interconnection is never congested, the situation is as described in section 3.6.1. Aggregate equilibrium capacities are denoted $K_i^{CU}$, where the superscript $^C$ refers to Cournot competition, and the superscript $^U$ refers to the unconstrained case. Aggregate cumulative capacity $K_2^{CU}$ is the unique solution of

$$\Psi_N(K_2^{CU}, c_2) = r_2,$$

while aggregate baseload capacity $K_1^{CU}$ is the unique solution to

$$\Psi_1(K_1^{CU}, c_1) - \Psi_1(K_1^{CU}, c_2) = r_1 - r_2.$$

Suppose the interconnection is congested from market 1 into market 2. Power flows and prices are illustrated in figure 7.3. Each firm $n$ receives the local market price for its production in each market plus the FTR payment: $(p_2(\theta) - p_1(\theta))\frac{\Phi}{N}$, hence its operating profit in state $\theta$ is thus

$$\begin{aligned}
\pi^n =& q_1^n(p_1 - c_1) + q_2^n(p_2 - c_2) + \frac{\Phi}{N}(p_2 - p_1)\\
=& q_1^n\left(P\left(\frac{Q_1 - \Phi}{\alpha_1}, \theta\right) - c_1\right) + q_2^n\left(P\left(\frac{Q_2 + \Phi}{\alpha_2}, \theta\right) - c_2\right)\\
&+ \frac{\Phi}{N}\left(P\left(\frac{Q_2 + \Phi}{\alpha_2}, \theta\right) - P\left(\frac{Q_1 - \Phi}{\alpha_1}, \theta\right)\right)\\
=& \alpha_1\frac{q_1^n - \frac{\Phi}{N}}{\alpha_1}\left(P\left(\frac{Q_1(t) - \Phi}{\alpha_1}, \theta\right) - c_1\right)\\
&+ \alpha_2\frac{q_2^n + \frac{\Phi}{N}}{\alpha_2}\left(P\left(\frac{Q_2 + \Phi}{\alpha_2}, \theta\right) - c_2\right) + \frac{\Phi}{N}(c_2 - c_1).
\end{aligned}$$

Define the adjusted outputs $\gamma_1^n = \frac{q_1^n - \frac{\Phi}{N}}{\alpha_1}$, $\gamma_2^n = \frac{q_2^n + \frac{\Phi}{N}}{\alpha_2}$, $X^+ = \frac{\Phi}{\alpha_2}$ and $\Gamma_m = \sum_{n=1}^N \gamma_m^n$ for $m = 1, 2$. Then,

$$\pi^n = \alpha_1\gamma_1^n(P(\Gamma_1, \theta) - c_1) + \alpha_2\gamma_2^n(P(\Gamma_2, \theta) - c_2) + \alpha_2\frac{X^+}{N}(c_2 - c_1). \tag{7.2}$$

Adjusted output $\gamma_m^n$ is firm $n$'s decision variable in market $m$, which incorporates market size, the impact of imports (and exports), and the FTR payment. When the interconnection is congested, dynamics in each market are independent, thus firms optimize separately in each market. Equation (7.2) shows that the profit function is the sum of two "standard" Cournot profit functions, where adjusted output $\gamma_m^n$ replaces output $q_m^n$. The equilibrium of the congested spot-market game is therefore easily obtained.

The simplicity of the solution to the spot-market game is a result of the inclusion of the FTR payment in the profit function and the symmetry of generators. These assumptions are the main difference with Borenstein, Bushnell, and Stoft (2000) presented above. Since most electricity markets use FTRs, the first feature is realistic, while the second corresponds to the long-term equilibrium with free entry in each market.

Adjusted baseload capacity for producer $n$ is defined by $x_1^n = \frac{k_1^n - \Phi^+}{\theta_1}$, and the aggregate adjusted baseload capacity by $X_1 = \frac{K_1 - \Phi^+}{\theta_1}$.

As in section 6.4, two additional assumptions on the values of the parameters are made: (i) $\alpha_1 K_2^{CU} \le \alpha_2 K_1^U$, and (ii) $(\Phi^+ + \Phi^-) \ge \alpha_1 k_2^{CU}$ . If these assumptions do not hold, other cases need to be considered, although the economic intuition is unchanged.

Léautier (2014b) proves the equilibrium is summarized in the following:

**Proposition 7.1.** Equilibrium-generation mix $(K_1^C, K_2^C)$.

1. If $\Phi^+ \ge \alpha_2 K_1^{CU}$, the transmission line is never congested, $K_2^C = K_2^{CU}$ and $K_1^C = K_1^{CU}$.

2. If $\Phi^+ < \alpha_2 K_1^U$ and $(\Phi^+ + \Phi^-) \geq \alpha_1 K_2^{CU}$, the transmission line is congested from market 1 to market 2 for some states of the world. The cumulative installed capacity $K_2^C$ is the cumulative uncongested capacity:

$$K_2^C = K_2^{CU},$$

and the baseload capacity is the uncongested baseload capacity scaled down by its domestic market size $\alpha_1 K_1^U$, plus the interconnection capacity $\Phi^+$:

$$X_1^C = K_1^{CU} \Leftrightarrow K_1^C = \alpha_1 K_1^U + \Phi^+.$$

Proposition 7.1 closely mirrors Proposition 6.1: first, aggregate cumulative capacity is unchanged by the presence of a transmission constraint, since it is no longer binding on-peak. Second, in this very simple setting, baseload equilibrium capacity when constraints are binding can easily be deduced from baseload equilibrium capacity absent any transmission constraint.

### 7.4.2 Transmission-Constrained Oligopoly in a Two-Market Network

Again, to simplify the notation, we introduce the demand functions $D_m(p) = \alpha_m D(p)$ for $m = 1, 2$. Since Gilbert, Neuhoff, and Newbery (2004) consider a Cournot oligopoly, inverse demand functions are required: for $m = 1, 2$, define $P_m(Q) = P\left(\frac{Q}{\alpha_m}\right)$. We immediately verify that $P_m(D_m(p)) = p$. The interested reader may also consult Gilbert, Neuhoff, and Newbery's (2002) working paper, leading to the published article, Gilbert, Neuhoff, and Newbery (2004), which contains a much richer analysis.

#### 7.4.2.1 Import-constrained oligopoly

Suppose now market 1 is perfectly competitive, with constant short-term marginal cost $c_1$. $N$ identical firms compete à la Cournot in market 2, with constant short-term marginal cost $c_2 > c_1$. This extends to an oligopoly; see Joskow and Tirole's (2000) analysis of an import-constrained monopoly presented in section 7.3. The power flows are represented in figure 7.6.

**Spot-market equilibrium** Since the interconnection is congested from market 1 to market 2, the SO issues FTRs paying $(p_2 - p_1)$. The profit of producer $n$ located in market 2 who owns $m_2^n \Phi$ FTRs is

$$\pi_2^n = q_2^n (P_2(Q_2^s + \Phi) - c_2) + m_2^n \Phi (P_2(Q_2^s + \Phi) - c_1).$$

Gilbert, Neuholf, and Newbery (2004) describe the relationship between an FTR purchase and a forward sale: since she does not control the price in market 1, a producer located in market 2 who purchases FTRs ending in her market de facto purchases a (long) position in market 2's spot price; hence she increases her exposure to market 2's spot price. Owning FTRs raises the value of a high price in market 2, hence worsens (from a net surplus perspective) the exercise of market power.

Following the analysis presented in section 4.5, Gilbert, Neuhoff, and Newbery (2004) confirm that an increase in firm $n$'s FTR position leads to an aggregate output reduction, that is, $\frac{\partial Q_2^s}{\partial m_2^n} < 0$, hence a price increase, in market 2. This is the opposite of the standard Allaz and Vila (1993) effect. Similarly, producer $n$'s increase in her own FTR position leads the producer $n$'s to reduce her output, since she gains more from a price increase: $\frac{\partial q_2^n}{\partial m_2^n} < 0$. Since quantities are strategic substitutes, this leads other producers to increase their output at the equilibrium: $\sum_{l \neq n} \frac{\partial q_2^l}{\partial m_2^n} > 0$. However, this output increase is not sufficient to compensate producer $n$'s output reduction, and the aggregate output is reduced: $\frac{\partial Q_2^s}{\partial m_2^n} = \frac{\partial q_2^n}{\partial m_2^n} + \sum_{l \neq n} \frac{\partial q_2^l}{\partial m_2^n} < 0$. We therefore expect producers import-constrained purchase FTR to increase their profits.

Gilbert, Neuhoff, and Newbery (2004) examine the equilibrium in the FTR market. As in Joskow and Tirole (2000), they find that this equilibrium depends on the microstructure of this market, specifically whether the SO uses a uniform price or a pay-as-bid auction.

**Uniform price auction for FTRs** In this case, the equilibrium price for FTRs is their expected value. Joskow and Tirole (2000) find that a monopoly does not purchase FTRs in a uniform price auction. Gilbert, Neuhoff, and Newbery (2004) extend the analysis, and find that import-constrained oligopoly producers have, in fact, incentives to sell FTRs.

As in Allaz and Vila (1993), FTR purchase has no direct impact on producers' profits since no arbitrage implies that their purchase price is equal to their ex post value. Its only impact is through the change in equilibrium output, hence price:

$$\begin{aligned}
\frac{\partial \pi_2^n}{\partial m_2^n} &= \frac{\partial q_2^n}{\partial m_2^n}(P_2(Q_2^s + \Phi) - c_2) + q_2^n \partial_q P_2(Q_2^s + \Phi)\frac{\partial Q_2^s}{\partial m_2^n} \\
&= \partial_q P_2(Q_2^s + \Phi)\left(q_2^n \frac{\partial Q_2^s}{\partial m_2^n} - (q_2^n + m_2^n)\frac{\partial q_2^n}{\partial m_2^n}\right) \\
&= \partial_q P_2(Q_2^s + \Phi)\left(q_2^n\left(\frac{\partial Q_2^s}{\partial m_2^n} - \frac{\partial q_2^n}{\partial m_2^n}\right) - m_2^n \frac{\partial q_2^n}{\partial m_2^n}\right).
\end{aligned}$$

Consider first a monopoly, $N = 1$. Since $Q_2^s = q_2^n$, then $\frac{\partial Q_2^s}{\partial m_2^n} = \frac{\partial q_2^n}{\partial m_2^n}$ and

$$\frac{\partial \pi_2^n}{\partial m_2^n} = -\partial_q P_2(Q_2^s + \Phi) m_2^n \frac{\partial q_2^n}{\partial m_2^n} \leq 0.$$

The monopoly purchases no FTRs. The intuition is identical to Joskow and Tirole (2000): traders capture ex ante the value of the FTRs, hence the monopoly refrains from purchasing any. For $N > 1$, since $\frac{\partial Q_2^s}{\partial m_2^n} - \frac{\partial q_2^n}{\partial m_2^n} = \sum_{l \neq n} \frac{\partial q_2^l}{\partial m_2^n} > 0$, $\frac{\partial \pi_2^n}{\partial m_2^n} < 0$: producers have incentives to sell FTRs.

This result is closely related to Allaz and Vila (1993): in the latter, forward sales reduce producers' incentives to exercise their market power in the spot market; however they cannot refrain from selling forward and thereby limit their future profits. In this case, (forward) FTR purchases increase producers' incentives to exercise their market power in the spot market; however, they cannot refrain from selling FTRs, thus limiting their future profits.

Gilbert, Neuhoff, and Newbery (2004) find that allowing import-constrained oligopolists to participate in a uniform-price auction of FTRs ending in their market is procompetitive.

**Pay-as-bid auction** Gilbert, Neuhoff, and Newbery (2004) characterize the equilibrium of the pay-as-bid auction for oligopolists. Like Joskow and Tirole (2000), they also find that market participants play mixed strategies. However, they find that, if a sufficient condition is met, oligopoly producers on average purchase FTRs, hence increase their market power. Thus allowing import-constrained oligopolists to participate in a pay-as-bid auction of FTRs ending in their market is not procompetitive.

#### 7.4.2.2 Export-constrained oligopoly

Suppose that $N$ producers compete à la Cournot in market 1, while market 2 is perfectly competitive. Suppose also that the Cournot equilibrium price in market 1 is lower than the marginal cost $c_2$ in market 2, hence that the interconnection between markets 1 and 2 is congested. Oligopoly producers are limited by the congestion on the interconnection.

The profit of producer $n$ located in market 1 who has purchased the fraction $m_1^n \Phi$ FTRs is

$$\pi_1^n = q_1^n (P_1(Q_1^s - \Phi) - c_1) + m_1^n \Phi (c_2 - P_1(Q_1^s - \Phi)).$$

An export-constrained producer who has purchased FTRs is in the situation described by Allaz and Vila (1993): its exposure to the wholesale spot price is reduced, hence her equilibrium output is higher than if no FTRs had been purchased. Proposition 4.1 presented in section 4.3 applies and yields $\frac{\partial q_1^n}{\partial m_1^n} > \frac{\partial Q_1^s}{\partial m_1^n} > 0$.

Suppose FTRs are sold in a uniform-price auction. As before, their equilibrium price is their ex post value. The results of section 4.3 hold. If $N = 1$,

$$\frac{\partial \pi_1^n}{\partial m_1^n} = \partial_q P_1(Q_1^s - \Phi) m_1^n \Phi \frac{\partial Q_1^s}{\partial m_1^n} < 0:$$

the monopoly producer purchases no FTRs. If possible, she sells FTRs forward.

For $N > 1$, equilibrium FTR purchase (if it exists) is defined by

$$\frac{\partial \pi_1^n}{\partial m_1^n} = 0 \Leftrightarrow m_1^n \Phi = q_1^n \left( 1 - \frac{\frac{\partial Q_1^s}{\partial m_1^n}}{\frac{\partial q_1^n}{\partial m_1^n}} \right).$$

We verify that $m_1^n \Phi^+ \in (0, q_1^n)$: export-constrained producers purchase FTRs, which limit their exercise of market power in the wholesale spot market. The FTR market therefore mitigates market power exercised by export-constrained producers.

Depending on the value of the parameters, the aggregate demand for FTRs may exceed the physical capacity of the interconnection. Suppose $\alpha_1 = 1$. The first-order condition defining $q_1^n$ is

$$P(Q_1^s - \Phi) + (q_1^n - m_1^n \Phi)\partial_Q P(Q_1^s - \Phi) = c_1.$$

Summing over all producers and introducing $M_1 = \sum_{n=1}^{N} m_1^n$ yields

$$P(Q_1^s - \Phi) + \frac{Q_1^s - M_1 \Phi}{N}\partial_Q P(Q_1^s - \Phi) = c_1.$$

Suppose, for example, demand is linear with constant slope $P(Q) = a - bQ$. Algebraic manipulation yields

$$bQ_1^s = \frac{N(a - c_1) + (M_1 + N)b\Phi}{N+1}.$$

The analysis presented in section 4.3 yields $M_1 \Phi = \frac{N-1}{N} Q_1^s$. Combining both relations yields

$$b\frac{N}{N-1}M_1\Phi = \frac{N(a - c_1) + (M_1 + N)b\Phi}{N+1} \Leftrightarrow M_1 = \frac{N(N-1)}{N^2+1}\left(\frac{a - c_1}{b\Phi} + 1\right).$$

Therefore, $M_1 \leq 1$ if and only if

$$\frac{a - c_1}{b\Phi} \leq \frac{N+1}{N(N-1)}.$$

If this condition is not verified, export-constrained producers would like to purchase more FTRs than the physical capacity of the line.

### 7.4.3 Three-Market Network

As discussed in chapter 6, loop flows modify the analysis but not fundamentally the economic intuition. Joskow and Tirole (2000) and Gilbert, Neuhoff, and Newbery (2004) illustrate the rich interaction between transmission constraints and market power on a three-market network.

#### 7.4.3.1 Oligopoly on the "right" side of the constraint

Consider the three-market network presented in figure 7.7.

Producers are located in markets 1 and 2, and demand $D(p)$ is located in market 3. All interconnections have identical characteristics, except that interconnection $(1, 2)$ has limited capacity $\Phi$. Market 1 is perfectly competitive, at marginal cost $c_1$. Market 2 is an oligopoly of $N$ identical producers, with marginal cost $c_2 > c_1$.

Absent congestion, perfectly competitive producers in market 1 serve the entire demand, and producers in market 2 do not produce. Consider now that demand in market 3 is high enough that interconnection $(1, 3)$ is congested (specifically $\frac{2}{3}D(c_1) > \Phi \Leftrightarrow P\left(\frac{3\Phi}{2}\right) > c_1$)

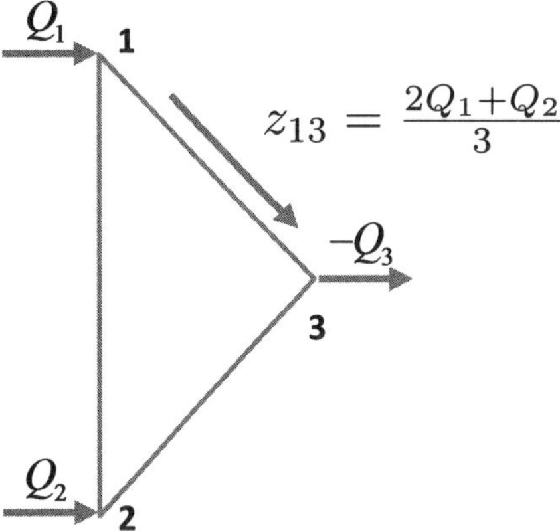


**Figure 7.7**
Three-market network with asymmetric transmission constraint. Producers are located in markets 1 and 2, and demand is located in market 3. All interconnections have identical characteristics, except that interconnection $(1, 3)$ has limited capacity $\Phi$.

and producers in market 2 are called to produce. Their situation is complex: they are exporting into market 3 and are protected by the constraint on an interconnection, which neither originates nor terminates in market 2 (and is potentially distant). The impact of FTRs on their behavior is a priori ambiguous. As it turns out, FTRs increase their incentive to exercise market power.

To prove this result, start from the SO's optimal dispatch program:

$$max_{\{Q_1^s, Q_2^s, Q_3^d\}}[U(Q_3^d) - c_1 Q_1^s - p_2 Q_2^s]$$
$$st: \begin{cases} Q_1^s + Q_2^s = Q_3^d & (\mu_e) \\ \frac{2Q_1^s + Q_2^s}{3} \leq \Phi & (\eta) \end{cases}.$$

The associated Lagrangian is

$$\mathcal{L} = [U(Q_3^d) - c_1 Q_1^s - p_2 Q_2^s] + \mu_e(Q_1^s + Q_2^s - Q_3^d) + \eta\left(\Phi - \frac{2Q_1^s + Q_2^s}{3}\right),$$

hence the nodal prices are

$$\begin{cases} p_3 = U'(Q_3^d) = P(Q_3^d) = \mu_e \\ p_1 = c_1 = p_3 - \frac{2\eta}{3} \\ p_2 = p_3 - \frac{\eta}{3} \end{cases}.$$

Market 3 is taken as the reference, hence $p_3$ is the value of the marginal MWh, equal to $\mu_e$, the shadow price of the global energy balance constraint. To export into market 3, producers in market 1 pay a "congestion tax" equal to $\frac{2}{3}$ of the shadow price of the constraint $\eta$, while producers in market 2, who use less of the interconnection, pay only $\frac{\eta}{3}$ as a "congestion tax."

We now compute $p_2$ as a function of $Q_2^s$. We first compute the value of $\eta$:

$$c_1 = p_3 - \frac{2\eta}{3} \Leftrightarrow \eta = \frac{3}{2}(p_3 - c_1).$$

Then,

$$p_2 = p_3 - \frac{\eta}{3} = \frac{p_3 + c_1}{2}$$

Then, combining the energy balance $Q_3^d = Q_1^s + Q_2^s$ with the flow on the interconnection

$$2Q_1^s + Q_2^s = 3\Phi \Leftrightarrow Q_1^s = \frac{3\Phi - Q_2^s}{2}$$

yields

$$Q_3^d = \frac{3\Phi - Q_2^s}{2} + Q_2^s = \frac{3\Phi + Q_2^s}{2}.$$

Therefore,

$$p_3 = P\left(\frac{3\Phi + Q_2^s}{2}\right) \textit{ and } p_2 = \frac{P\left(\frac{3\Phi + Q_2^s}{2}\right) + c_1}{2}:$$

producers in market 2 face an inverse demand similar to import-constrained producers.

These producers may purchase FTRs from market 2 to market 3. The total value of these FTRs is

$$\mathcal{F}(Q_2^s) = (p_3 - p_2)\Phi = \frac{\eta}{3}\Phi = \left(\frac{P\left(\frac{3\Phi + Q_2^s}{2}\right) - c_1}{2}\right)\Phi.$$

Thus purchasing FTRs increase market 2 producers' incentives to exercise their market power. Even though they are exporters in market 3, they behave as importers. The result does not change if producers purchase FTRs from market 1 to market 3.

#### 7.4.3.2 Oligopoly on the wrong side of the constraint

Suppose now interconnection (2, 3) is potentially congested and that all other elements remain unchanged. This situation is represented in figure 7.8.

Congestion occurs when $\frac{1}{3}D(c_1) > \Phi \Leftrightarrow P(3\Phi) > c_1$. The nodal prices are

$$\begin{cases} p_3 = P(Q_3^d) = \mu_e \\ p_1 = c_1 = p_3 - \frac{\eta}{3} \\ p_2 = p_3 - \frac{2\eta}{3} \end{cases}.$$

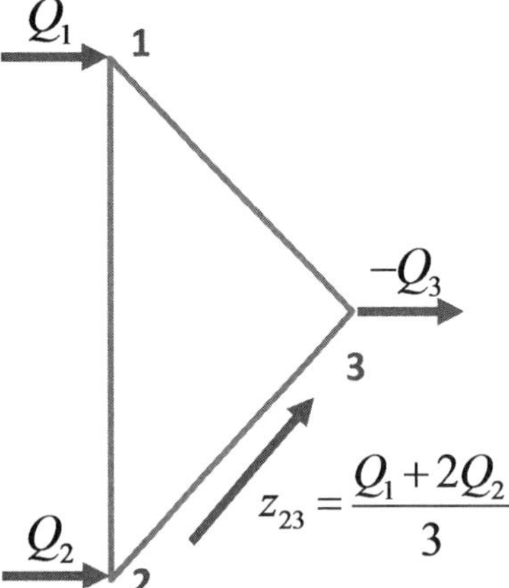


**Figure 7.8**
Three-market network with asymmetric transmission constraint. Producers are located in markets 1 and 2, and demand is located in market 3. All interconnections have identical characteristics, except that interconnection (2, 3) has limited capacity $\Phi$.

The shadow price of the transmission constraint is $\eta = 3(p_3 - c_1)$: a unit increase in transmission capacity enables the production of three more units in market 1 and their consumption in market 3.

Then,

$$p_2 = p_3 - \frac{2\eta}{3} = 2c_1 - p_3.$$

Then, combining the energy balance $Q_3^d = Q_1^s + Q_2^s$, with the flow on the interconnection $Q_1^s + 2Q_2^s = 3\Phi$ yields $Q_3^d = 3\Phi - Q_2^s$. Therefore,

$$p_3 = P(3\Phi - Q_2^s) \ and \ p_2 = 2c_1 - P(3\Phi - Q_2^s).$$

The oligopolists profit is

$$\pi_2^n = q_2^n(2c_1 - P(3\Phi - Q_2^s) - c_2),$$

hence

$$\left.\frac{\partial \pi_2^n}{\partial q_2^n}\right|_{q_2^n=0} = 2c_1 - P(3\Phi - Q_2^s) - c_2 = -(c_2 - c_1 + P(3\Phi - Q_2^s) - c_1) < 0,$$

since $c_2 > c_1$ by assumption and $P(3\Phi - Q_2^s) \geq P(3\Phi) > c_1$ since the interconnection is congested. Producers in market 2 are so severely export-constrained that they produce nothing.

However, they may start producing if they hold FTRs on interconnection (2, 3). The value of these FTR is

$$\mathcal{F}(Q_2^s) = (p_3 - p_2)\Phi = \frac{2\eta}{3}\Phi = 2(P(3\Phi - 2Q_2^s) - c_1)\Phi.$$

By increasing their output, producers in market 2 increase the value of their FTRs. If this increase exceeds the margin lost on their output, they end up producing. This can be verified analytically. Suppose producer $n$ purchases a fraction $m_2^n$ of these FTRs,

$$\left.\frac{\partial \pi_2^n}{\partial q_2^n}\right|_{Q_2=0} = 2c_1 - P(3\Phi) - c_2 - 4m_2^n \Phi \partial_Q P(3\Phi).$$

The first term on the right-hand side is negative, since $c_1 < P(3\Phi)$ and $c_1 < c_2$, while the second term is positive. The sum may be positive.

# IV POLICY ISSUES

# 8 Gone with the Wind (and the Sun): Renewable Energy Sources Entry into Electricity Markets

Take a good look my dear. It's an historic moment you can tell your grandchildren about—how you watched the Old South fall one night.
—Rhett Butler

## 8.1 Introduction

As mentioned in chapter 1, slowing climate change requires massive decarbonization of the electricity industry. This chapter explores various aspects of this decarbonization challenge and is structured as follows. Section 8.2 presents a few facts on climate change and describes their implication for the electric power industry, mostly the high and rapid penetration of low-carbon renewable energy sources (RES). Section 8.3 reviews the policy instruments available to fight climate change. Sections 8.4 to 8.6 present an analytic model of the power market when RES entry is mandated by policy makers, which draws heavily on Green and Léautier (2017).

## 8.2 The Energy-Transition Challenge

This section first presents a few issues regarding climate change, then discusses implications for the power industry, including non-$CO_2$-emitting production technologies. Of course, it does not do justice to such an important topic. Interested readers should start with the Intergovernmental Panel on Climate Change (IPCC) reports, whose executive summary is clear and nontechnical. In addition, the best book I have read on this subject is *Sustainable Energy—Without the Hot Air,* by the late David MacKay (2008), a mathematician and physicist and regius professor of engineering at the University of Cambridge, which offers a wonderfully clear perspective on the challenge and possible answers.

### 8.2.1 Climate Change

The discussion about climate change starts from facts: the concentration of greenhouse gases (GHG) in the atmosphere and the earth's average temperature have increased since the beginning of the Industrial Age. For example, in its fifth assessment report, published during the winter of 2013–2014, the IPCC (2014) concluded that the period 1983–2012 was likely to have been the warmest 30-year period of the last 1,400 years in the Northern Hemisphere, and as a result unprecedented ice melting has already occurred. A very vivid illustration of temperature increase can be found in NASA (2018).

Most scientists agree that this joint increase is not coincidence and that the former causes the latter: temperature rises owing to the emissions of anthropogenic GHG, chief among which is carbon dioxide. GHG also include methane, nitrous oxide, ozone, and chloro- and hydro-fluorocarbon. Their average global warning potential is measured in $CO_2e$ equivalent, denoted $CO_2e$. The theory of global warming is that GHG trap incoming heat from the sun around the earth, in the similar way a greenhouse traps heat to make plants grow faster. Using this theory and climate models, scientists estimate the impact of continued GHG emissions. Their main conclusion is that GHG emissions must stop by 2050 if we are to maintain temperature increase at an acceptable level (defined as 2° Celsius above the preindustrial temperature average).

#### 8.2.1.1 Climate and economic models

Climate scientists hypothesize a linear relationship between the stock of GHG in the atmosphere and average global temperature. Using this relationship, for any path of GHG emissions, they can compute the expected global temperature increase. They also use most sophisticated models to estimate the temperature in different points of the globe, taking various factors into effect. The IPCC (2014) report anticipates that, if current trends continue, (a) global average temperature increase compared with 1986–2005 levels could range from 0.3 to 4.8°C by the turn of this century, (b) oceans temperature will continue to increase during the twenty-first century, potentially affecting ocean circulation, and (c) rising sea level compared with the end of the twentieth century could reach 26 to 82 cm by 2100, leading to massive population displacements.

Economists then estimate how increased temperature, hence GHG, impacts economic activity. This type of analysis usually requires computer simulations. These models are called integrated assessment models (IAMs). One example is the DICE model developed by Richard Nordhaus (1994) from Yale University. These models are then used to compute the true social cost of emitting one ton of $CO_2e$. For example, the Stern review (2007) estimates the cost of climate change somewhere between 2 percent and 35 percent of world GDP by the year 2200.

Climate science is relatively new, and the physics of climate (hence climate change) are complex. For example, all scientists involved in climate modeling agree that oceans play a significant role, but we still know very little about them. Furthermore, forecasting the future is of course complex. Models are often extremely sensitive to input parameters, about which little is known today.

Thus predictions about climate change are imperfect at best. This has facilitated the emergence of climate skeptics.

#### 8.2.1.2 Climate skeptics

Since this book examines both the microeconomics and the political economy of the power industry, a word on climate skeptics is warranted. Climate skeptics make some valid points: for example, temperature increase on earth appears to have slowed down recently. (Climate scientists explain this phenomenon by an increase in temperature in the oceans.)

Climate skeptics also argue that climate change advocates are often its beneficiaries. In the same way that oil companies are skeptic about the negative long-term impact of burning oil, wind turbines developers are adamant that climate Armageddon is near and that decisive action is required, starting with subsidies for wind power.

It is hard to disagree with these observations. The science of climate change is relatively recent, hence its predictions do not perfectly match the data. This does not mean that we should dismiss existing scientific evidence; rather, more science is required. It is also obvious that an organization's (or individual's) stated positions on the severity of climate change are often influenced by the economic position. Coal mining and oil producing companies are less convinced by the science of climate change than photovoltaic panels manufacturers.

Still, since there appears to exist a scientific consensus on the reality of man-made climate change, few policy makers are overtly climate skeptics, and public opinion is broadly aligned with the need to fight climate change. As a result, numerous climate change fighting policies are adopted today with public opinion support, and more are expected to come.

### 8.2.2 Implications for the Electricity Industry

The power industry is essential if we are to mitigate climate change. For example, David MacKay (2008) proposes the following plan to reduce $CO_2$ emissions: (a) electrify transport (i.e., replace combustion engines by electric-powered vehicles); (b) electrify heating (i.e., replace combustion of gas and oil in boilers by heat pumps) and reduce heating requirement through insulation; and (c) produce low $CO_2$ electricity. Most other proposals follow essentially the same steps, although there is a debate on the role of natural gas, which emits less $CO_2$ per MWh than coal, as a transition fuel to produce electricity. The justification is simple: we can produce electricity without emitting $CO_2$, while combustion of fossil fuel always produces $CO_2$.

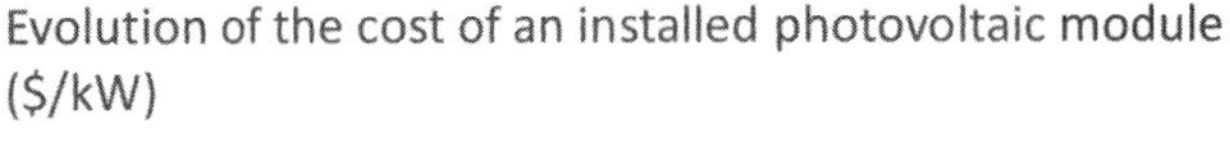

Evolution of the cost of an installed photovoltaic module
($/kW)


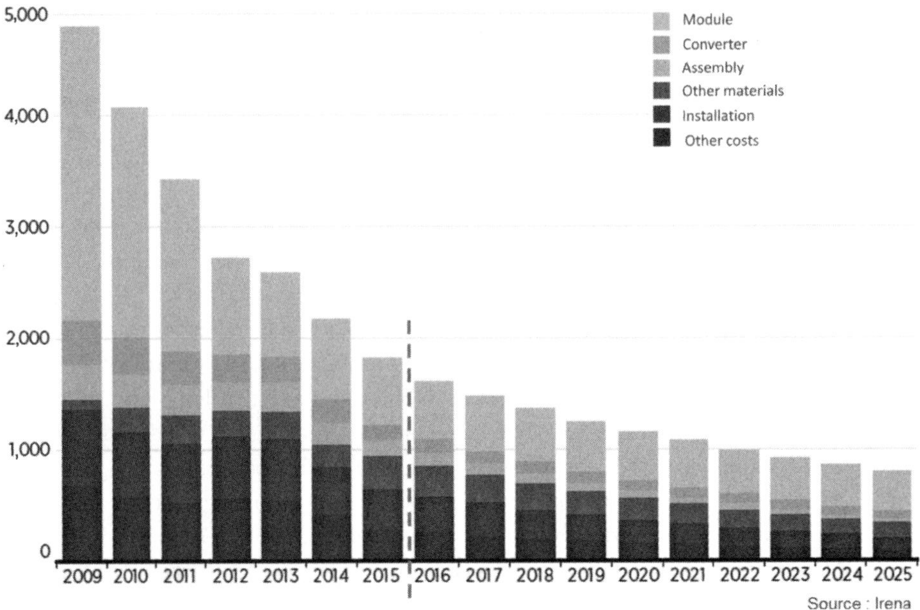


**Figure 8.1**
Learning curve for photovoltaic modules.
*Source*: IRENA (2016), figure E-S1, page 13.

### 8.2.2.1 Electric power demand

As discussed in the introduction, demand for electrical power is expected to grow very slowly in OECD, member countries: the impact of these new usages will be more or less offset by energy efficiency. On the other hand, demand is expected to grow rapidly in developing countries (mostly China and India), although energy efficiency will play a role.

### 8.2.2.2 Renewable energy sources

The rapid and unexpected decrease in the cost of renewable energy sources (RES) is the most significant industrial advance of the past ten years. As illustrated in figure 8.1, the cost of photovoltaic modules in 2016 was less than half the 2009 cost and is expected to further decrease to reach 20 percent of the 2009 cost by 2025. This is not Moore's law, which stipulates that the power of computers doubles every eighteen months, but nevertheless extremely impressive. Furthermore, it is completely unexpected: no one anticipated such a rapid reduction.

This cost decrease translates into price decrease. In 2015 the world was stunned when a contract to produce power from photovoltaic panels in the Arabian desert was awarded at \$40/MWh. In 2016 the world was stunned again when a similar contract was awarded at \$30/MWh. Another shock came in the same year when a contract to provide electricity from offshore wind in the North Sea was awarded at €72/MWh. In 2017 Orsted (previously Dong Energy) agreed to build and operate an offshore wind farm located in the North Sea at market price, that is, with no subsidy.

A word of caution is in order: first, there may be little profit for the developers in these contracts. For example, analysts suspect national pride played a role in inducing investors to take risks in the second photovoltaic contract. Second, these low costs may be attributable to specific circumstances that may not be replicable. The developers of the offshore wind field seem to have benefited from extremely low steel price and shipping rates. The sun conditions in the Arabian desert are extremely favorable and not to be found in Scotland, for example. Third, the offshore wind farm contracts are options: should economic conditions change, Orsted, for example, could decide not to build the wind farm and pay a break-up fee.

Still, this cost decrease is significant. It is attributable to large and sustained demand for RES, fueled by subsidies. Confident that they could sell their product, manufacturers and developers have constantly innovated and built large factories to capture economics of scale. There is also a so-called Chinese effect: manufacturing overcapacity in China has pushed the sales price of photovoltaic modules close to their variable production cost. Thus German taxpayers, who foot a very large RES subsidy bill, and Chinese taxpayers, who may be liable for bailouts, are to thank for this remarkable decrease in cost.

Still, RES are now (almost) competitive in advanced economies and offer a reasonable hope for developing countries. Africa has leapfrogged standard telephony by moving rapidly to wireless. For remote areas, local solutions power by RES may offer a faster way to electricity than expanding the network. This is extremely important and should not be overlooked. This is great news for the billions of humans who lack reliable access to electricity.

#### 8.2.2.3 Electricity storage

A side effect of the dramatic reduction (and largely unexpected, as illustrated in box 8.1) in RES costs is the reduction in storage costs. One particular issue with RES is that their production is variable (e.g., a solar panel does not produce at night) and unpredictable. Thus a power system with a high share of photovoltaic power, for example, needs another energy source or significant demand response to produce every night and to react rapidly when a cloud arrives. One obvious solution is storing energy. Renewable energy sources have increased the market value of storage. For example, the International Renewable Energy Agency in a report published in October 2017 (IRENA 2017) predicts that “electricity storage capacity in energy terms will need to grow from 4.67 Terawatt-hours in 2017 to 11.89–15.72 Terawatt-hours if the share of renewable energy in the energy system is to be doubled by 2030.”

**Box 8.1**
Energy storage: The next technological surprise?

Two personal anecdotes are revealing on the recent and unanticipated progress of electricity storage. In the early 2000s, I discussed electricity storage with Les Silverman, who at the time was leading McKinsey and Company's electric power and natural gas practice. Les was principal deputy assistant secretary for policy and evaluation at the U.S. Department of Energy from 1979 to 1981, hence he has a long experience in the power industry. He fondly remembered various attempts at storing electricity, in particular through inertia, and concluded laconically that this were doomed efforts.

Ten years later, in the early 2010s, I discussed batteries with Bill Hogan, who many consider the father of the current architecture of electricity markets in the United States. Bill is a professor at Harvard University and was also assistant secretary of energy in the 1970s. I mentioned a journal article presenting exciting prospects for batteries. Bill put his hand on my shoulder, and told me: "My poor child. Batteries are still dozens of years away from being economic."

Les and Bill are two of the most knowledgeable persons I know about the power industry. Their knowledge certainly was as good as it got. Yet entrepreneurs from many countries are betting they were wrong.

Multiple storage technologies are available. When feasible, energy is stored in hydro-reservoirs, for example, pump-storage plants. Alternatively, energy can be stored in batteries. Different technologies have different usages. Lithium-ion batteries are used for daily storage and can sustain hundreds of charge-discharge cycles. Their cost is decreasing rapidly through the standard effects: economies of scale and learning by doing. To capture these two effects, Tesla's Gigafactory aims to produce 35 Gigawatt-hours of batteries annually. IRENA (2017) finds that "the cost of Li-ion batteries have fallen by as much as 73% between 2010 and 2016 for transport applications. Benefitting from the growth in scale of Li-ion battery manufacturing for Electric Vehicles, the cost could decrease in stationary applications by another 54-61% by 2030."

In parallel, other firms are developing other types of batteries. For example, sulfur-ion batteries are explored for seasonal storages. Other firms are using hydrogen to store energy: electricity is used to split water into hydrogen and oxygen. The hydrogen then stores the energy until a fuel cell or engine converts it back to electricity. Hydrogen can be injected in the natural gas network or also be recombined with captured $CO_2$ to produce a synthetic natural gas that can be used in power plants or transportation applications.

#### 8.2.2.4 Carbon capture and storage

Carbon capture and storage (CCS) is the great white hope of gas and coal companies. The concept is extremely simple: first, install a large net on top of a power plant's chimney, which traps $CO_2$ molecules. Second, pipe the captured molecules into permanent storage areas, for example, depleted oil or gas fields, or salt mines. The challenge lies in the execution. While the concept and the science behind it are clear, and a few prototypes are in operation, no one knows how to turn CCS into a commercially viable activity. Over the years, knowledgeable and competent experts have assured me that CCS is just around the corner, while equally knowledgeable and competent experts have assured me it will never occur.

#### 8.2.2.5 Nuclear generation

Nuclear generation stands out among low $CO_2$ electricity production technologies and is the subject of heated debate. On the one hand, nuclear operators point out that nuclear is the only low-$CO_2$-generation technology available today that is both affordable and controllable and has a large development potential. Hydro is also low in $CO_2$ and controllable, but the availability of new hydro sites is limited. Thus decarbonization of power generation will require an increased share of nuclear generation. Furthermore, some analysts (e.g., David MacKay) observe that there is simply not enough space to produce enough electricity using only wind turbines and photovoltaic panels: we need to produce a large number of Megawatt-hours per square meter of occupied space. Today, nuclear is the only technology that does so. These arguments supported the nuclear "renaissance" that occurred in the 2000s, which ended dramatically and tragically with the accident at the Fukushima Daiichi plant in Japan on March 11, 2011.

On the other hand, most "green" NGOs, which are very active in promoting the energy transition, are also opposed to the nuclear industry and envision a nuclear-free world.

**The nuclear power conundrum** To discuss nuclear generation, it is useful to distinguish between two levels of decision. The first is whether to include nuclear technology in the electricity-generation mix. This is a complex decision, which goes well beyond economics. On the one hand, nuclear power produces low-$CO_2$ electricity reliably at a reasonable cost. On the other hand, the construction and operation of nuclear power plants requires specific technical skills but also robust governance. The consequences of a nuclear accident are frightening, however rare it might be. Nuclear waste has a life span of millions of years. The decisions taken by our generation will also involve (practically) all future generations.

Various countries adopted nuclear electricity production in the 1960s. The United States was a pioneer. In 1946 the legendary admiral Hyman Rickover established a program to use nuclear reactors to propel the U.S. Navy submarines and aircraft carriers, considerably improving their performance during the Cold War. The USS *Nautilus,* the first U.S. Navy nuclear submarine, was launched in 1954. American manufacturers then used the pressurized water technology developed by the Navy to produce civil electricity. A similar approach was used in the USSR, Great Britain, and, to a lesser extent, Canada (which focused on civil nuclear power).

France followed the same approach. In the 1950s, the military nuclear program allowed France to join the very exclusive club of nuclear powers. This was clearly a political decision. In the 1970s and 1980s, the civil nuclear program allowed France to reduce its dependency on oil imported from the Middle East. Again, it was a political and strategic decision, well beyond mere economic efficiency. For the interested reader, Gabrielle Hecht (2009) provides a fascinating account of the history of the French nuclear industry.

When the British government decided to renew nuclear power in 2013, it was largely for geostrategic reasons: North Sea gas was starting to run out, and electricity production using natural gas was leading to a strong dependency on foreign imports, particularly from Russia.

Economists can contribute to this first level of decision by producing cost-benefit analyses, but, as the previous examples illustrate, the final decision is made using broader criteria than simply economic ones.

The second level of decision concerns implementation of the policy. After a country decides to include nuclear power as part of its generation mix, it has to define the economic and financial conditions. This involves solving issues of efficiency, incentives, and risk sharing, for which economists have a significant contribution to make.

**New nuclear generation and markets** Nuclear generation is perceived to be incompatible with the electricity market. This view is supported by two observations. First, existing plants in countries where markets have replaced monopolies, for example, Germany and the United States, are closing down, unable to operate profitably. This is true, but more complex than it may appear. First, French nuclear power plants are not retired. On the contrary, their owner, EOF, is investing tens of billions of euros to increase their operating life, since, under most scenarios, the wholesale power price exceeds the capital and operating cost of these assets. Second, the decision not to extend the life of the German nuclear power plants and to anticipate the closure of some of them is purely political as clearly demonstrated in Hansen and Percebois (2017). We do not know how these plants would have fared economically. Finally, the rules supporting RES are often unfavorable to nuclear plants.

Another observation supporting the incompatibility between markets and nuclear production is that almost all new constructions take place in countries where regional (or national) monopolies still operate the power system, for example, China. One variation on this theme is that construction of nuclear generation is so expensive and so risky that government guarantees are required. For example the Tory government in Great Britain guaranteed a price of £92,5/MWh, largely above the prevailing market price at the time, for all the output of the new nuclear power plant to be built at Hinkley Point.

This is true, but the causality also goes in the other direction: it is precisely because the nuclear industry has always been supported by governments that it finds it difficult to operate in a market. One of the fascinating facts about the nuclear industry is that costs always go up: contrary to all other industries, there is no learning curve. A portion of this increase can be explained by the ever-increasing safety requirements. New plants are safer, hence more expensive to build, than older ones. Another portion of the increase can be explained by the culture of the industry, which does not emphasize cost controls. Public purchasers are not known for being tough negotiators. In addition, who would object to cost increases when safety—or national pride—is at stake? The relative sizes of these portions is a matter of debate.

As a result, nuclear technology no longer has a clear cost advantage among low-$CO_2$ power production technologies. Comparing costs alone is misleading, since nuclear production is controllable, while other low-$CO_2$ power production technologies, in particular wind and solar, are not. Still, the argument that nuclear is the cheapest way to produce low-$CO_2$ power no longer holds today. This explains—at least partially—why new nuclear plants are difficult to finance in markets. Another important factor is that other low-$CO_2$ power production technologies have been heavily subsidized and are still advantaged in many markets.

Setting aside the issue of the social desirability of nuclear power, I see no reason why nuclear plants could not be built in a market, provided profound changes to their design and construction methods are made. At the minimum, design-to-cost of new nuclear plants would significantly improve their economic performance, without sacrificing on their safety performance. Alternative designs, such as smaller plants, or alternative fuels, such as thorium instead of uranium, could also be tested and prove more economic. Nuclear renaissance—if it occurs—will have to be a true renaissance.

## 8.3 The Tools to Fight Climate Change

Most economists agree that the first policy to be deployed is putting a price on $CO_2$ and other GHG. However, to most economists' chagrin, this recommendation has not been embraced by politicians at the time of this writing. Instead, they have opted to support RES.

### 8.3.1 Carbon Pricing

The economic argument supporting a $CO_2$ price is extremely robust. The underlying economic logic is presented in figure 8.2.

Consider, for example, a power producer burning coal to produce one Megawatt-hour coal and emitting approximately one ton of $CO_2$. This generates a direct marginal cost, covering the extraction of the fuel and its combustion, called the private marginal cost, since it is directly incurred by the power producer. The market equilibrium is such that the value derived from the marginal ton of $CO_2$ is exactly equal to this private marginal cost.

Emission of one ton of $CO_2$ also creates costly future environmental damages, called externalities, since the power producer today does suffer from these future damages. The social marginal cost is the sum of the private marginal cost and the cost of the negative

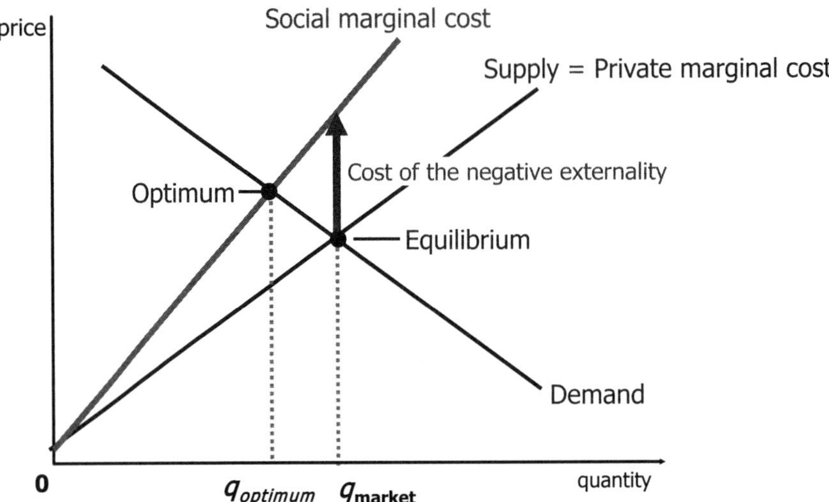


**Figure 8.2**
Equilibrium and optimum $CO_2$ emissions.

externality. The optimum is such that the value derived from the marginal ton of $CO_2$ is exactly equal to this social marginal cost. Since the latter exceeds the private marginal cost, users emit too much compared to the socially desirable level: as illustrated in figure 8.2, the equilibrium quantity exceeds the optimal one. To correct this over consumption, users must pay a price that includes the external effect.

While this logic is intellectually compelling, it so far failed to convince policy makers. Figure 8.3 shows the fraction of emissions that are actually priced in the world. Europe is clearly the most committed region, pricing 60 percent of its $CO_2$ emissions. On the contrary, some regions subsidize $CO_2$ emissions, for example the Middle East, which produces oil, and in 2013 subsidized 63 percent of its $CO_2$ emissions. Interestingly, in China and the United States, $CO_2$ emissions are simultaneously priced and subsidized. Even the COP 21, hailed as a great success for climate control, has stopped short of recommending adoption of a $CO_2$ price.

#### 8.3.1.1 Political economy of carbon pricing

The low penetration of carbon pricing and the failure to set a global uniform carbon price described in figure 8.3 can be explained by political economy and game theory analyses (see Tirole 2017). Three interrelated factors explain the situation.

First, most citizens, including elected officials and voters, are not physically experiencing global warming. Statistics often tell us that each year is the warmest on record, but, owing to a variety of cognitive biases, these statistical facts fail to impress us.

In addition, while no one seriously questions the reality of the increase in average temperature since the 1850s, some voters and their elected officials question whether

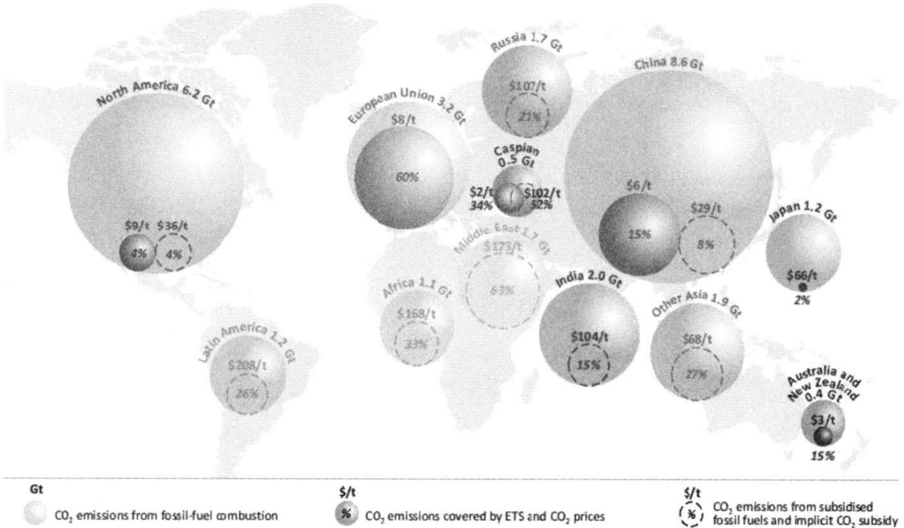


**Figure 8.3**
Worldwide $CO_2$ emissions, subsidies and prices.
*Source*: International Energy Agency (2015).

human GHG emissions are responsible for it. Their number may be small, but they are located in the United States, one of the most important contributors to GHG emissions and an essential architect of any global action.

A second factor, related to the first, is that, even if they are convinced of the long-term damage caused by today's GHG emissions, elected officials who decide on carbon pricing understand well that the costs of carbon pricing are immediate and are paid by their current voters, while the benefits accrue to future generations, who do not vote today. They are therefore loath to increase the carbon price in the economy to a level where it would have a real impact on investment decisions precisely because it would have an impact on investment decisions and jobs.

The third factor magnifies the second: countries vary in their perceived cost associated with future global warming and with the cost of today's actions. Some countries are vitally in danger (e.g., Maldives Islands), some are indifferent, and might even benefit from a temperature increase (e.g., Russia and Canada), while others are unsure. On the other hand, some countries are adamantly opposed to any action leading to reduced use of fossil fuels (e.g., Saudi Arabia and Venezuela). Developing countries like China and India trade-off differently than advanced economies the future losses from global warming against today's costs to reduce or alleviate it. As a result, countries differ in the cost they are willing to bear. If country B implements a carbon price while country A refuses to do so, it runs the

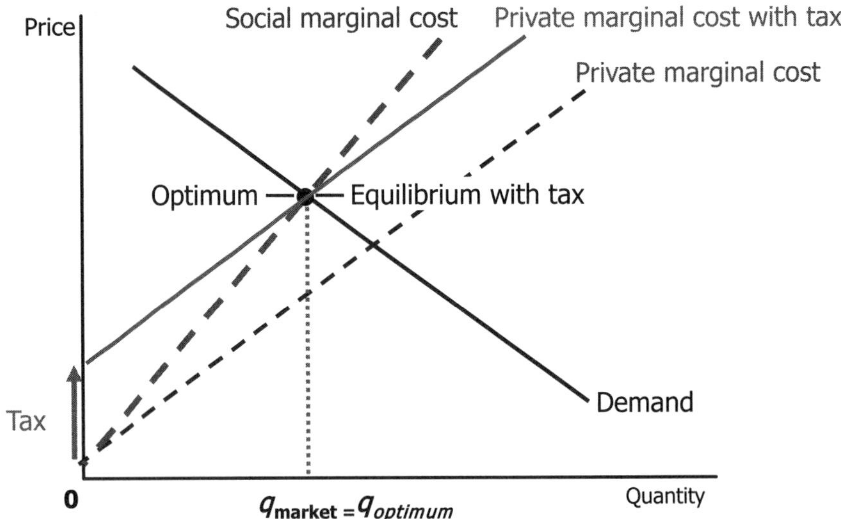


**Figure 8.4**
Equilibrium $CO_2$ emissions are optimal if a Pigovian tax is imposed.

risk of outsourcing its production—and its emissions—to country A, a phenomenon called "carbon leakage."

As a result of these three factors, elected officials in countries B (a loose characterization of which could be "advanced economies") collectively agree that the best course of action is to do nothing or, more exactly, to commit to ambitious goals without specifying the tools to achieve them. In doing so, they are joined by countries A ("emerging economies").

The European Union is a semioutlier to this narrative. Following the signature of the Kyoto protocol in 1992, it has created the first and largest $CO_2$ market in the world. The political economy of this decision is idiosyncratic. The European Commission is always in search of a federal project. Large members were in favor: France because of its strong nuclear power industry, Germany because of its strong industry. Great Britain and Nordic countries may have joined for (good) ideological reasons. Thus was born the European Emissions Trading Scheme (ETS), later called the European Union Emissions Trading System (EU ETS).

#### 8.3.1.2 Market versus tax

Two approaches are available to put a price on carbon. The first one, illustrated in figure 8.4, is to put a tax on every ton of $CO_2$ emitted, set at the best estimate of the indirect cost. If the tax is perfectly equal to the indirect cost, the market equilibrium is also the optimum. Economists call this a Pigovian tax, named after Arthur Cecil Pigou, the Cambridge economist who studied the issue of externalities and its solution in his text *The Economics of Welfare*, published in 1920.

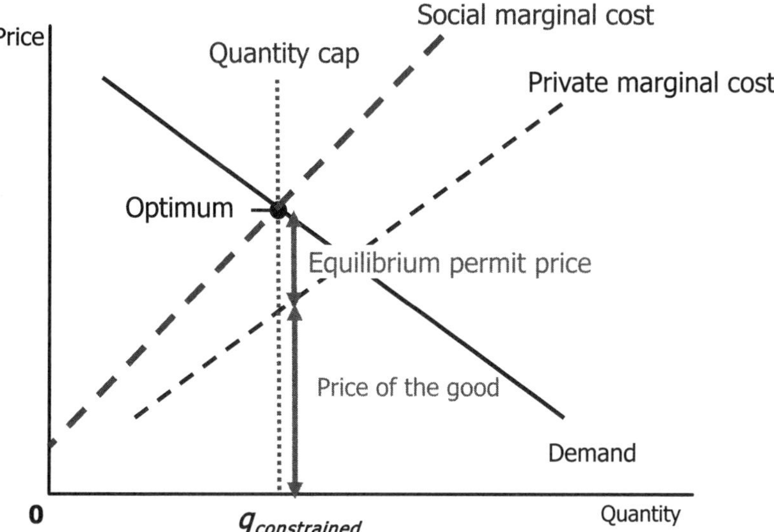


**Figure 8.5**
By construction equilibrium $CO_2$ emissions are optimal if the volume cap is set optimally.

The second approach relies on tradable property rights and was developed by the Chicago economist Ronald Coase in his article "The Problem of Social Cost," also published in 1920.[1] This approach is illustrated in figure 8.5. Policy makers determine a total emissions volume for a given period and issue permits (or credits) for that exact amount. Producers are allowed to emit exactly the amount of permits they possess. If their emissions exceed their permits, they pay a hefty penalty. If a firm anticipates its emissions will be lower than the amount of permits in its possession, it can resell them to another firm.

By construction, equilibrium emissions are equal to or lower than the permits issued. If policy makers issue the optimal amount of permits, equilibrium permit price is equal to the marginal cost of the externality at the optimum quantity.

The previous descriptions are voluntarily vague, since I want to focus on the principles: a Pigovian tax is a price intervention: it adds a tax to every ton of $CO_2$ emitted and lets users determine the quantity actually emitted. By contrast, a carbon market is a quantity intervention, it sets a maximum quantity of $CO_2$ to be emitted, and lets users determine the price at which permits (i.e., tons of $CO_2$) are traded.

The natural question is thus: Which approach is preferable, and why?

1. Even though he was born and spent the first half of his life in Great Britain, I classify Coase as a Chicago economist, since his insights on externalities and property rights are one of the intellectual foundations of many subsequent works by Chicago economists.

**Common challenges** First, observe that both approaches face significant implementation difficulties. One important one is the ability to effectively measure $CO_2$ emissions by user. As of 2016, only the total stock of $CO_2$ in the atmosphere is measured; hence the emissions per period are computed by difference. Each country calculates its $CO_2$ emissions and reports them to the United Nations, following a common and approved procedure: measure data about activities that generate emissions (e.g., how much gasoline is consumed to move vehicles, how much coal is burned to produce power), and compute the emissions per activity, using formulas and models. It is reasonable to assume one can compute the emissions of coal-fired power plant. Much more complex would be the task of computing the emissions of all cars using gasoline.

Linking aggregate emissions thus calculated to single users to assess a tax or allocate permits will require significant work and multiple hypotheses.

Second, there are significant uncertainties. The social cost of a ton of $CO_2$ is the value in today's money of the cost of future damages caused by this ton of $CO_2$. This requires three key ingredients: first, a physical model linking $CO_2$ emissions to temperature; second, climate and economic model linking temperature to economic activity, and third, a discount rate to evaluate today's value of future costs. While all of these ingredients exist, no scientist claims that there are perfect and error-proof tools.

**Technical argument** If policy makers had perfect foresight, they could compute the optimal tax and the optimal amount of permits to issue, and both approaches would be technically equivalent. In reality, there is significant uncertainty on the marginal benefit of abatement (i.e., reducing $CO_2$ emissions by one ton) and on the marginal cost of doing so. The question is: Which approach is more robust to uncertainty?

The answer was derived more than forty years ago by Martin Weitzman, an economics professor at Harvard University. Weitzman's (1974) seminal analysis was conducted in a simple setting. Later work has expanded it in multiple directions, for example, a dynamic version, or inclusion of a business cycle (Grodecka and Kuralbayeva 2015). The conclusion still holds: in the case of $CO_2$ emissions, price instruments (carbon taxes) are more robust to uncertainty than quantity restrictions (caps on emission), hence are preferable.

The argument is illustrated in figure 8.6: the marginal cost and benefit of abatement are presented on the vertical axis as a function of abated quantities, presented on the horizontal axis. Figure 8.6 is related to figures 8.2, 8.4, and 8.5. The marginal benefit of abating a ton of $CO_2$ in figure 8.6 is the marginal social cost of a ton of $CO_2$ in figures 8.2, 8.4, and 8.5, with a twist: a ton abated is a negative ton produced, hence, since the marginal social cost of $CO_2$ increases, the marginal benefit of abatement decreases. Similarly, the marginal cost of abatement is the demand for emissions (i.e., the marginal value of a ton of $CO_2$).

Emissions of $CO_2$ are a stock pollutant: a marginal ton adds a small amount to the stock, hence the marginal benefit of abatement is close to constant. On the other hand, the marginal cost of abatement increases faster. Weitzman (1974) first proves that, if the

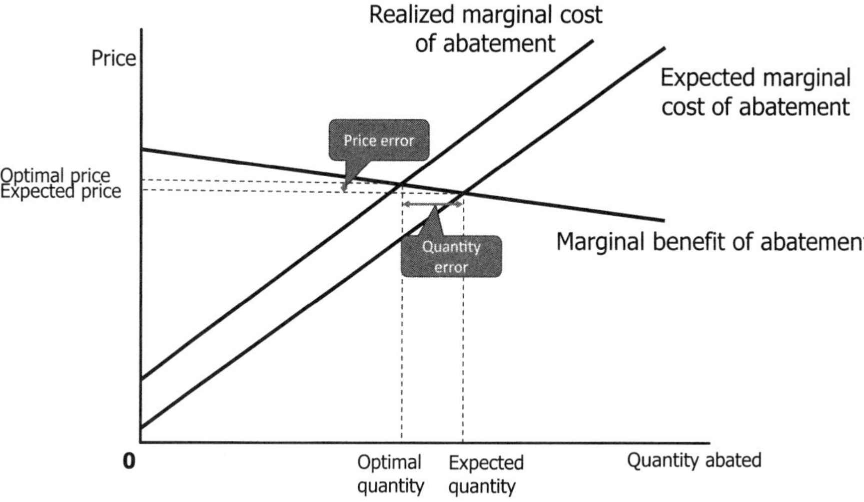


**Figure 8.6**
Marginal cost and marginal benefit of abatement, as a function of the cumulative abated quantity. Realized marginal cost is higher than expected, hence ex post optimal abated emissions lower than expected.

uncertainties on marginal cost and marginal benefits of abatement are not correlated, only the former matters.

The optimal abatement quantity is such that the marginal benefit of the last unit abated equals its marginal cost. Suppose that the marginal cost of abatement turns out to be higher than expected. The ex post optimal abated quantity is lower than the ex ante expected quantity. Because the slope of the marginal benefit curve is lower than the slope of the marginal cost curve, the quantity error is larger than the price error, as illustrated in figure 8.6. This implies that the net surplus loss is higher when setting a wrong quantity than when setting a wrong price, as illustrated in figure 8.7.

**Political economy arguments** A global and uniform tax is extremely simple, and the analysis presented above suggests it is more robust to uncertainty. National carbon taxes have been implemented in many countries. On the other hand, emissions markets are complex to set up, require heavy monitoring (as all markets do), and produce an unpredictable price. For these reasons, a carbon tax is embraced by many stakeholders as a practical solution. However, this simplicity is deceptive.

To be effective, a $CO_2$ tax has to be global. A significant national $CO_2$ tax would mostly penalize local producers and consumers; hence it is unlikely to be implemented (except in Scandinavia). However, taxes have always been at the core of a government authority. The ability to levy a tax in exchange for protection was the contract linking peasants to their feudal lord, and it is today the social contract linking citizens to their elected governments. It is therefore highly unlikely that governments willingly give up that essential prerogative

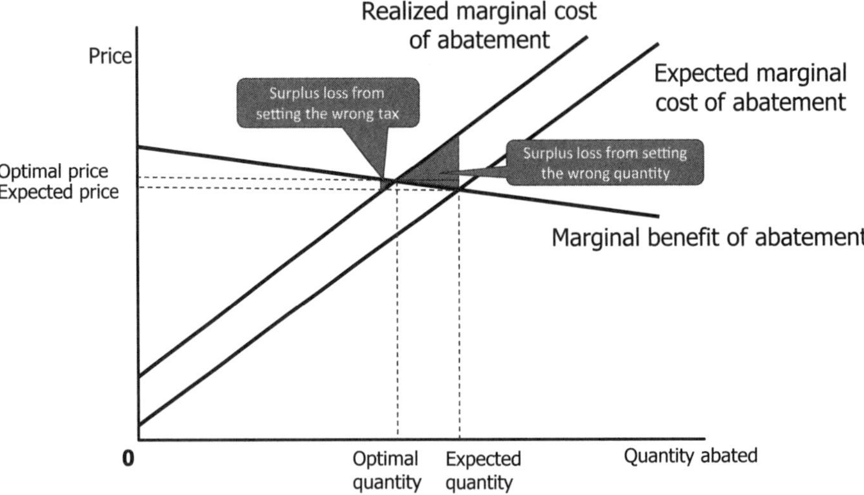


**Figure 8.7**
Net surplus loss from setting the wrong price and the wrong quantity.

to a supranational authority. Even the European Union, which has many responsibilities in Europe, does not have fiscal authority, which stays firmly with member states.

I therefore believe that political economy arguments favor the market for emissions solution and will take precedence over the economic arguments although convergence of national taxes may occur. Economists are split on the issue (Gollier and Tirole 2015; Tirole 2017), although I suspect that most economists started with a tax and are slowly migrating toward markets.

In the emission tax versus allowance market discussion, I give precedence to political economy arguments, while when advocating for peak-load pricing, I give precedence to microeconomic arguments. This apparent inconsistency can be explained by a simple reason: demand response can make peak-load pricing politically acceptable, while I believe a supranational tax is far from ever being politically acceptable.

#### 8.3.1.3 The European Union Emissions Trading System

The European Union Emissions Trading System (EU ETS) is the oldest $CO_2$ market and the most ambitious to date. The price of a ton of $CO_2$ on the EU ETS market hovered around €5 for most of 2017 and rose to around €8 as of January 2018. As a comparison, the penalty for noncompliance on the EU ETS market is €100/ton, most articles put the social cost of carbon around $37/ton, and the carbon price required to switch from coal to gas is €30/ton. Thus all observers agree that the EU ETS produces a $CO_2$ price too low to matter.

All policy makers and most economists agree that the market has thus failed and needs to be reformed. I have a different perspective on this point. I expand on it because I believe

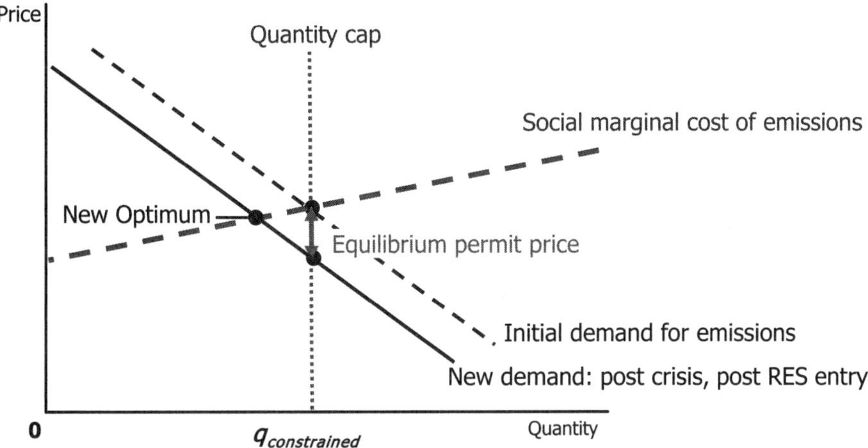


**Figure 8.8**
Equilibrium $CO_2$ emissions post economic crisis and RES entry.

it is a perfect illustration of the main theme of this book: imperfect regulation contributes to imperfect markets.

The EU ETS price is the correct opportunity cost of $CO_2$, given the rules in place, in particular, the quantity of permits that were issued. To my best knowledge, there is no evidence of market manipulation by participants. There was an early issue of value-added tax fraud, but it has been resolved. Thus the problem is not with the operation of the market per se but with the determination of the quantity of permits. This distinction may appear theoretical, but it is essential if we are to devise sound public policies.

The equilibrium price in the EU ETS market is much lower than our estimate of the social cost of $CO_2$ because supply of permits exceeds demand, which has shifted to the left for two reasons: first, the quantity of permits issued did not anticipate the permanent reduction in emissions caused by the 2008 financial, then economic, crisis. Second, as they enter the market, RES substitute for $CO_2$-emitting electricity generation technology, hence they reduce the demand for $CO_2$ permits. In addition, the initial allocation was overly generous, that is, too many permits were issued. This is illustrated in figure 8.8.

Since demand has shifted to the left and the quantity cap has not been adjusted, the equilibrium permit price is much lower than the cost of the externality. This is a perfect illustration of Martin Weitzman's (1994) result presented above. Furthermore, equilibrium emissions are by construction equal to the quantity cap and higher than the new optimum.

How should we solve this problem? First, most observers believe that the demand reduction is permanent and not transitory. Second, policy makers use conflicting instruments: by subsidizing RES entry, they reduce the demand for $CO_2$ permits, hence ceteris paribus, reducing the equilibrium permit price, and thereby reducing the impact of the $CO_2$ market on investment decisions. Policy makers should therefore reduce the quantity of $CO_2$ permits by the impact of RES entry and the economic crisis.

The European solution to reduce $CO_2$ permits is to create a permit stabilization fund to adjust the quantity of $CO_2$ permits so that the equilibrium price of $CO_2$ permits remains within a preagreed band. This is a practical solution to a politically complex (although economically simple) issue. Policy makers have a target price for a ton $CO_2$. The direct solution would be to set a European tax at this level. However, it has proven so far outside the realm of the politically feasible. Furthermore, is inconsistent with the official policy objectives, which are set in terms of reduction in $CO_2$ emissions, not in terms of $CO_2$ permits price. Thus policy makers adjust the rules of the $CO_2$ permits market to achieve their policy objective.

This approach has produced results: as of October 2018, the price of a ton of $CO_2$ stands slightly above €20. However, it is not consistent with the policy objectives as officially stated, which is confusing. Furthermore, its implementation is fraught with perils: as we have seen in chapter 3, increasing the complexity of market rules increases the potential for manipulations, either when the rules are designed or when they are implemented.

For these reasons, I believe a better approach would have been to adopt a more ambitious target for the speed of $CO_2$ reduction, which is (a) possible, since science tells us (with the usual caveats) that $CO_2$ emissions should be equal to zero by 2050, and (b) less costly for the economy, since demand for $CO_2$ permits has been permanently reduced.

### 8.3.2 Renewable Energy Sources Support Mechanisms

While elected officials have resisted increasing the price of $CO_2$ in the economy, they have enthusiastically supported subsidizing renewable energy sources. There is sound economic justification for a public subsidy to a new technology, but the design of the RES support mechanisms in the 2000s has been uneconomical, and has led to the creation and capture of a large economic rent.

#### 8.3.2.1 Renewable support approaches

Three main approaches have been implemented to supporting renewable generation. One is to require the local grid company (or some other agency) to purchase all the output from a renewable generator at a fixed price and to sell it on to retailers and ultimately to consumers. The cost of these purchases is added to consumers' bills.

The fixed price can be set administratively or be market based. The first situation is usually called a feed-in tariff (FiT), and it has been the primary support mechanism used in Europe. A FiT gives price security to the generator but has often created an open-ended promise to pay this price to any generator that meets the eligibility conditions. A more market-based approach is to have the fixed price arise from a competitive outcome. For example, the UK auctions off contracts for differences (CfDs) for RES, which pay the difference between the fixed price resulting from the auction and the wholesale spot price.

Thus the contract payment plus the wholesale spot price together give a predictable income stream, equivalent to an FiT. However, RES compete for these CfDs.

A second approach requires the generator to sell its power in the wholesale market (or via a contract) but tops up this revenue with additional payments. In Europe, this is typically in the form of a fixed payment derived from electricity consumers, a feed-in premium (FiP) and many countries, such as Spain and Germany, already use both FiTs and FiPs. In its renewable support guidelines, the European Commission is recommending a move towards FiP (European Commission 2014). In the United States, a renewable tax credit gives a rebate on corporate taxes for each MWh generated and comes at the expense of taxpayers.

The third approach concentrates on the quantity of renewable power to be procured, rather than its price. The U.S. Renewable Portfolio Standards requires retailers to procure output from renewable generators but need not lay down any conditions on how this is done. In Europe, tradable green certificate schemes require retailers to acquire certificates equal to a set proportion of their sales or to pay a buy-out fee. Generators are given certificates for each unit of renewable power they produce, which they can sell to retailers; they also have to sell their power in the wholesale market or through long-term contracts.

#### 8.3.2.2 Economic justification for subsidies

**Learning externalities** The most common argument for supporting RES in the presence of a carbon price arises from the magnitude of the learning curve associated with the development of the renewable technologies. The first units deployed cost more than the conventional technology, but as more renewable generators use new technologies, learning by doing and the chance of obtaining economies of scale means that future units will cost less.

The learning curve effect is real, as illustrated in figure 8.1. It is a valid argument, which can be rephrased in terms of externality. Developing the first unit generates know-how that can be used by other producers. For example, firm B learns by observing firm A's new design or improved installation procedure. There lies the externality. The first producer cannot appropriate all the know-how, hence his incentives to invest are lower than socially optimal. This provides an economic justification for a subsidy. As illustrated in figure 8.9, the subsidy to a marginal Megawatt of RES capacity is the difference, when positive, between the cost of installing the asset and the market value of the energy produced by this asset, which is decreasing almost linearly as RES capacity increases, hence is represented by a straight line. The marginal subsidy is no longer required when the cost of installing of Megawatt of RES capacity falls below the market value of the energy produced by the asset.

The cumulative subsidy, that is, the subsidy to all installed RES capacity eligible for a subsidy, is the surface between the cost curve of RES capacity and the market value of energy given installed RES capacity.

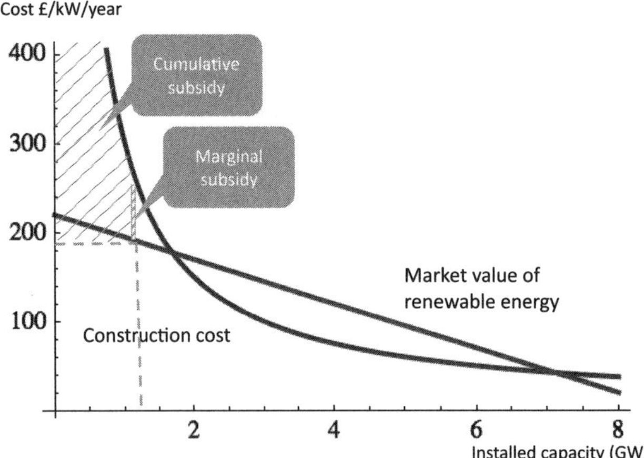


**Figure 8.9**
Marginal and cumulative RES subsidies.

**Risk mitigation** Other arguments have been used, with less economic might. The first one is that developing a new technology is risky, hence government support reduces the risk, hence the cost of capital, hence the cost of developing the new technology. I am personally not comfortable with this argument. Investors routinely back risky projects (in 2016, investors flocked to buy shares in Tesla, that is, backed electric cars and Gigafactories for batteries). Financial markets—like all markets—are imperfect, hence investors may not correctly price the risk of new technologies, although this is not immediately obvious.

What is obvious, on the other hand, is that this argument opens the door to rent seeking. Companies invest significantly in convincing governments that their project or their technology needs to be subsidized, maybe more than in developing the technology itself.

**Industrial policy** Proponents of industrial policy argue that supporting RES can create jobs in manufacturing wind turbines or solar panels. This is most likely to be a good policy where a strong exporting industry can be established, as with the Danish wind turbine company Vestas. Making RES purely for the domestic market will create jobs in that sector, but the resulting higher price of power risks destroying jobs in energy-using sectors, and it is unclear whether the net gain is positive.

Germany provides a clear example of industrial policy that misfired: one of the assumption underlying the Energiewende was that German industrial firms would lead the world in new electric power technologies in the twenty-first century as they had in the emerging electric power technologies at the beginning of the twentieth century. This assumption proved wrong. In 2016 Chinese firms were market leaders in manufacturing solar panels.

**Energy security** Another argument for supporting RES is that increased use of domestically generated renewable power can reduce the amount of fuel that must be imported from abroad, with benefits for energy security. This is true, but the availability of many RES depends on the weather, which creates a generation adequacy risk (as discussed in chapter 2), unless adequate backup is available.

**Concluding thoughts** The existence of an economic justification is not sufficient reason to subsidize an industry. Learning externalities are present in almost every industry and do not automatically lead to subsidies. Economists argue that R&D should be subsidized, not production. Subsidizing RES is therefore a political decision, not simply an economic one.

#### 8.3.2.3 The design of these subsidies has been uneconomic

Support mechanisms for RES were poorly designed. In fact, I believe they are the textbook example of a public policy designed without a careful and thoughtful impact analysis.

In Europe, the main mechanism for RES support was the feed-in tariff: RES were guaranteed a fixed price, computed by a government agency, for every Megawatt-hour they produced. They were also granted physical dispatch priority, that is, the system operator was forced to dispatch power produced by RES. This led to negative wholesale prices in some hours. Furthermore, in many instances, RES were also not responsible for the difference between their day-ahead predictions and their actual production.

These provisions (detailed below) were clearly favorable to RES. As a result, we have observed massive RES entry, much higher and much faster than was anticipated: between 2004 and 2016, the total RES investment in Europe was $890 billion, with a peak in 2011 at $120 billion (Frankfurt school-UNEP 2017). At the same time, the amount of subsidies ballooned: in 2016 a German retail customer paid around €30/MWh for the wholesale cost of energy and around €60/MWh for the RES subsidy. Since Germany has granted the largest subsidies to RES over the past decade, we should all be grateful to German rate payers for their massive contribution to the reduction in RES costs.

German consumers have so far been less sanguine about this subsidy than as one might expect. Two explanations are possible, not mutually exclusive. First, customers do not know why their bill increases. Second, many customers are also investors in local RES, hence are also on the receiving end of the subsidies.

In the words of a senior French civil servant. "We did not realize that leaving significant rents on the table would attract such entry." Bureaucrats did not realize that markets were able to spot and capture risk-free profits.

**Administrative determination of prices** The fixed price was determined by the government, not by competition. To compute the fixed price, government officials estimated the cost of RES using publicly available data and conducting interviews with manufacturers and developers. It is reasonable to assume that all sources, well aware of the purpose of the exercise, overestimated their costs.

Furthermore, the interval between changes in feed-in tariffs was too long, for the usual reasons. First, administrative processes are by construction slow. Bureaucracies are not designed for speed. Second, once a favorable rate has been set, stakeholders lobby hard for it not to be rapidly reset. Thus new plants with cheaper costs benefit from the old rate, computed for older plants with higher costs.

It is hard to put an exact figure on how much rent has been created, but one can make a reasonable guess. In April 2014, the French Energy Regulatory Commission (CRE) issued a report on the profitability of RES in mainland France. It found that the average return on capital invested for solar in calls for tenders is 2 percent below the return when the price is set administratively. Similarly, the return on capital for onshore wind energy seems to be an average of 2 percent higher than a just and reasonable rate of return on capital. Assuming that this extra return on invested capital represents an unwarranted economic rent, and applying to capital invested between 2004 and 2014 (i.e., assuming that new RES support approaches discussed below eliminate the rent from 2015 onward), it totaled $82 billion from 2004 to 2014. This approximate calculation suggests European taxpayers may have transferred somewhere between $50 billion and $100 billion of unjustified economic rent to the RES industry during that period.

**Physical dispatch priority** Physical dispatch priority was given to RES. This provided them with assurance that the Megawatt-hours they produce will actually be sold. In other words, it protects them against SOs who could be less than enthusiastic at the thought of dispatching intermittent (in SO parlance) generation sources. While this may sound reasonable, it creates another problem: negative prices.

This situation may appear to be a logical puzzle from "Alice in Wonderland," and it deserves a bit of explanation. All generation units have start-up and shut-down costs. For some technologies, for example, coal and nuclear, these are very high. For this reason (and their cost structure, as discussed in chapter 2), their owners aim to operate them as baseload plants, that is, have them running uninterrupted. When market price falls below the variable cost of production for a few hours, they may prefer to run and earn a negative operating margin rather than pay the shut-down and start-up costs. For example, in the late 1990s in the Midwest of the United States, coal-fired plants, the marginal cost of which was around $14–16/MWh, were willing to sell at $10/MWh overnight rather than shut down at night and restart the following morning.

The advent of RES has made the gap even more important. Since RES are given physical dispatch priority, and since their revenue is independent of spot-market price, they produce at their maximum capacity. For some hours, demand net of RES output exceeds the capacity of the baseload technology, hence the price falls below the marginal cost of the baseload technology. If RES output is small, that effect is negligible. If RES output is large, that effect can no longer be neglected, and price may have to become negative to balance supply and demand. In other words, producers pay to produce and consumers are paid to consume.

Negative prices are a problem: producing a Megawatt-hour the price of which is negative reduces net surplus. In other words, not producing this Megawatt-hour would increase net surplus. Policy makers should strive to avoid negative prices.

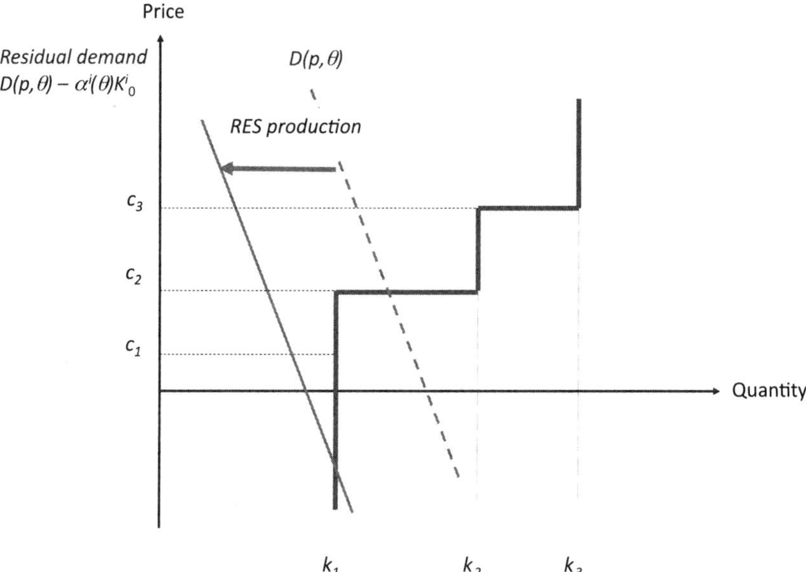


**Figure 8.10**
Negative prices occur when residual demand exceeds inflexible generation capacity.

This is illustrated in figure 8.10. The supply curve is different from that in figure 2.11: since the baseload technology is inflexible, it can never produce less than installed capacity, denoted $k_1$. The totality of RES potential production is actually accepted by the SO: RES producers sell as much as they can produce, since they are indifferent to the actual wholesale spot price; and the SO must give physical priority to RES production. Total demand in state $\theta$ is $D(p, \theta)$, and residual demand, which is total demand minus the entire RES production, is $D(p, \theta) - \alpha^i(\theta)K_0^i$. The notation will be discussed later in this chapter. Since residual demand intersects with the supply curve at a negative price, the wholesale spot-market price is negative.

This is not a theoretical possibility. In Texas, negative prices are commonly observed. They also occur in Germany, and in France (albeit rarely).

#### 8.3.2.4 The future of RES support

While the cost of developing RES is decreasing fast, and faster than was anticipated, support mechanisms are likely to be required for the next few years. How should this support be structured to minimize its impact on the public purse?

**Pricing $CO_2$** The first step of course is to put a price to $CO_2$. This recommendation, made by all economists, is worth repeating, at the risk of being tedious. The political economy of $CO_2$ pricing is challenging, to say the least. However, I believe a silver lining may be on the horizon. Firms in different industries realize that some form of $CO_2$ legislation will be enacted in the next ten to twenty years. The only thing corporations dislike more than a cost increase is uncertainty. Most corporate chieftains prefer a carbon price to piecemeal and ever-changing rules and regulations—even though some may build a business on exploiting these rules. During the COP 21 held in Paris in December 2015, a group of firms led by the CEO of Engie, a large European energy firm, strongly argued for a $CO_2$ price.[2]

**Changes to existing RES support mechanisms** A few simple changes to these mechanisms will dramatically reduce their cost. Some of them are already implemented. First, as is currently the case in Great Britain, policy makers (or their agents) should use competitive tendering for their chosen RES volumes instead of FiTs. Investors in RES would still receive a guaranteed price for their output, but it would be determined by a competitive process instead of bureaucratic fiat. This will provide two benefits: first, the price will be set competitively, not administratively; second, the government can control the volume of entry, hence the amount of subsidies granted to RES, hence retail price increase.

One of the drawbacks of this approach, which explains the popularity of feed-in tariffs, is that responding to a tender requires commercial sophistication. The early phase was politically supported because communities and individuals could participate. It is now likely that the next phase of RES expansion will be undertaken by larger firms.

Economists would prefer the tenders to be technology neutral, to avoid governments picking winners. I am skeptical that policy makers will agree.

Second, RES should be given balancing responsibility, that is, made accountable for the difference between their day-ahead production plan and their actual output. This will lead them to invest in better prediction. Balancing responsibility will not be an issue for large RES producers. It is likely that RES aggregators will emerge to offer that service to small RES producers.

**Feed-in premium** The third change is more complex. As previously discussed, policy makers are moving from a guaranteed fixed price (determined administratively or competitively) to a feed-in premium (FiP), that is, for every Megawatt-hour they produce, RES producers receive the wholesale spot price plus a premium, usually determined competitively. The FiPs approach provides complete transparency on the subsidy, plus it transfers the market price risk to RES producers, which seems fair now that the technology has progressed.

2. Interestingly, even some climate skeptics recognize the merits of $CO_2$ pricing. In researching for this chapter, I found a highly entertaining and articulate climate skeptic blog. As a disclaimer, I do not endorse all of its economic analyses, but I found the argument for $CO_2$ pricing particularly well constructed: whether or not global warming is occurring with the magnitude we fear, a stable $CO_2$ price is preferable to subsidies targeted to political allies (Meyer 2016).

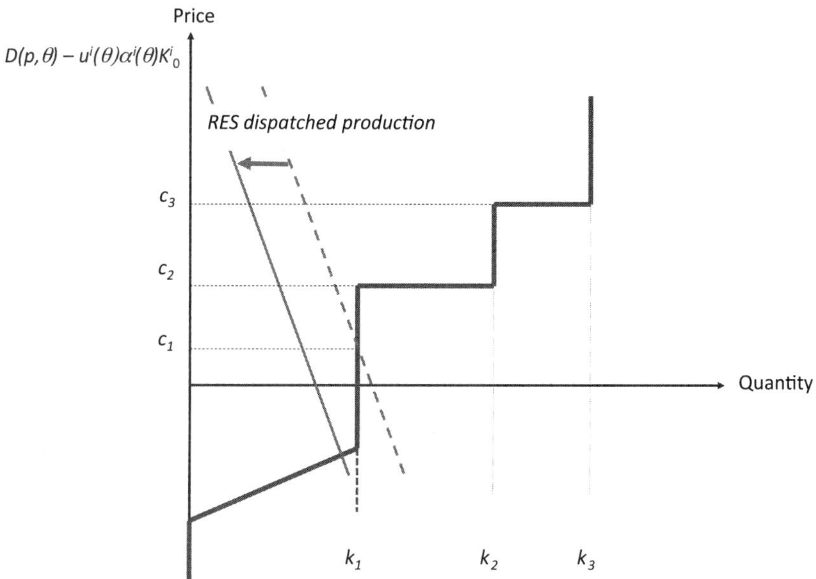


**Figure 8.11**
Supply curve with FiT.

In addition, if the market uses locational marginal prices for energy, discussed in chapter 6, FiPs provide information on which location is most economic: an RES producer locating in a constrained-off zone will require a higher FiP.

However, FiPs have drawbacks. First, to compute her revenues, an RES producer who bids for an FiP must estimate the wholesale spot price for duration of the FiP, typically twenty years. This requires some sophistication, which rules out smaller producers. Second, transferring the price risk to RES producers is fair but also favors large, well-funded, and diversified producers over small ones.

Finally, FiPs do not resolve completely the problem of negative prices. An RES producer who has received an FiP will produce and sell in the market as long as the wholesale spot price exceeds the opposite of the FiP. The supply curve will therefore be as presented in figure 8.11. RES producers will progressively reduce their output, as price become more and more negative. If price becomes smaller than the opposite of the largest FiP awarded, no RES producer will sell.

**Financial insurance** Should policy makers feel the provision of a guaranteed fixed price is required for RES, one solution to avoid—or at least seriously reduce the occurrence of—negative prices is to physically curtail RES when prices become negative while paying for the energy that could have been produced.

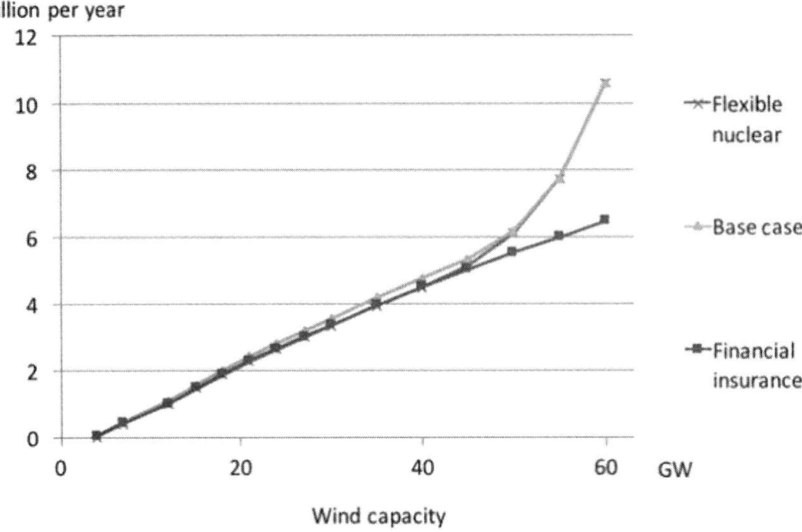


**Figure 8.12**
Evolution of net surplus loss as RES capacity increases for different cases.

Perfectly implementing this policy requires (a) the SO to be able to determine precisely the available production from each renewable facility, even if it is not actually producing, and, (b) the producers to be able to shut down their facilities remotely. These conditions appear more likely to be met for large wind farms than for individual solar panels.

With my colleague Richard Green, who teaches at Imperial College in London, we have examined the impact of increasing RES capacity on the power market in Great Britain (Green and Léautier 2017). The analysis is detailed later in this chapter. In particular, we have computed the loss of net surplus caused by subsidizing RES. The findings are presented in figure 8.12. The installed RES capacity (composed of onshore and offshore wind turbines) is presented on the horizontal axis, the net surplus loss on the vertical axis. For example, for 30 GW of renewables, the cumulative loss in net surplus is £3.4 billion per year. For low-RES capacity, financial insurance and physical dispatch priority generate the same surplus loss, however, financial insurance performs better for high-RES capacity.

Financial dispatch insurance does not modify expected revenues for RES producers, nor does it expose them to additional risk, hence it does not increase their cost of capital compared with physical dispatch insurance. If negative prices occur frequently, financial insurance can be substituted to physical dispatch insurance for installed RES assets.

### 8.3.3 Impact of RES Support Mechanisms

#### 8.3.3.1 Impact on electricity prices

**Impact on average wholesale price** Over the past few years, the rapid rise in renewable capacity in Europe has depressed market prices. Since the marginal cost of wind (or solar)

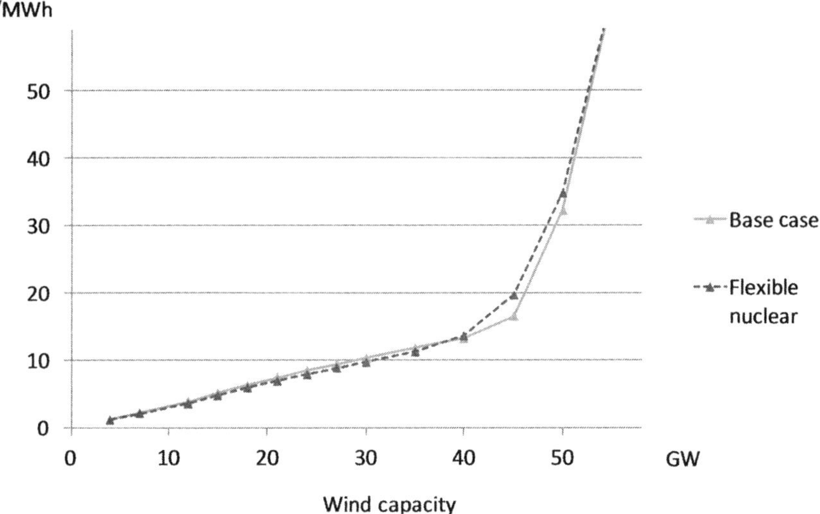


**Figure 8.13**
Evolution of the unit tax levied on all Megawatt-hours to finance the cumulative subsidy as RES capacity increases.

generation is effectively zero, when RES produce they push the industry's supply curve to the right, so that this intersects with demand at a lower equilibrium price, a feature named the "merit-order" effect.

The merit-order effect is a disequilibrium phenomenon, however, because the lower prices mean that some (or all) conventional stations will be unable to recover their full economic costs. This may lead to retirements, or at the very least a shortage of new investment, and so the industry's conventional capacity should fall over time. The time-weighted average price of power in a long-run equilibrium which will tend to the average cost of a baseload station does not depend on the amount of renewable capacity. The demand-weighted average price might change. If there is more renewable output, on average, at times of high demand (e.g., solar power in a system with summer-peaking demands), then the demand-weighted price will be reduced as renewable capacity grows.

**Impact on retail prices** While the wholesale spot price decreases in the short run, as new zero-variable-cost RES enter the market, the retail price increases, owing to the subsidy. The academic analysis of RES penetration in Great Britain previously discussed also estimates the evolution of the unit levy required to cover the subsidy, presented in figure 8.13. The levy increases almost linearly up to £13/MWh when 40 Gigawatts of RES capacity is installed. Then, as more RES capacity is installed, nuclear is pushed out of the market, and the levy increases much more steeply.

The levy is much lower than the current levy of €60/MWh retail customers pay in Germany for two main reasons. First, RES technology is more mature, hence cheaper.

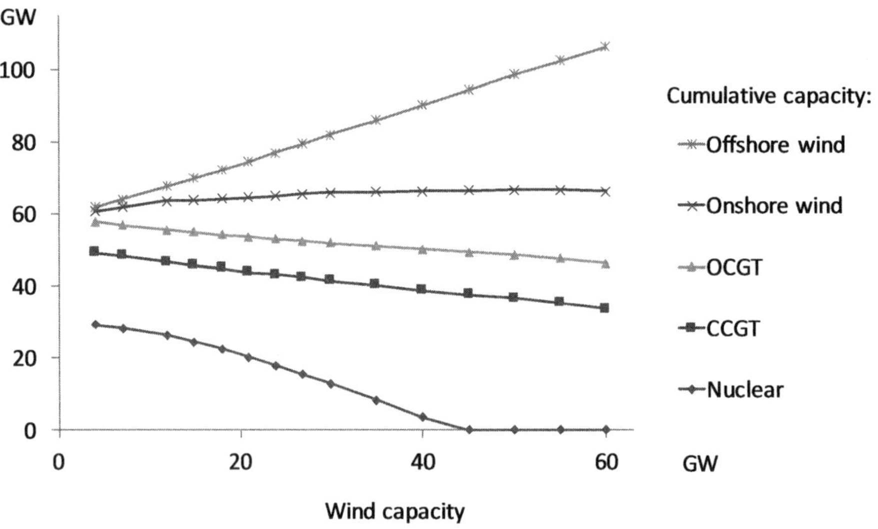


**Figure 8.14**
Evolution of the generation mix as RES capacity increases.
*Source*: Green and Léautier (2017).

The analysis uses today's wind turbines, which are much cheaper than the solar panels built over the past fifteen years in Germany. Second, the analysis assumes procurement is optimal, that is, the subsidy covers the true cost of installed RES capacity and leaves no rent to developers. By contrast, feed-in tariffs were used in Germany, which left significant rents on the table.

### 8.3.3.2 Impact on incumbent generation

When RES production capacity increases, residual demand, that is, total demand net of renewable production, decreases. In the long term, output of incumbent technologies, mainly thermal and nuclear power, is therefore reduced, as is their installed capacity. This is, of course, the goal of the energy transition: more renewable production, less conventional production, more wind farms and fewer coal plants.

Most observers and politicians agree that a decarbonized power sector will include both renewable and nuclear energy. Some would prefer to see a higher proportion of renewables, while others lean in favor of nuclear production, but the overall consensus seems to be that both will coexist by 2050–2100. As the following discussion illustrates, reality is more complex.

**The case of Great Britain** Figure 8.14 presents the evolution of the generation mix as RES capacity increases, extracted from the analysis of the British system Richard Green and I conducted. Wind capacity is presented on the horizontal axis, cumulative capacities up to different technologies are presented on the vertical axis. For example, nuclear capacity is represented by the surface below the lowest line, while combined-cycle gas turbine capacity is represented by the difference between the lowest line and the next one. Total cumulative capacity (the upper line) increases by almost 50 GW, from 60 to 110 GW, since RES have a utilization factor around 30 percent.

The starting point is the long-term equilibrium assuming a carbon tax set at £70 per ton, which is the British government target. For this reason, coal does not appear in the generation mix, which includes nuclear, and combined-cycle and open cycle gas turbines. With 30 GW of wind capacity, corresponding to 25 percent of electricity being produced from renewables, the level implied by the more ambitious UK targets for the early 2020s, the capacity of profitable nuclear stations would decrease to around 13 GW. The capacity of CCGT stations would increase to 29 GW, while that of OCGT plant would increase to 10 GW. Nuclear power disappears from the economically efficient generation mix when production from onshore and offshore wind power reaches 45 percent of demand.

Achieving a nuclear and RES generation mix may be possible. However, it would be very costly, as British citizens would have to subsidize both generation sources. In the not-too-distant future, policy makers will have to acknowledge this reality.

The substitution of natural gas for nuclear production raises the question of the evolution of $CO_2$ emissions. Since we assume a carbon price of £70 per ton, even with very low levels of RES output we obtain emissions of 38 million tons of $CO_2$, compared with actual emissions of 147.9 million tons of $CO_2$ equivalent in 2013 (Department of Energy and Climate Change 2015). Adding RES to this low-carbon system has the counter intuitive effect of increasing emissions at first, because they displace more nuclear output than they generate themselves. By the time nuclear stations are fully crowded out of the market, with 45 GW of RES, emissions have risen to 76 million tons of $CO_2$. Beyond this point, however, the wind output is entirely crowding out gas-fired generation, and with our maximum deployment of 60 GW, emissions have fallen back to 60 million tons of $CO_2$.

**Underlying mechanism** A simplified example illustrates the impact of RES entry on incumbent generation. Formal derivations are presented later in this chapter. Suppose the year divided into two periods of 4,000 hours each. One period corresponds to on-peak consumption, during which the hourly demand is 90 Gigawatts. The other period is off-peak hours, with an hourly demand of 30 Gigawatts. Total consumption is therefore (30 + 90) x 4,000 = 480,000 Gigawatt-hours, or 480 Terawatt-hours. Two conventional production technologies are required: a baseload technology that produces during every hour of the year (e.g., coal, or nuclear if the price of carbon is very high in the power industry) and a peaking technology that produces only during peak hours. It therefore takes 30 Gigawatts of baseload capacity to satisfy the off-peak demand and 60 Gigawatts (90 – 30) of peaking technology to satisfy on-peak demand. This situation is illustrated in figure 8.15.

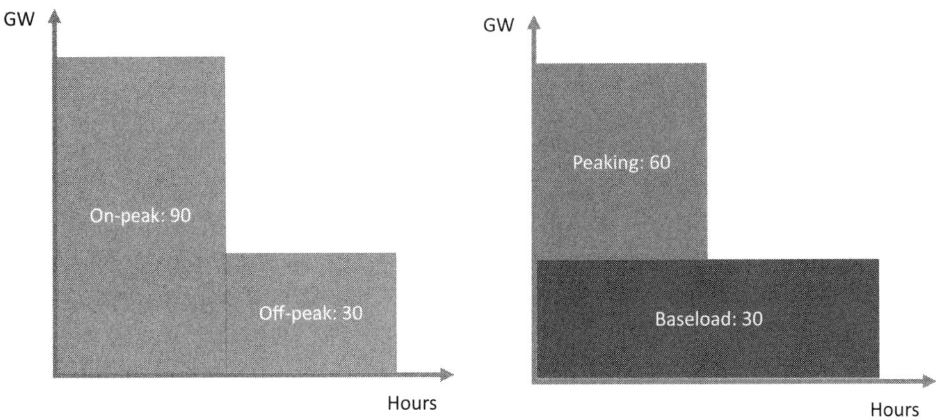


**Figure 8.15**
Two-period market (left panel): on-peak demand is 90 GW, off-peak demand is 30 GW. Optimal generation mix (right panel): baseload generation capacity serves the entire off-peak demand (30 GW), peaking technology meets remaining on-peak demand (60 GW).

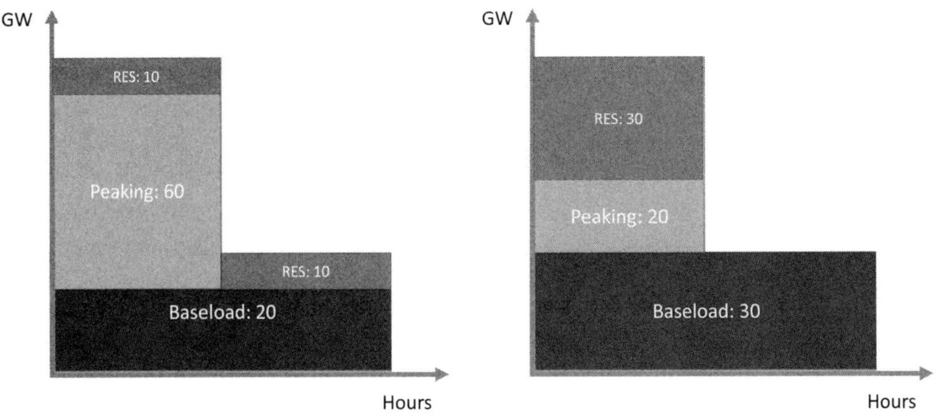


**Figure 8.16**
RES entry if (a) RES output is not correlated with demand (left panel) and (b) RES produces on-peak only (right panel).

We now introduce RES. Assume that RES production is constant during both periods. For example, the wind blows with the same intensity year-round, winter afternoons and summer mornings alike. This is not a bad representation of reality in Great Britain, which enjoys different wind regimes. In technical terms, RES production is not correlated with demand.

How do incumbent production facilities adapt if renewable production is 10 Gigawatts per hour? Residual demand is 20 Gigawatts $(30-10)$ during off-peak hours and

80 Gigawatts (90 – 10) during on-peak hours. Baseload capacity is therefore 20 Gigawatts, and peak capacity remains unchanged at 60 Gigawatts (80 – 20). This is illustrated on the left panel of figure 8.16.

If renewable production rises to 30 Gigawatts per hour, off-peak residual demand, and therefore baseload capacity, disappears. Residual demand during on-peak hours remains unchanged, however, at 60 Gigawatts (90 – 30). Thus RES replace the baseload technology.

What production levels do renewable sources have to reach for this change to occur? Renewables would have to produce 30 Gigawatts per hour: $30 \times 8,000 = 240,000$ Gigawatt-hours, or 240 Terawatt-hours. In this simplified example, RES replace the baseload technology when they reach 50 percent of demand.

Renewable energy sources are not always incompatible with the baseload technology. Assume now that RES produce only during peak hours. This is (almost) the case of photovoltaic panels in areas where air-conditioning accounts for a significant percentage of peak demand, such as the southwestern United States: when the sun shines, the panels produce and the temperature rises, leading to high demand for air-conditioning. In the previous example, if RES production is 30 Gigawatts at peak hours and nil at off-peak hours, baseload capacity will be unchanged. However, on-peak capacity will be reduced to 30 Gigawatts (90 – 30 – 30). RES replace the peaking technology, since they produce only during on-peak hours. This is illustrated on the right panel of figure 8.16.

A similar example illustrates that, if RES produce only off-peak, then baseload capacity is reduced while peaking capacity increases.

The analysis presented above shows that RES substitute for incumbent technologies according to the correlation between RES production and demand, not policy makers directives.

#### 8.3.3.3 Impact on existing RES

**General discussion** As RES-installed capacity increases, in the long-term equilibrium, the value of a marginal RES and the value of all installed RES may decrease.

An example illustrates this result. Consider one wind turbine installed in a market. To simplify, consider the turbine produces equally at all hours. The value of the energy it produces, per Megawatt of installed capacity, is therefore the average hourly wholesale price times 8,760 hours.

Suppose now a second turbine is installed right next to the first one. Both turbines produce exactly at the same hours. During these hours, output slightly increases, while demand remains unchanged. Wholesale spot price therefore decreases (possibly slightly); hence the value of the second turbine is lower than the value of the first turbine when it was alone. Furthermore, the value of the first turbine is now lower.

This example assumes that they turbines are colocated, hence produce at exactly the same hours. This is not always the case if the wind turbines are exposed to different wind regimes, or if a solar panel is introduced after a wind turbine. Derivations presented later

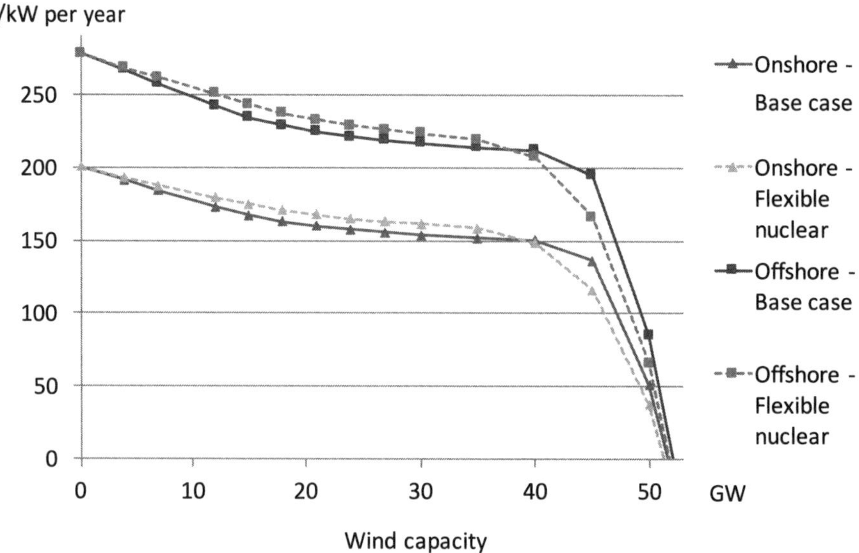


**Figure 8.17**
Value of the marginal RES unit as RES capacity increases.

in this chapter prove that the short-term price impact is proportional to the expectation of the product of the availabilities of both RES.

What about the long-term impact? In the long term, incumbent-generation capacity will be retired as a result of RES entry. Ceteris paribus, this capacity reduction increases the price. Derivations presented later in this chapter prove that the capacity reduction is proportional to the expected availability of the second turbine. The value of the first RES is thus increased by the expected availability of the first RES times the expected availability of the second RES.

Combining both terms, the change in value of the first RES is the expectation of the product of the availabilities minus the product of the expected availability. Readers who remember their probability course will have recognized the covariance between availabilities.

If both RES availabilities are positively correlated, entry by the second reduces the value of the first in the long-term equilibrium. In some instances, the correlation could be negative: for example, the sun shines during the day while the wind blows at night. In that case, the value of the first RES in the long-term equilibrium is increased by the entry of the second RES, through the impact on incumbent-generation capacity.

**The example of Great Britain** As shown in figure 8.17, the market value of the first increment of onshore wind capacity is £200/kW per year, decreasing to £150/kW per year for 30 GW renewable capacity, and £280/kW per year decreasing to £220/kW per year for offshore wind capacity.

This decrease is almost linear. The average slope is proportional to the average variance-covariance matrix of the availabilities on the vertical segments of the supply curve:

$$\begin{bmatrix} 1.6 & 1.5 \\ 1.5 & 3.2 \end{bmatrix}.$$

Thus a 1 GW increase in onshore wind capacity decreases the marginal value of onshore wind by £1.6/kW per year, and the marginal value of offshore wind by £1.5/kW per year, while a similar increase in offshore wind capacity decreases the marginal value of offshore wind by £3.2/kW per year, and the marginal value of onshore wind by £1.5 /kW per year.

The marginal value decreases sharply after 40 Gigawatts of RES is installed, when nuclear generation disappears from the mix.

### 8.3.4 Two RES Paradoxes

Over the years, participants in seminars I have given on this topic have raised apparent paradoxes arising from increased RES penetration. Two are discussed below.

**The paradox of RES producing off-peak** Observers sometimes worry about the following situation. Consider the previous example and suppose RES produce only during the off-peak hours, that is, do not produce on-peak. Suppose 30 Gigawatts of RES production capacity is installed (i.e., capacity that produces 30 Gigawatts every off-peak hour). Suppose also the total cost of RES is lower than the variable production cost of both incumbent technologies.

The paradox is as follows: off-peak, the price is zero since output from RES is sufficient to serve the entire market. On-peak, price is set by the incumbent technologies, but RES do not capture this price since they do not produce on-peak. Thus RES are not profitable, even though they are a cheaper technology. How can that be?

The answer is to be found in the microeconomics of power market. Since RES capacity is precisely equal to off-peak demand, the value of the marginal Megawatt of RES is precisely equal to zero, thus, the value of all RES capacity is also equal to zero. If only 29 Gigawatts of RES capacity was installed, off-peak price would be the variable cost of the incumbent technologies, which exceeds the fixed cost of RES by assumption. Thus RES would be profitable.

This example is highly simplified, hence produces a simple and striking result: the value of RES is positive for any capacity strictly lower than 30 Gigawatts and equal to zero if RES capacity is equal to 30 Gigawatts. In reality, matters are more complex. The intuition, however, is identical: if too much RES capacity is installed, it is no longer economical. It is an application of a more general result: if too much capacity is installed, it does not cover its capacity cost.

**The paradox of all RES market** Observers also worry about an all-RES market. In that case, the price would always be equal to zero. How could RES cover their fixed cost without a subsidy?

The answer requires applying peak-load pricing to this market. As we have seen in chapter 2, if only one technology is present, price is set at the variable cost of production off-peak, and at the value of the marginal Megawatt-hour consumed on-peak. In this case, the variable cost of production is precisely equal to zero. RES cover their fixed capacity cost on-peak. If RES capacity is optimal, on-peak revenues cover exactly RES fixed capacity cost. If this does not occur, RES capacity is higher than optimal.

## 8.4 A Model of the Electric Power Market with RES

The impact of the penetration of RES on wholesale power markets has been studied extensively (see the excellent survey by Wiser et al. 2017). Most analyses rely on numerical simulation. The rest of this chapter presents the only analytical model of electricity markets with subsidized RES, that I am aware of. It follows closely Green and Léautier (2017). Since it builds on the model introduced in chapter 2 and used throughout this text, only differences are highlighted.

### 8.4.1 Demand

As in chapter 2, all customers are homogenous, and inverse demand $P(Q, \theta)$ satisfies Assumption 2. Different from chapter 2, a per-unit levy $\tau \geq 0$ is levied to cover the cost of subsidizing renewable generators. Individual demand in state of the world $\theta$ is thus $D(p + \tau, \theta)$, where $p$ is the wholesale electricity price.

The unit levy distorts consumption, hence reduces surplus. A similar analysis could be conducted if the subsidy was financed through general taxation. We would then introduce the shadow cost of public funds, which captures the distortion to the economy caused by taxes.

### 8.4.2 Supply

$(N + I)$ generation technologies are available: $N$ incumbent technologies indexed by $n \in [1, N]$, and $I$ RES corresponding to $n = 0$ and indexed by $i \in [1, I]$. For technology $n \geq 1$, $c_n$ is the variable cost per unit and $r_n$ is the hourly capacity cost per unit (i.e., annual capital cost plus fixed annual operating and maintenance costs expressed in £/MW/year divided by 8,760 hours per year). Both are assumed constant and expressed in £/MWh. Without loss of generality, incumbent generation technologies are ordered by increasing operating cost: $c_n > c_m \; \forall \; n \geq m$. Consistent with the rest of this text, $k_n$ is the installed capacity of incumbent technology $n$, and $K_n = \sum_{m=1}^{n} k_m$ is the cumulative installed capacity up to incumbent technology $n$.

As seen in chapter 2, there is a trade-off between capital and operating costs: if a technology has lower fixed cost, it then produces at higher variable cost, that is, $r_n < r_m \ \forall \ n \geq m$. We consider all available incumbent-generation technologies are present in the long-term equilibrium before RES are introduced.

All generation technologies have start-up cost and operating limits, which reduces their start-up and shut-down flexibility. As discussed in chapter 2, full representation of these start-up costs requires complex modeling. Green and Vasilakos (2011) propose a simplifying approach: inflexibility is included only for the baseload technology, which in practice is the most inflexible. It is represented by a minimum production level $m_1 k_1$. This minimum production constraint creates a virtual cost of shutting down, which plays an economic role similar to the real shut-down and start-up costs.

This chapter follows the same approach. The base case is no baseload flexibility, that is, $m_1 = 1$. This chapter also models the case of full flexibility, that is, $m_1 = 0$, with no change in costs, even though in reality full flexibility is costly. Since reality is somewhere between the two extremes considered, the results presented below provide upper and lower bounds of the real impact.

### 8.4.3 Renewable Energy Sources

We consider $I$ RES, for example, onshore wind, offshore wind, and photovoltaic panels. For $i = 1, \ldots, I$, $K_0^i$ is the installed capacity of RES $i$ and $\mathbf{K}_0 \in \mathbb{R}^I$ the vector of installed RES capacities $K_0^i$. $\mathbf{K}_0$ is determined by policy makers, not by a free-entry condition. Denote $r_0^i(K_0^i)$ the marginal capacity cost of RES $i$ when capacity $K_0^i$ is installed, $R_0^i(K_0^i) = \int_0^{K_0^i} r_0^i(x)dx$ the cumulative capacity cost of RES $i$ up to capacity $K_0^i$, and $R_0(\mathbf{K}_0) = \sum_{i=1}^{I} R_0^i(K_0^i)$ the aggregate cumulative capacity cost of RES capacities $\mathbf{K}_0$.

For all RES, variable operating cost is negligible, that is, $c_0^i = 0$. Learning by doing and economies of scale imply that $r_0^i(.)$ is decreasing in $K_0^i$: installing an additional offshore wind turbine reduces the cost of the next offshore wind turbine. As a first approximation, we assume that there are no cross-technology externalities from learning or economies of scale: installing an additional onshore wind turbine does not significantly reduce the cost of the next offshore wind turbine. Thus

$$R_0(\mathbf{K}_0) = \sum_{i=1}^{I} \int_0^{K_0^i} r_0^i(x)dx \Leftrightarrow \frac{\partial R_0}{\partial K_0^i} = r_0^i(K_0^i).$$

RES production is often variable. Availability of renewable technology $i$ in state $\theta$ is $\alpha^i(\theta) \in [0, 1]$, hence available production from RES $i$ in state $\theta$ is $\alpha^i(\theta) K_0^i$. This could reflect two dimensions of variability. First, RES can be predictably unavailable, for example, the sun does not shine at night.

Second, renewables may be unpredictably unavailable to an extent that requires the SO to procure additional flexibility. This aspect is not included in the model, and constitutes a fertile avenue for further research.

### 8.4.4 Wholesale Market Structure and Equilibrium

We suppose the market is centralized, that is, an SO receives bids from all producers and consumers, selects the optimal dispatch (defined later), and declares a unique market price. This constitutes an adequate description of U.S. markets. European markets are decentralized, hence buyers and sellers transact either in power exchanges or bilaterally, then communicate their negotiated transactions to the SO, who takes any actions needed to ensure a feasible and secure dispatch. Since this analysis is focussed on the impact of RES support policy, competition is assumed to be perfect, hence both market structures are equivalent. We also abstract from transmission constraints discussed in chapter 6.

In every state of the world, the SO assigns the dispatch rate $u_n(\theta) \in [0, 1]$ to each technology $n \geq 0$. Production from technology $n \geq 1$ is $u_n(\theta)k_n$, and production from RES $i$ is $u_0^i(\theta)\alpha^i(\theta)K_0^i$.

Inflexible technology 1 is slightly different, for it faces significant start-up costs. We first suppose that, to avoid shutting down, producer 1 is willing to accept a price lower than $c_1$ for a few hours. Therefore, $u_1(\theta) = 1$. This assumption is relaxed later.

RES receive physical dispatch insurance. They have dispatch priority, that is, $u_0^i(\theta) = 1$. This is the case in most jurisdictions, where the SO can cut off RES only when the operational security of the system is threatened, a situation we do not model.

The economic optimum would be for the SO to set the retail price (which determines consumption) equal to the marginal cost of power. The wholesale price would then be the retail price minus the tax, lower than the short-term marginal cost $c_n$ when technology $n$ is marginal. This is unrealistic. Producers are unwilling to participate in a market that guarantees price lower than $c_n$ when they are marginal. To account for the constraint $p(\theta) \geq c_n$ when producer $n$ is dispatched, the SO adds the unit tax $\tau$ to the opportunity cost of power.

Taking into account the constraint $u_1(\theta) = 1$ and the unit tax $\tau$, the SO dispatch program is

$$\max_{p(\mathbf{K}_0,\theta),\{u_n(\theta)\}_{n\geq 2}} \mathbb{E}[S(p(\theta)+\tau,\theta) - \textstyle\sum_{n=1}^N (c_n+\tau)u_n(\theta)k_n]$$
$$st: D(p(\theta)+\tau,\theta) \leq \textstyle\sum_{n=1}^N u_n(\theta)k_n + \sum_{j=1}^I \alpha^j(\theta)K_0^j \lambda(\theta),$$

where $p(\mathbf{K}_0, \theta)$ is the long-run equilibrium wholesale price in state $\theta$ when $\mathbf{K}_0$ has already been installed. The Lagrangian is

$$\mathcal{L}=\mathbb{E}\left[S(p(\theta)+\tau,\theta)-\sum_{n=1}^{N}(c_n+\tau)u_n(\theta)k_n\right.$$
$$\left.+\lambda(\theta)\left(\sum_{n=1}^{N}u_n(\theta)k_n+\sum_{j=1}^{I}\alpha^j(\theta)K_0^j-D(p(\theta)+\tau,\theta)\right)\right].$$

The first-order condition for price is

$$\frac{\partial\mathcal{L}}{\partial p(\mathbf{K}_0,\theta)}=(p(\mathbf{K}_0,\theta)+\tau-\lambda(\theta))\frac{\partial D(p(\mathbf{K}_0,\theta)+\tau,\theta)}{\partial p}=0\Leftrightarrow p(\mathbf{K}_0,\theta)+\tau=\lambda(\theta):$$

price paid by consumers is equal to the opportunity cost of power.

For $n\geq 2$, the first-order derivative with respect to dispatch rate is

$$\frac{\partial\mathcal{L}}{\partial u_n}=(\lambda(\theta)-(c_n+\tau))k_n=(p(\mathbf{K}_0,\theta)-c_n)k_n,$$

which is consistent with the producers' participation constraint.

When technology $n\geq 2$ is marginal, the wholesale price is $p(\mathbf{K}_0,\theta)=c_n$ and $u_n(\theta)>0$ is determined to balance supply and demand:

$$K_{n-1}+u_n(\theta)k_n+\sum_{j=1}^{I}\alpha^j(\theta)K_0^j=D(c_n+\tau,\theta).$$

When technology $n\geq 1$ produces at capacity and technology $(n+1)$ does not produce, the wholesale price is determined by the intersection of the vertical supply curve and the demand curve:

$$K_n+\sum_{j=1}^{I}\alpha^j(\theta)K_0^j=D(p(\mathbf{K}_0,\theta)+\tau,\theta)\Leftrightarrow p(\mathbf{K}_0,\theta)=P(K_n+\sum_{j=1}^{I}\alpha^j(\theta)K_0^j,\theta)-\tau.$$

This includes states of the world for which the price is lower than $c_1$.

The supply curve is the "staircase" presented in chapter 2, shifted upward by the unit tax $\tau$: on the horizontal portions, the wholesale price is the marginal cost of the marginal technology producing; on the vertical portions, the marginal technology produces at capacity, and the wholesale price is set by the intersection of the demand curve and the vertical supply curve, minus the tax.

Recalling that $c_0=0$ and using the convention $c_{N+1}\to+\infty$, the steps of the staircase are formally defined for $0\leq n\leq N$ by $v_n=\{\theta:c_n<p(\mathbf{K}_0,\theta)<c_{n+1}\}$ and $h_n=\{\theta:p(\mathbf{K}_0,\theta)=c_n\}$. The sets $v_n$ and $h_n$ are functions of $\mathbf{K}_0$. To simplify the notation, the reference is omitted.

While technology 1 earns a negative operational margin when $p(\mathbf{K}_0, \theta) < c_1$, it still produces since it cannot reduce its output. Invested capacity in technology 1 is determined to precisely balance the positive and negative margins. When baseload generation is inflexible, $h_1 = \emptyset$ since price is never set at $c_1$. Similarly, when renewables receive physical dispatch insurance, $h_0 = \emptyset$ since price is never set at 0. This is illustrated in figure 8.10.

Finally, in a competitive equilibrium, investors in nonrenewable generators invest until their marginal profit is equal to zero, which yields

$$\mathbb{E}[(p(\mathbf{K}_0, \theta) - c_n)u_n(\theta)] = r_n, \quad for\ n \geq 1. \tag{8.1}$$

In the long-term equilibrium, the conventional generation mix optimally adapts to RES capacities $\mathbf{K}_0$ mandated by policy makers

### 8.4.5 Renewable Support Policy

The market value of energy produced by renewable technology $i$ is $\mathbb{E}[\alpha^i(\theta)p(\mathbf{K}_0, \theta)]$. As discussed in section 8.3 and illustrated in figure 8.9, in many cases, it is less than the generator's cost, and it would not be viable without some kind of state support. Consistent with practices presented in section 8.3, we assume that policy makers purchase all available RES production at a constant rate, which can be exactly adjusted as capacity is added and RES costs fall, so that the fixed price $f^i(K_0^i)$ for the marginal investor is the minimum amount required to precisely cover the marginal capital cost:

$$f^i(K_0^i)\mathbb{E}[\alpha^i(\theta)] = r_0^i(K_0^i) > \mathbb{E}[\alpha^i(\theta)p(\mathbf{K}_0, \theta)].$$

This approach constitutes a first-best benchmark. For example, the SO (or a government agency) runs a series of calls for tenders until cumulative capacity is $\mathbf{K}_0$ and competition among developers is perfect, hence the price is driven down to the cost. For installed renewable capacity $K_0^i$, the cumulated expected revenues from the fixed-price contracts cover exactly the cumulative capital cost

$$\int_0^{K_0^i} f^i(x)\mathbb{E}[\alpha^i(\theta)]dx = \int_0^{K_0^i} r_0^i(x)dx = R_0^i(K_0^i).$$

The subsidy required by a marginal unit of technology $i$ when $\mathbf{K}_0$ has been installed, denoted $\varphi^i(\mathbf{K}_0)$, is the difference, when positive, between the marginal cost and capacity and the marginal value of the energy produced:

$$\varphi^i(\mathbf{K}_0) = \max(r_0^i(K_0^i) - \mathbb{E}[\alpha^i(\theta)p(\mathbf{K}_0, \theta)], 0). \tag{8.2}$$

The marginal subsidy is the dark shaded rectangle in figure 8.9. Even with a carbon tax, the first MW of renewable production must be subsidized: $\varphi^i(0) > 0$ for all $i$. However, it is widely believed that $r_0^i(.)$ is decreasing sufficiently rapidly that $\varphi^i(.)$ is decreasing, and there exists $\bar{\mathbf{K}}_0 > 0$ such that $\varphi^i(\bar{\mathbf{K}}_0) = 0$, hence subsidies will no longer be required for all $K_0^i \geq \bar{K}_0^i$. Expression (8.2) illustrates that this common wisdom may not stand up to rigorous economic analysis. There may exist a $\bar{\mathbf{K}}_0 > 0$ such that $\varphi^i(\bar{\mathbf{K}}_0) = 0$. However, as $K_0^i$ increases, $p(\mathbf{K}_0, \theta)$ decreases (as proved below), and so $\varphi^i(\mathbf{K}_0)$ may become negative again. This situation is illustrated in figure 8.9.

In the remainder of this chapter, we assume that costs are such that all technologies must be subsidized, that is, $\varphi^i(\mathbf{K}_0) > 0$. This simplifies the notation without altering the economic insights. It is verified empirically on the examples we consider.

As illustrated in figure 8.9, for technology $i$, the cumulative subsidy up to $K_0^i$, denoted $\Phi^i(\mathbf{K}_0)$, is the difference between the fixed payments to producers and the revenues from sale of renewable energy:

$$\Phi^i(\mathbf{K}_0) = R_0^i(K_0^i) - \mathbb{E}[\alpha^i(\theta)p(\mathbf{K}_0, \theta)]K_0^i.$$

This relation can be aggregated over all renewable technologies. The cumulative subsidy up to $\mathbf{K}_0$ is the difference between the fixed payments to producers and the revenues from sale of renewable energy:

$$\Phi(\mathbf{K}_0) = R_0(\mathbf{K}_0) - \sum_{i=1}^{I} \mathbb{E}[\alpha^i(\theta)p(\mathbf{K}_0, \theta)]K_0^i. \tag{8.3}$$

This subsidy is financed through a unit levy on the retail power price of $\tau$ paid by all users. Denoting the expected demand by $\bar{D}(\mathbf{K}_0) = \mathbb{E}[D(p(\mathbf{K}_0, \theta) + \tau(\mathbf{K}_0), \theta)]$, the unit levy is determined by

$$\tau(\mathbf{K}_0)\mathbb{E}[D(p(\mathbf{K}_0, \theta) + \tau(\mathbf{K}_0), \theta)] = \Phi(\mathbf{K}_0). \tag{8.4}$$

In practice, the realized availability rate and demand may be higher or lower than expected, and the levy may adjust for the previous year's out-turn. We abstract from this issue, as we ignore potential risk aversion.

We present a static model: policy makers set a target renewable capacity $\mathbf{K}_0$, and perfectly adjust the subsidy to cover the marginal investment cost $r_0^i(x)$ for all $x \leq K_0^i$. Thus we ignore the temporal dimension: all periods are collapsed into one. Extending the model to different periods would simply make the notation more complex and lead to the same economic intuition.

## 8.5 Marginal Impact of Renewables

This section derives the marginal impact of renewable capacity on conventional capacities, subsidies, levies, and net surplus. The cost of grid enhancements required to accommodate renewables is included in the grid rate, hence not covered by this analysis.

Two mutually exclusive situations are possible: (a) baseload technology is present at the long-term equilibrium, or (b) RES entry is so large that no baseload technology is present at the long-term equilibrium. The economic intuition is identical, but the details of the analysis are slightly different for each case. For ease of exposition, we examine each in turn. Finally, we extend the results to flexible baseload.

### 8.5.1 Baseload Technology Present at the Long-Term Equilibrium

Throughout this subsection, renewable capacity is assumed to be small enough that baseload technology is present at the equilibrium. We first establish the following:

**Lemma 8.1.** The expected price on the vertical segments of the supply curve does not vary with installed renewable capacity. Specifically, for all $i \geq 1$

$$\mathbb{E}\left[\frac{\partial p(\mathbf{K}_0,\theta)}{\partial K_0^i}\,|p(\mathbf{K}_0,\theta) < c_2\right] = 0,$$

and for all $n \geq 2$,

$$\mathbb{E}\left[\frac{\partial p(\mathbf{K}_0,\theta)}{\partial K_0^i}\,|v_n\right] = 0.$$

The time-weighted average price is constant:

$$\frac{\partial}{\partial K_0^i}\mathbb{E}[p(\mathbf{K}_0,\theta)] = 0.$$

***Proof.*** At the long-run equilibrium, the expected price on the vertical segments of the supply curve is set to yield profits equal to the capital cost of the marginal technology, and hence does not depend on the renewable capacity. Similarly, since the baseload technology cannot be turned off, it produces all the time. At the long-run equilibrium, the time-weighted average price is equal to the long-run marginal cost of the baseload technology, hence does not depend on the renewable capacity.

Formally, for $n \geq 2$, equation (8.1) can be rewritten as:

$$\mathbb{E}[(p(\mathbf{K}_0,\theta) - c_n)\mathbb{I}_{\{p(\mathbf{K}_0,\theta)\geq c_n\}}] = r_n w.$$

where $\mathbb{I}_{\{x\geq 0\}}$ is the indicator function that takes the value 1 if $x \geq 0$, and 0 otherwise. For $n = N$, differentiation yields

$$\frac{\partial}{\partial K_0^i}\mathbb{E}[(p(\mathbf{K}_0,\theta) - c_N)\mathbb{I}_{\{p(\mathbf{K}_0,\theta)\geq c_N\}}] = \mathbb{E}\left[\frac{\partial p}{\partial K_0^i}\mathbb{I}_{\{p(\mathbf{K}_0,\theta)\geq c_N\}}\right] = 0$$

since by construction the integrand is equal to zero at the lower bound: $p(\mathbf{K}_0,\theta) = c_N$. This proves the result for $n = N$.

For $1 < n < N$, subtraction yields

$$\mathbb{E}[(p(\mathbf{K}_0, \theta) - c_n)\mathbb{I}_{\{p(\mathbf{K}_0,\theta)\geq c_n\}}] - \mathbb{E}[(p(\mathbf{K}_0, \theta) - c_{n+1})\mathbb{I}_{\{p(\mathbf{K}_0,\theta)\geq c_{n+1}\}}] = r_n - r_{n+1}.$$

Differentiating with respect to $K_0^i$ yields

$$\mathbb{E}\left[\frac{\partial p}{\partial K_0^i}\mathbb{I}_{\{p(\mathbf{K}_0,\theta)\geq c_n\}}\right] - \mathbb{E}\left[\frac{\partial p}{\partial K_0^i}\mathbb{I}_{\{p(\mathbf{K}_0,\theta)\geq c_{n+1}\}}\right] = \mathbb{E}\left[\frac{\partial p}{\partial K_0^i}\mathbb{I}_{\{c_n \leq p(\mathbf{K}_0,\theta)< c_{n+1}\}}\right]$$

$$= \mathbb{E}\left[\frac{\partial p}{\partial K_0^i}\,|v_n\right] \times \Pr(v_n) = 0.$$

The same argument applies for n = 1 on the vertical

$\theta$: $p(\mathbf{K}_0, \theta) < c_2$ □

#### 8.5.1.1 Marginal impact on conventional capacity

**Proposition 8.1.** Installed conventional capacity changes as renewable capacity increases for two reasons: demand changes through the change in unit levy, and renewable capacity substitutes for conventional capacity. Specifically,

$$\frac{\partial K_1}{\partial K_0^i} = \frac{\frac{\partial \tau}{\partial K_0^i} - \mathbb{E}[P_q \alpha^i(\theta)\,|\,p(\mathbf{K}_0, \theta) < c_2]}{\mathbb{E}[P_q\,|\,p(\mathbf{K}_0, \theta) < c_2]},$$

and for $n \geq 2$,

$$\frac{\partial K_n}{\partial K_0^i} = \frac{\frac{\partial \tau}{\partial K_0^i} - \mathbb{E}[P_q \alpha^i(\theta)\,|v_n]}{\mathbb{E}[P_q\,|v_n]}. \tag{8.5}$$

***Proof.*** For $n \geq 2$, $\mathbb{E}\left[\frac{\partial p}{\partial K_0^i}\,|v_n\right] = 0$ yields

$$\mathbb{E}\left[\left(P_q(K_n + \sum_{j=1}^{I}\alpha^j(\theta)K_0^j, \theta) \times \left(\frac{\partial K_n}{\partial K_0^i} + \alpha^i(\theta)\right) - \frac{\partial \tau}{\partial K_0^i}\right)|v_n\right] = 0.$$

Rearranging yields equation (8.5). The same argument applies for $n = 1$ on the vertical $\{\theta : p(\mathbf{K}_0, \theta) < c_2\}$.

Intuition for equation (8.5) is easier to obtain when assuming inverse demand is linear with constant slope, $P(Q, \theta) = a(\theta) - bQ$, in which case it simplifies to

$$\frac{\partial K_n}{\partial K_0^i} = -\frac{1}{b}\frac{\partial \tau}{\partial K_0^i} - \mathbb{E}[\alpha^i(\theta)\,|v_n]$$

for $n \geq 2$, and

$$\frac{\partial K_1}{\partial K_0^i} = -\frac{1}{b}\frac{\partial \tau}{\partial K_0^i} - \mathbb{E}[\alpha^i(\theta)\,|\,p(\mathbf{K}_0, \theta) < c_2].$$

The change in $K_n$ is the sum of two effects. First, the levy usually increases, hence demand decreases, and so does $K_n$. Second, all incumbent technologies up to $n$ are replaced by RES. Increasing $K_0$ by 1 reduces cumulative capacity $K_n$ by the expectation of $\alpha^i(\theta)$, conditional on $K_n$ producing at capacity.

If demand is not linear with constant slope, these substitution effects are weighted by the slope of the demand function.

Proposition 8.1 provides the formal proof of figure 8.16. The evolution of capacity of peaking technology is given by

$$\frac{\partial k_2}{\partial K_0^i} = \frac{\partial K_2}{\partial K_0^i} - \frac{\partial K_1}{\partial K_0^i} = \mathbb{E}[\alpha^i(\theta)\,|v_2] - \mathbb{E}[\alpha^i(\theta)\,|\,p(\mathbf{K}_0, \theta) < c_2].$$

On the left panel of figure 8.16, $\mathbb{E}[\alpha^i(\theta)\,|\,p(\mathbf{K}_0, \theta) < c_2] = \mathbb{E}[\alpha^i(\theta)\,|v_2]$, thus $\frac{\partial k_2}{\partial K_0^i} = 0$: the peaking technology is unchanged. On the right panel, $\mathbb{E}[\alpha^i(\theta)\,|\,p(\mathbf{K}_0, \theta) < c_2] = 0$, hence $\frac{\partial k_1}{\partial K_0^i} = \frac{\partial K_1}{\partial K_0^i} = 0$ since the tax impact is not included in the example.

If renewable capacity is very large, we may reach a point where $K_n(\mathbf{K}_0) = 0$, which is discussed later. Proposition 8.1 is illustrated in figure 8.14.

Finally, Proposition 8.1 enables us to determine the impact of renewable capacity on expected demand $\bar{D}(\mathbf{K}_0)$:

**Corollary 8.1.** The marginal impact of $K_0^i$ on expected demand $\bar{D}(\mathbf{K}_0)$ is

$$\frac{\partial \bar{D}}{\partial K_0^i} = -\frac{1}{B}\frac{\partial \tau}{\partial K_0^i} + \Gamma^i, \tag{8.6}$$

where

$$\frac{1}{B} = -\sum_{n=2}^{N}\left(\mathbb{E}\left[\frac{\partial D(c_n+\tau, \theta)}{\partial p}\,|h_n\right] \times \Pr(h_n) + \frac{\Pr(v_n)}{\mathbb{E}[P_q\,|v_n]}\right) - \frac{\Pr(p < c_2)}{\mathbb{E}[P_q\,|\,p < c_2]},$$

and

$$\Gamma^i = \sum_{n=2}^{N}(\mathbb{E}[\alpha^i(\theta)\,|v_n] - \frac{\mathbb{E}[\alpha^i(\theta)P_q\,|v_n]}{\mathbb{E}[P_q\,|v_n]})\Pr(v_n)$$
$$+\left(\mathbb{E}[\alpha^i(\theta)\,|\,p < c_2] - \frac{\mathbb{E}[\alpha^i(\theta)P_q\,|\,p < c_2]}{\mathbb{E}[P_q\,|\,p < c_2]}\right)\Pr(p < c_2).$$

***Proof.*** Start with

$$\begin{aligned}\frac{\partial}{\partial K_0^i}\mathbb{E}[D(p+\tau,\theta)] &= \mathbb{E}\left[\frac{\partial}{\partial K_0^i}D(p+\tau,\theta)\right]\\ &= \mathbb{E}\left[\frac{\partial D}{\partial p}\left(\frac{\partial p}{\partial K_0^i}+\frac{\partial \tau}{\partial K_0^i},\theta\right)\right]\\ &= \sum_{n=2}^{N}\left(\frac{\partial \tau}{\partial K_0^i}\mathbb{E}\left[\frac{\partial D}{\partial p}\,|h_n\right]\Pr(h_n)+\mathbb{E}\left[\left(\frac{\partial K_n}{\partial K_0^i}+\alpha^i(\theta)\right)|v_n\right]\Pr(v_n)\right)\\ &\quad+\mathbb{E}\left[\left(\frac{\partial K_1}{\partial K_0^i}+\alpha^i(\theta)\right)|p<c_2\right]\Pr(p<c_2)\end{aligned}$$

since price is constant on $h_n$ and $D(p+\tau,\theta)=K_n+\sum_{j=1}^{I}\alpha^j(\theta)K_0^j$ on $v_n$. For $n\geq 2$

$$\frac{\partial K_n}{\partial K_0^i}+\alpha^i(\theta)=\frac{\frac{\partial \tau}{\partial K_0^i}+\alpha^i(\theta)\mathbb{E}[P_q\,|v_n]-\mathbb{E}[\alpha^i(\theta)P_q\,|v_n]}{\mathbb{E}[P_q\,|v_n]},$$

thus

$$\mathbb{E}\left[\frac{\partial K_n}{\partial K_0^i}+\alpha^i(\theta)\,|v_n\right]=\frac{\partial \tau}{\partial K_0^i}\frac{1}{\mathbb{E}[P_q\,|v_n]}+\left(\mathbb{E}[\alpha^i(\theta)\,|v_n]-\frac{\mathbb{E}[\alpha^i(\theta)P_q\,|v_n]}{\mathbb{E}[P_q\,|v_n]}\right).$$

Deriving a similar expression for $n=1$ and summing over all intervals yields equation (8.6).

Suppose again inverse demand is linear with constant slope $P(Q,\theta)=a(\theta)-bQ$. Then, $B=b$, $\Gamma^i=0$, and equation (8.6) simplifies to

$$\frac{\partial \bar{D}}{\partial K_0^i}=-\frac{1}{b}\frac{\partial \tau}{\partial K_0^i}.$$

An increase in $K_0^i$ leads to a change in levy. If demand is linear with constant slope, this leads to a proportional change in expected demand. Under reasonable assumptions, $\frac{\partial \tau}{\partial K_0^i}\geq 0$: the levy increases to finance an increase in RES target capacity. The levy effect is thus negative: as renewable capacity increases, so does the levy, and demand decreases.

#### 8.5.1.2 Marginal impact on the value of RES capacity

We now determine the impact of the level of RES capacity on its marginal value:

**Proposition 8.2.** The marginal impact of $K_0^i$ on the marginal value of RES technology $j$ is

$$\frac{\partial}{\partial K_0^i}\mathbb{E}[\alpha^j(\theta)p(\theta,\mathbf{K}_0)]=\mathbb{E}\left[\alpha^j(\theta)\frac{\partial p}{\partial K_0^i}\right]=-\Gamma^j\frac{\partial \tau}{\partial K_0^i}-E^{ij}, \tag{8.7}$$

where

$$E^{ij}=\sum_{n=2}^{N}\left(\frac{\mathbb{E}[P_q\alpha^i\,|v_n]\mathbb{E}[P_q\alpha^j\,|v_n]}{\mathbb{E}[P_q\,|v_n]}-\mathbb{E}[P_q\alpha^i\alpha^j\,|v_n]\right)\Pr(v_n)$$
$$+\left(\frac{\mathbb{E}[P_q\alpha^i\,|p<c_2]\mathbb{E}[P_q\alpha^j\,|p<c_2]}{\mathbb{E}[P_q\,|p<c_2]}-\mathbb{E}[P_q\alpha^i\alpha^j\,|p<c_2]\right)\Pr(p<c_2).$$

***Proof.*** We have:

$$\mathbb{E}\left[\alpha^j(\theta)\frac{\partial p}{\partial K_0^i}\right]=\sum_{n=2}^{N}\mathbb{E}\left[\alpha^j(\theta)\left(P_q\times\left(\frac{\partial K_n}{\partial K_0^i}+\alpha^i(\theta)\right)-\frac{\partial\tau}{\partial K_0^i}\right)|v_n\right]\Pr(v_n).$$
$$+\mathbb{E}\left[\alpha^j(\theta)\left(P_q\times\left(\frac{\partial K_1}{\partial K_0^i}+\alpha^i(\theta)\right)-\frac{\partial\tau}{\partial K_0^i}\right)|p<c_2\right]\Pr(p<c_2)$$

For $n\geq 2$,

$$\mathbb{E}\left[P_q\times\left(\frac{\partial K_n}{\partial K_0^i}+\alpha^i(\theta)\right)\alpha^j(\theta)\,|v_n\right]$$
$$=\frac{1}{\mathbb{E}[P_q\,|v_n]}\left(\begin{array}{c}\frac{\partial\tau}{\partial K_0^i}\mathbb{E}[P_q\alpha^j\,|v_n]-\mathbb{E}[P_q\alpha^i\,|v_n]\mathbb{E}[P_q\alpha^j\,|v_n]\\+\mathbb{E}[P_q\alpha^i\alpha^j\,|v_n]\mathbb{E}[P_q\,|v_n]\end{array}\right)$$
$$=\frac{\mathbb{E}[P_q\alpha^j\,|v_n]}{\mathbb{E}[P_q\,|v_n]}\frac{\partial\tau}{\partial K_0^i}+\left(\mathbb{E}[P_q\alpha^i\alpha^j\,|v_n]-\frac{\mathbb{E}[P_q\alpha^i\,|v_n]\mathbb{E}[P_q\alpha^j\,|v_n]}{\mathbb{E}[P_q\,|v_n]}\right).$$

A similar derivation obtains for $p(K_0,\theta)<c_2$. Summing over all vertical segments of the supply curve yields

$$\mathbb{E}\left[a^j(\theta)\frac{\partial p}{\partial K_0^i}\right]=-\Gamma^j\frac{\partial\tau}{\partial K_0^i}-E^{ij},$$

which is equation (8.7).

If demand is linear with constant slope,

$$E^{ij}=b\left(\sum_{n=2}^{N}cov[\alpha^i,\alpha^j\,|v_n]\Pr(v_n)+cov[\alpha^i,\alpha^j\,|p<c_2]\Pr(p<c_2)\right)$$
$$=b\widehat{cov}_{\mathbf{K}_0}[\alpha^i(\theta),\alpha^j(\theta)],$$

hence equation (8.7) simplifies to

$$\mathbb{E}\left[\alpha^j(\theta)\frac{\partial p}{\partial K_0^i}\right] = -b\widehat{cov}_{\mathbf{K}_0}[\alpha^i(\theta), \alpha^j(\theta)]. \tag{8.8}$$

An increase in $K_0^i$ has two impacts on the amount of capacity producing on $v_n$: it increases renewable output by $\alpha^i(\theta)$ in state $\theta$, and it reduces cumulative conventional capacity by $\mathbb{E}[\alpha^i(\theta)\,|v_n]$. Multiplying by $\alpha^j(\theta)$ and taking the expectation yields the covariance. The result follows since inverse demand is linear with constant slope. Proposition 8.2 is illustrated in figure 8.17 and justifies the approximation using the variance-covariance matrix.

If demand is not linear with constant slope, additional terms corresponding to the variation of the slope are added to equations (8.7).

#### 8.5.1.3 Subsidy for the marginal unit

Proposition 8.2 leads to the following:

**Proposition 8.3.** If demand is linear with constant slope, the subsidy required by a marginal unit of RES $i$ may increase or decrease as RES capacity $i$ increases, and increases as RES capacity $j$ increases if and only if availabilities on the vertical segments of the supply curve are positively correlated .

***Proof.*** The subsidy to the marginal unit is

$$\varphi^i(\mathbf{K}_0) = r_0^i(K_0^i) - \mathbb{E}[\alpha^i(\theta)p(\mathbf{K}_0, \theta)],$$

hence

$$\frac{\partial\varphi^i(\mathbf{K}_0)}{\partial K_0^i} = \frac{d}{dK_0^i}r_0^i(K_0^i) + b\widehat{var}_{\mathbf{K}_0}[\alpha^i(\theta)].$$

The first term is negative since marginal cost is decreasing, while the second is positive. This proves the first point. Then,

$$\frac{\partial\varphi^i(\mathbf{K}_0)}{\partial K_0^j} = b\widehat{cov}_{\mathbf{K}_0}[\alpha^j(\theta), \alpha^i(\theta)]$$

proves the second point. □

The first point of Proposition 8.3 illustrates the race between falling costs and falling prices. For low capacity, significant learning effects are present, hence falling costs most likely outweigh falling prices. As capacity increases, costs fall much more slowly, and maybe not sufficiently to compensate the price decrease. The subsidy will then increase. This point is illustrated in figure 8.18, from Green and Léautier (2017).

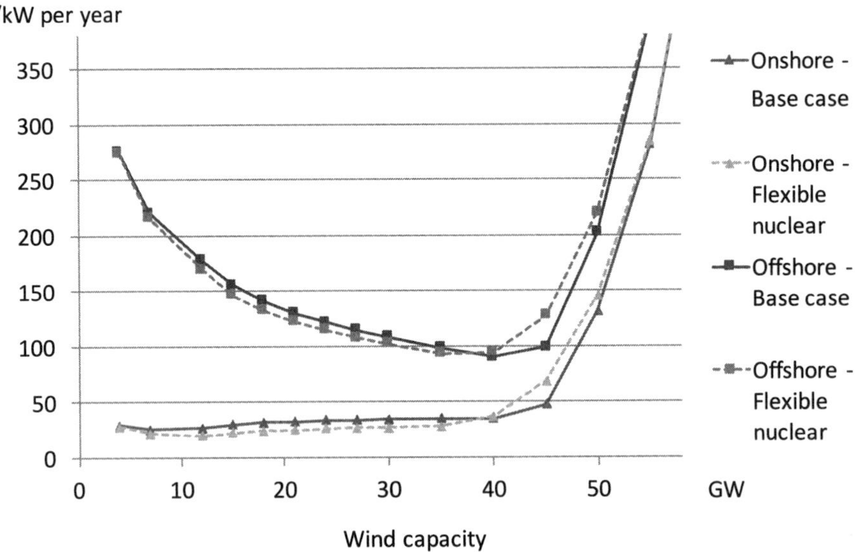


**Figure 8.18**
Evolution of the marginal RES subsidy as RES capacity increases.

Given the structure of our learning curve (discussed later), the cost of the first kW of renewable capacity, hence the subsidy, is infinite. We therefore start computing the marginal subsidy from the 2009 values: 3 GW onshore and 1 GW offshore. Over this interval, the marginal subsidy to onshore wind remains constant around £50/kW per year: the cost reduction from learning is not sufficient to compensate the reduction in value of the energy produced. For offshore wind, the subsidy decreases from £350/kW per year to £150/kW per year, suggesting that the gains from learning outweigh the loss of value of the energy produced.

The second point of Proposition 8.3 illustrates the complementarity and substitutability of technologies: if the outputs from two technologies are positively correlated, increasing the capacity of one increases the supply and hence reduces the price available to the other one, and so increases the required subsidy.

### 8.5.1.4 Marginal impact on levy

Differentiation of equation (8.4) with respect to $K_0^i$ yields

$$\frac{\partial \tau}{\partial K_0^i}\bar{D}(\mathbf{K}_0)+\tau\frac{\partial \bar{D}}{\partial K_0^i}=r_0^i(K_0^i)-\mathbb{E}[\alpha^i(\theta)p(\mathbf{K}_0,\theta)]-\sum_{j=1}^{I}\mathbb{E}\left[\alpha^j(\theta)\frac{\partial p}{\partial K_0^i}\right]K_0^j$$

$\Leftrightarrow$

$$\frac{\partial \tau}{\partial K_0^i}\bar{D}(\mathbf{K}_0) = \varphi^i(K_0^i) - \tau \frac{\partial \bar{D}}{\partial K_0^i} - \sum_{j=1}^{I} \mathbb{E}\left[\alpha^j(\theta)\frac{\partial p}{\partial K_0^i}\right] K_0^j. \tag{8.9}$$

To finance incremental renewable capacity, the gross levy receipts $\frac{\partial \tau}{\partial K_0^i}\bar{D}(\mathbf{K}_0)$ must change to cover the subsidy to the marginal unit $\varphi^i(K_0^i)$ and the reduction in levy receipts, owing to the demand reduction $\tau \frac{\partial \bar{D}}{\partial K_0^i}$. If increasing renewable capacity $i$ decreases the value of all inframarginal units, a third cost is added: the reduction in market value of all inframarginal renewables benefiting from the feed-in tariff.

When granting a fixed-price contract, policy makers commit the customers to a fixed payment to a renewable producer, hence their net liability is this payment minus the market value of this renewable capacity. As more fixed-price contracts are granted, the market value of renewable energy (usually) decreases, and the liability (usually) increases.

Using previous results, we now establish the following:

**Proposition 8.4.** The marginal impact of $K_0^i$ on the unit levy necessary to finance RES is

$$\frac{\partial \tau}{\partial K_0^i} = \frac{B\left(\varphi^i(\mathbf{K}_0) + \sum_{j=1}^{I} E^{ij} K_0^j + \tau(\mathbf{K}_0)\Gamma^i\right)}{B\bar{D}(\mathbf{K}_0) - \tau(\mathbf{K}_0) + B\sum_{j=1}^{I}\Gamma^j K_0^j}. \tag{8.10}$$

***Proof.*** Inserting equations (8.6) and (8.7) into expression (8.9) yields

$$\frac{\partial \tau}{\partial K_0^i}\left(\bar{D} - \frac{\tau}{B} + \sum_{j=1}^{I}\Gamma^j K_0^j\right) - \Gamma^i\tau + \mathbb{E}[p(\mathbf{K}_0,\theta)\alpha^i(\theta)] - \sum_{j=1}^{I} E^{ij} K_0^j = r_0^i(K_0^i),$$

which leads to equation (8.10).

If demand is linear with constant slope, equation (8.10) yields

$$\bar{D}(\mathbf{K}_0)\frac{\partial \tau}{\partial K_0^i} = \frac{1}{1 - \frac{\tau(\mathbf{K}_0)}{b\bar{D}(\mathbf{K}_0)}}\left(\varphi^i(\mathbf{K}_0) + b\sum_{j=1}^{I}\widehat{cov}_{\mathbf{K}_0}[\alpha^i(\theta),\alpha^j(\theta)]K_0^j\right),$$

which illustrates the marginal impact of $K_0^i$ on levy. First, the levy must increase to cover the marginal subsidy $\varphi^i(\mathbf{K}_0)$. Second, the levy must change to cover the changes in the value of all other renewable capacities. Finally, this change is magnified by the factor $\frac{1}{1-\frac{\tau(\mathbf{K}_0)}{b\bar{D}(\mathbf{K}_0)}} > 1$ to account for the deadweight loss from levies.

#### 8.5.1.5 Marginal impact on net surplus

We now compute the change in net surplus caused by a marginal increase in renewable capacity. Since this analysis focuses on net surplus, and not overall welfare, it ignores distributional issues. The net surplus is the standard net surplus from consumption derived in chapter 2, adding the distortion caused by the levy:

$$W(\mathbf{K}_0) = \mathbb{E}[S(p(\mathbf{K}_0,\theta)+\tau,\theta) - (p(\mathbf{K}_0,\theta)+\tau)D(p(\mathbf{K}_0,\theta)+\tau,\theta,)].$$

**Proposition 8.5.** The marginal loss of hourly net surplus resulting from a marginal increase in $K_0^i$ is equal to the marginal subsidy plus the marginal demand reduction, valued at the levy rate. It can be expressed as

$$\frac{\partial W}{\partial K_0^i} = -\left(\varphi^i(\mathbf{K}_0) + \tau(\mathbf{K}_0)\Gamma^i + \frac{\tau(\mathbf{K}_0)\left(\varphi^i(\mathbf{K}_0) + \sum_{j=1}^I E^{ij}K_0^j + \tau(\mathbf{K}_0)\Gamma^i\right)}{B\bar{D}(\mathbf{K}_0) - \tau(\mathbf{K}_0) + B\sum_{j=1}^I \Gamma^j K_0^j}\right). \tag{8.11}$$

***Proof.*** The envelope theorem yields

$$\frac{\partial W}{\partial K_0^i} = -\left(\mathbb{E}\left[\left(\frac{\partial p}{\partial K_0^i} + \frac{\partial \tau}{\partial K_0^i}\right) D(p(\mathbf{K}_0,\theta)+\tau,\theta)\right]\right).$$

Since $\frac{\partial p}{\partial K_0^i} \neq 0$ only when each technology produces at capacity,

$$\begin{aligned}
&\mathbb{E}\left[\frac{\partial p}{\partial K_0^i} D(p(\mathbf{K}_0,\theta)+\tau,\theta)\right] \\
&= \sum_{n=2}^N \mathbb{E}\left[\frac{\partial p}{\partial K_0^i} \times \left(K_n + \sum_{j=1}^I \alpha^j(\theta)K_0^j\right) | v_n\right] \Pr(v_n) \\
&\quad + \mathbb{E}\left[\frac{\partial p}{\partial K_0^i}\left(K_1 + \sum_{j=1}^I \alpha^j(\theta)K_0^j\right) | p < c_2\right] \Pr(p < c_2) \\
&= \sum_{n=2}^N \left(\mathbb{E}\left[\frac{\partial p}{\partial K_0^i} | v_n\right] K_n + \sum_{j=1}^I \mathbb{E}\left[\frac{\partial p}{\partial K_0^i}\alpha^j(\theta) | v_n\right] K_0^j\right) \Pr(v_n) \\
&\quad + \left(\mathbb{E}\left[\frac{\partial p}{\partial K_0^i} | p < c_2\right] K_1 + \sum_{j=1}^I \mathbb{E}\left[\frac{\partial p}{\partial K_0^i}\alpha^j(\theta) | p < c_2\right] K_0^j\right) \Pr(p < c_2)
\end{aligned}$$

$$= \sum_{j=1}^{I} \left( \begin{array}{c} \sum_{n=2}^{N} \mathbb{E}\left[ \frac{\partial p}{\partial K_0^i} \alpha^j(\theta) \,|v_n \right] \Pr(v_n) \\ + \mathbb{E}\left[ \frac{\partial p}{\partial K_0^i} \alpha^j(\theta) \,|p < c_2 \right] \Pr(p < c_2) \end{array} \right) K_0^j = \sum_{j=1}^{I} \mathbb{E}\left[ \frac{\partial p}{\partial K_0^i} \alpha^j(\theta) \right] K_0^j,$$

since $\mathbb{E}\left[ \frac{\partial p}{\partial K_0^i} \,|v_n \right] = \mathbb{E}\left[ \frac{\partial p}{\partial K_0^i} \,|p < c_2 \right] = 0$. Thus,

$$\frac{\partial W}{\partial K_0^i} = -\left( \frac{\partial \tau}{\partial K_0^i} \bar{D}(K_0) + \sum_{j=1}^{I} \mathbb{E}\left[ \frac{\partial p}{\partial K_0^i} \alpha^j(\theta) \right] K_0^j \right)$$

$$= -\varphi(K_0) + \tau \frac{\partial \bar{D}}{\partial K_0^i}$$

by inserting equation (8.9). Observing that

$$\frac{\partial \bar{D}}{\partial K_0^i} = -\frac{1}{B} \frac{\partial \tau}{\partial K_0^i} + \Gamma^i = -\frac{\varphi^i(\mathbf{K}_0) + \sum_{j=1}^{I} E^{ij} K_0^j + \tau \Gamma^i}{B\bar{D}(\mathbf{K}_0) - \tau + B\sum_{j=1}^{I} \Gamma^j K_0^j} + \Gamma^i$$

yields

$$\frac{\partial W}{\partial K_0^i} = -\left( \varphi^i(\mathbf{K}_0) + \tau \frac{\varphi^i(\mathbf{K}_0) + \sum_{j=1}^{I} E^{ij} K_0^j + \tau \Gamma^i}{B\bar{D}(\mathbf{K}_0) - \tau + B\sum_{j=1}^{I} \Gamma^j K_0^j} - \tau \Gamma^i \right)$$

which is equation (8.11). □

Numerically, the term $\tau \frac{\partial \bar{D}}{\partial K_0^i}$ is small, hence the marginal surplus loss is almost equal to the marginal subsidy.

### 8.5.2 Conventional Technologies Disappearing at the Long-Term Equilibrium

As indicated by Proposition 8.1, conventional capacity is likely to decrease as renewable capacity increases. How are the previous results modified when technology $n$ is no longer present at the long-term equilibrium?

**Proposition 8.6.** If technology $n \geq 2$ is no longer present at the long-term equilibrium, the above results still hold, with the convention that $v_n = h_n = \emptyset$.

If technology $n = 1$ is no longer present at the equilibrium, an additional term is included in all expressions. For example, the general expression for the marginal impact of renewables on expected price is

$$\frac{\partial}{\partial K_0^i}\mathbb{E}[p(\mathbf{K}_0,\theta)] = \mathbb{E}\left[\left(P_q\alpha^i(\theta) - \frac{\partial \tau}{\partial K_0^i}\right)\mathbb{I}_{\{p(\mathbf{K}_0,\theta)<c_2\}}\right]\mathbb{I}_{\{K_1=0\}}.$$

The general expression for the slope of the marginal value of renewable capacity is

$$\begin{aligned}\mathbb{E}\left[a^j(\theta)\frac{\partial p}{\partial K_0^i}\right] &= \Gamma^j\frac{\partial \tau}{\partial K_0^i} - E^{ij}\\ &+ \left(\frac{\mathbb{E}[P_q\alpha^j(\theta)\,|p<c_2]}{\mathbb{E}[P_q\,|p<c_2]}\mathbb{E}\left[\left(P_q\alpha^i(\theta) - \frac{\partial \tau}{\partial K_0^i}\right)\mathbb{I}_{\{p(\mathbf{K}_0,\theta)<c_2\}}\right]\right)\mathbb{I}_{\{K_1=0\}}.\end{aligned} \tag{8.12}$$

Marginal value is continuously differentiable everywhere (i.e., $\frac{\partial}{\partial K_0^i}\mathbb{E}[\alpha^j(\theta)p(\theta,\mathbf{K}_0)]$ is continuous for $\mathbf{K}_0$ such that $K_n(\mathbf{K}_0)=0$).

***Proof.*** The proof follows the steps of the previous ones, and is presented in Green and Léautier (2017).

The first result confirms the intuition that there is nothing unique about the technologies included in the dispatch. In other words, more (or fewer) technologies can be included, without modifying the expressions.

A baseload technology that always runs at full capacity is different. When $K_1 > 0$, the decrease in $K_1$ mitigates the supply effect of increasing $K_0^i$. When $K_1 = 0$, this mitigating effect disappears, and the price decrease is larger by a factor $\mathbb{E}\left[\left(P_q\alpha^i(\theta) - \frac{\partial \tau}{\partial K_0^i}\right)\mathbb{I}_{\{p(\mathbf{K}_0,\theta)<c_2\}}\right]$. Continuity of the derivatives when one technology disappears arises because the supply curve is continuous.

### 8.5.3 Flexible Baseload Technology

One could conclude from the previous analysis that the simplicity of the expressions previously obtained is mostly attributable to the assumption that the baseload technology is completely inflexible. This is not accurate, as shown below:

**Proposition 8.7.** Suppose the baseload technology is flexible. The average price is no longer constant:

$$\frac{\partial}{\partial K_0^i}\mathbb{E}[p(\mathbf{K}_0,\theta)] = \mathbb{E}\left[\left(P_q\left(\sum_{j=1}^{I}\alpha^j(\theta)K_0^j,\theta\right)\alpha^i(\theta) - \frac{\partial \tau}{\partial K_0^i}\right)\mathbb{I}_{\{p(\mathbf{K}_0,\theta)<c_1\}}\right].$$

The slope of the value of renewable technology $j$ is:

$$\mathbb{E}\left[a^j(\theta)\frac{\partial p}{\partial K_0^i}\right] = -\Gamma^j\frac{\partial \tau}{\partial K_0^i} - E^{ij} + \frac{\mathbb{E}\left[\alpha^j(\theta)P_q\,|p(\mathbf{K}_0,\theta)<c_1\right]}{\mathbb{E}[P_q\,|p(\mathbf{K}_0,\theta)<c_1]}$$

$$\times\,\mathbb{E}\left[\alpha^j(\theta)\left(P_q\alpha^i(\theta)-\frac{\partial\tau}{\partial K_0^i}\right)\mathbb{I}_{\{p(\mathbf{K}_0,\theta)<c_1\}}\right]. \qquad (8.13)$$

***Proof.*** The proof follows the steps of the previous ones, and is presented in Green and Léautier (2017).

When the baseload technology is inflexible, average price is constant, equal to the long-run marginal cost of this baseload technology. When the baseload technology is flexible, it stops producing if $p < c_1$. On this set, increasing renewable capacity (usually) reduces price. Hence, average price (usually) decreases as renewable capacity increases. The same effect explains why the marginal value of renewable technology $j$ decreases faster when $p < c_1$.

The flexible baseload technology is equivalent to the other technologies, hence, once the new definition of the supply curve is used, the expression of $\frac{\partial K_1}{\partial K_0^i}$ is formally identical to any $\frac{\partial K_n}{\partial K_0^i}$.

In the case of Great Britain, fully flexible nuclear remains economic for higher RES penetration. For example, if RES capacity is 40 GW, fully flexible nuclear long-term equilibrium capacity is 13 GW (assuming full flexibility does not increase capital or operating cost, which is optimistic), while it is only 4 GW if nuclear is inflexible. This is illustrated in figure 8.19.

Flexibility does not keep nuclear technology present in the long-term equilibrium: a subsidized baseload technology replaces an unsubsidized one.

## 8.6 Application to the Case of Great Britain

To illustrate these effects in practice, we present a simulation of the electricity industry in Great Britain, calibrated for the 2020s.

### 8.6.1 Data

#### 8.6.1.1 Wind data

The values of $\alpha^i(\theta)$ are clearly vital for the numerical results that we obtain; Staffell and Green (2014) use wind speed estimates from NASA's MERRA data set to simulate the hourly output of every wind farm in Great Britain. They we use eighteen years of their data and match it to the actual hourly demands over the same period. This gives us 157,680 observations from which to calculate the relationship between the level of demand and the

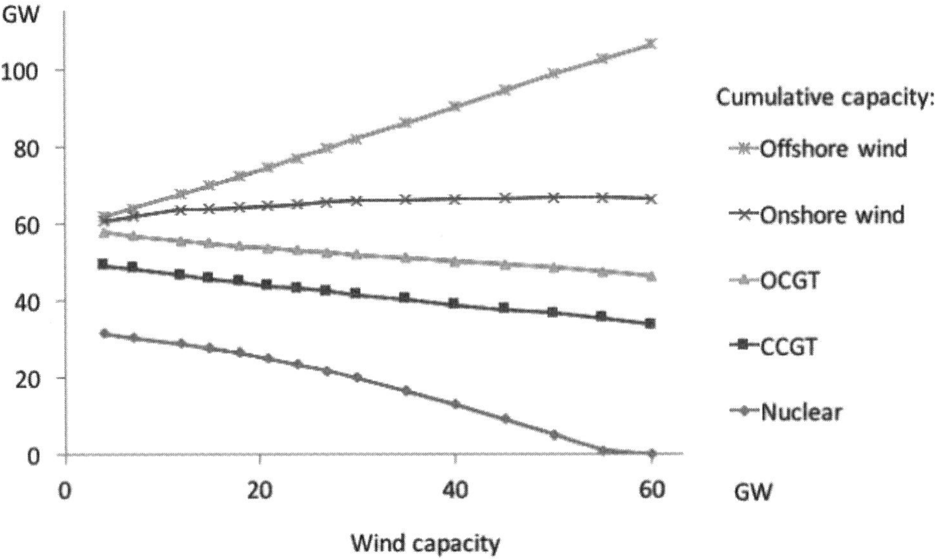


**Figure 8.19**
Evolution of the generation mix as RES capacity increases, assuming nuclear is fully flexible.

load factors of onshore and offshore wind stations, the two renewable technologies that we consider.

The MERRA data set estimates the wind speed at several heights above ground at a grid of points covering the entire globe, using computer modeling to match observations from satellites and weather stations. The virtual wind turbine model (Staffell and Green 2014) interpolates between these points to the location of any chosen wind farm and extrapolates the wind speed to the height of its turbines. The manufacturer's power curve for the type of turbine used at the farm (where known) gives the relationship between the estimated wind speed and the station's output. The hourly output for a given wind farm is estimated with a degree of error, but the monthly output for a farm, or the hourly output for a fleet of turbines spread across Great Britain, is remarkably close to actual values.

We assume that these load factors do not change as more capacity is added, even though it would be natural to expect the best sites to be developed first (an effect that might be offset by technical progress and the gradual move to larger turbines on taller masts, which capture more of the wind). In the figures below, many of our results are plotted against the total amount of wind capacity installed. Our minimum capacity is 4 GW, made up of 3 GW onshore and 1 GW offshore (the situation in 2009). By 2014, 8 GW of onshore wind and 4 GW of offshore capacity had been installed. As capacity grows beyond this point, we assume that more will be added offshore than onshore, so that 30 GW would consist of 14 GW onshore and 16 GW offshore. The highest level that we consider is

60 GW, made up of 20 GW onshore and 40 GW offshore, a "round-numbers" variant of the medium case for the level of deployment in 2030, as reported by the Department of Energy and Climate Change (2011). This is enough to generate 52 percent of the industry's total output.

#### 8.6.1.2 Demand data

We matched these estimated hourly load factors to eighteen years of actual demand data. Because electricity demand has grown significantly over this period, each year's observations were scaled to a common level of underlying demand, 350 TWh a year. This was done by multiplying every hourly observation by the ratio of the year's weather-corrected demand (published by National Grid) to the underlying level. In other words, if the weather-corrected demand in 1995 was 280 TWh, every observation for that year was multiplied by 1.25 when creating our scaled demand. If 1995 had, in fact, been a cold year with more above-average demands than usual, this would be preserved in our data set, and the modeled demand would exceed 350 TWh. The demand is scaled to give an annual total of 350 TWh, with a peak of around 60 GW and a minimum of 25 GW. Demand is assumed to be linear, with constant slope $b = 100$ £/MWh per GW change in consumption.

To make our modeling calculations more tractable, rather than using all 157,680 observations, we grouped our data points into bins of equal width. Each hour was allocated to one of twenty bins for demand, ten for the onshore load factor and ten for the offshore load factor. This gave 2,000 possible states of the world, each represented by the average demand and load factors for the hours within that set of bins. In practice, the probability of many of these states was zero—it would be an extraordinary weather pattern that gave a load factor of less than 10 percent for onshore wind farms and one of more than 90 percent for offshore farms at the same time, for example. The model therefore used the 1,268 combinations of bins that actually arose over the period as our states of the world, with probabilities based on their relative frequencies.

#### 8.6.1.3 Wind turbine costs

The marginal cost of wind turbines (excluding connection cost) is derived from a learning-curve model. As of 2016, around 10 GW of wind turbines are installed in Great Britain. Their marginal cost is £210/kW per year for onshore, and £455/kW per year for offshore. Assuming exponential learning, the marginal cost of wind turbines is

$$r_0^i(K_0^i) = r_0^i(\bar{K}_0^i) \times \left(\frac{\bar{K}_0^i}{K_0^i}\right)^{\beta}$$

where $\beta$ is a measure of the learning rate, and $\bar{K}_0^i$ the current renewable capacity. Learning is typically measured by the reduction in costs achieved for a doubling of production. The DECC study mentioned previously (DECC 2011, page 20) observes that a doubling of onshore wind turbine capacity leads to a 7 percent reduction in cost. This leads to

$$\frac{r_0(2K_0)}{r_0(K_0)} = \left(\frac{1}{2}\right)^{\beta} = 0.93 \Leftrightarrow \beta = -\frac{\ln(0.93)}{\ln(2)} = 0.104.$$

For offshore wind, the same study proposes an estimate of a 12 percent reduction in costs with each doubling of UK capacity (DECC 2011, page 42), which gives $\beta = 0.184$.

A word of caution concerning the estimates of the learning rates is warranted.

On the one hand, this analysis was conducted in 2015. Since then, the cost of wind turbines has fallen much faster than expected: two developpers have offered to build off-shore wind farms in the North Sea without any subsidies, incorporating future cost reduction into their bids. This suggests the learning rate used in this analysis is probably too low.

On the other hand, Nordhaus (2014) cautions that simple estimates of learning rates are probably biased, as observed cost reduction includes both learning and exogenous technological change.

In addition, procurement conditions have changed. Earlier turbines were developed under feed-in tariffs, which did not provide strong incentives for cost reduction. Auctions are now used to procure new wind turbines, which provides very strong cost reduction incentives. I believe we moved from one gentle learning curve to another steeper one.

Therefore, the significant cost reduction we observe is the result of exogenous technological progress, learning by doing, but also a discontinuous change in incentives. Disentangling these effects requires more data than we currently have. I have thus decided to use the original learning rates. The evolution of incumbent technologies is not affected by this decision, nor is the logic of a race between declining cost of RES and the declining value of the energy they produce. Should the learning rate prove higher than our estimate, the marginal subsidy may no longer be required.

#### 8.6.1.4 Incumbent-generation technologies

We consider three investment options apart from wind power: nuclear power, combined-cycle gas turbines, and open-cycle gas turbines for peaking use.

We assume carbon is priced at £70/ton, the level that the government's carbon price support is expected to reach in 2030. In looking for a long-run equilibrium, we disregard existing capacity, observing that the UK's remaining coal and oil stations will be uneconomical with such a carbon price. The cost estimates are taken from a report on generation costs prepared by the Department of Energy and Climate Change (2013). Inflexible nuclear stations have a fixed cost of £575/kW/year and a variable cost of £8/MWh. We also model flexible nuclear technology, with the same costs, as discussed above. Combined-cycle gas turbine stations have a fixed cost of £106/kW/year and a variable cost of £73/MWh, while the peaking OCGT stations have a fixed cost of £50/kW/year and a variable cost of £109/MWh. It should be noted that these costs are based on the value of fuel and carbon prices over the station's lifetime, according to DECC's central scenario, rather than predictions for a particular year in the 2020s. This lifetime perspective is the appropriate one when considering investment decisions, but the resulting electricity prices are greater than those likely to be seen in the near future.

Given these costs, nuclear stations are the most effective option if they can operate for at least 8,000 hours a year. Open-cycle gas turbine stations are the cheapest way of meeting demands that last for fewer than 1,700 hours a year. With no wind stations, the optimal mix of thermal capacity contains 30 GW of nuclear stations, 21 GW of CCGTs, and 8 GW of OCGT peaking plant. The time-weighted wholesale electricity price is equal to £81/MWh, while the demand-weighted price is £87/MWh.

### 8.6.2 Closing the Model for Any Vector of Renewables Capacities

As previously discussed, the expressions simplify significantly if inverse demand is linear with constant slope $P(Q,\theta)=a(\theta)-bQ$. The marginal impact of $K_0^i$ on demand, which is given by equation (8.6), simplifies to

$$\frac{\partial \bar{D}}{\partial K_0^i}=-\frac{1}{b}\frac{\partial \tau}{\partial K_0^i} \Leftrightarrow \bar{D}\left(\mathbf{K}_0\right)-\bar{D}(\mathbf{0})=-\frac{1}{b}\tau\left(\mathbf{K}_0\right)$$

since $\tau(\mathbf{0})=0$.

Inserting this expression of tax in combining with equation equations (8.4) and (8.3) then yields

$$\tau(\mathbf{K}_0)\left(\bar{D}(\mathbf{0})-\frac{1}{b}\tau(\mathbf{K}_0)\right)=R_0(\mathbf{K}_0)-\sum_{j=1}^{I}\mathbb{E}[\alpha^j(\theta)p(\mathbf{K}_0,\theta)]K_0^j.$$

Since $p(\mathbf{K}_0,\theta)$ also depends on $\tau(\mathbf{K}_0)$ on the vertical segments of the supply curve, the above equation is a fixed-point problem, which is solved iteratively using the following simple algorithm, equivalent to one described by Borenstein (2005). For a given tax rate, we first set the cumulative capacity of all $N$ generators. Profits are monotonically decreasing in capacity, and so if the peaking generators are making an economic profit, the industry's capacity must be increased. Once the profits of OCGT stations are zero, we adjust the cumulative capacity of generators 1 to $(N-1)$ (in this case, nuclear and CCGT stations) until the profits of type $(N-1)$ are zero. We continue in the same way until all generators are making zero profits. We then check how much revenue the tax is raising, and if this is less than the cost of the renewable subsidy, we increase the tax rate and reoptimize the capacity levels. In practice, the model can be solved quickly in an excel spreadsheet using VBA macros.

As indicated above, exactly closing the model requires a robust and detailed long-term model of a power market. Alternatively, a simple linear approximation is available, presented below.

Suppose the conditional covariances are not affected by the level of renewable penetration: $\widehat{cov}_{\mathbf{K}_0}[\alpha^i(\theta), \alpha^j(\theta)] = \widehat{cov}[\alpha^i(\theta), \alpha^j(\theta)]$. Since most expressions are linear, it is helpful to use vector notation. For any matrix $\mathbf{M}$, denote $\mathbf{M}_{ij}$ the element located on line $i$, column $j$. Introduce $\mathbf{C} \in \mathbb{R}^I \times \mathbb{R}^I$ the matrix of covariances, that is, $\mathbf{C}_{ij} = \widehat{cov}[\alpha^i(\theta), \alpha^j(\theta)]$, and $\mathbf{V}(\mathbf{K}_0) \in \mathbb{R}^I$ the vector of marginal values, that is, $\mathbf{V}_i(\mathbf{K}_0) = \mathbb{E}[\alpha^i(\theta) p(\mathbf{K}_0, \theta)]$.

From equation (8.8), the vector of marginal values is a linear function of the vector of renewable capacities:

$$\mathbf{V}(\mathbf{K}_0) = \mathbf{V}(\mathbf{0}) - b\mathbf{C} \cdot \mathbf{K}_0.$$

The slope of the own effect is the opposite of the variance of availability, restricted to being on the vertical portions of the supply curve. The slope of the cross effect is the covariance between availabilities (also restricted to being on the vertical portions of the supply curve).

Then, the vector of marginal subsidies $\varphi(\mathbf{K}_0) \in \mathbb{R}^I$ is

$$\varphi(\mathbf{K}_0) = r_0(\mathbf{K}_0) - \mathbf{V}(\mathbf{0}) + b\mathbf{C} \cdot \mathbf{K}_0,$$

and the cumulative total subsidy $\Phi(\mathbf{K}_0) \in \mathbb{R}$ is

$$\Phi(\mathbf{K}_0) = R_0(\mathbf{K}_0) + \mathbf{K}_0^T \cdot (b\mathbf{C} \cdot \mathbf{K}_0 - \mathbf{V}(\mathbf{0})).$$

Since we obtain a closed-form expression of $R_0(\mathbf{K}_0)$, the linear approximation yields a closed form expression of $\Phi(\mathbf{K}_0)$. The marginal subsidy $\varphi(.)$ and the cumulative subsidy $\Phi(.)$ are respectively linear and a quadratic functions of the capacity vector $\mathbf{K_0}$.

We derive a closed-form solution for $\tau(\mathbf{K}_0)$. As previously,

$$\tau(\mathbf{K}_0)\left(\bar{D}(\mathbf{0}) - \frac{1}{b}\tau(\mathbf{K}_0)\right) = \Phi(\mathbf{K}_0).$$

With the linear approximation, $\Phi(\mathbf{K}_\theta)$ given above no longer varies with $\tau$, hence the fixed-point problem disappears. Thus

$$b\bar{D}(\mathbf{0})\tau - \tau^2 = b\Phi(\mathbf{K}_0) \Leftrightarrow \tau^2 - b\bar{D}(\mathbf{0})\tau + b\Phi(\mathbf{K}_0) = 0.$$

The discriminant of the quadratic equation is

$$\Delta(\mathbf{K}_0) = (b\bar{D}(\mathbf{0}))^2 - 4b\Phi(\mathbf{K}_0).$$

We assume (and shall verify later) that $\Delta(\bar{K}_0) > 0$. The quadratic equation admits two roots. We choose the root increasing in each $K_0^i$

$$\tau(\mathbf{K}_0) = \frac{b\bar{D}(\mathbf{0}) - \sqrt{\Delta(\mathbf{K}_0)}}{2} = \frac{b\bar{D}(\mathbf{0}) - \sqrt{(b\bar{D}(\mathbf{0}))^2 - 4b\Phi(\mathbf{K}_0)}}{2}.$$

The tax is an "approximately" linear function of the capacity vector $\mathbf{K_0}$, since it is the square root of a quadratic form.

This expression of $\tau(\mathbf{K}_0)$ enables us to obtain a "simple" expression for the marginal welfare change:

$$\frac{\partial H}{\partial K_0^i} = -\left(\varphi^i(\mathbf{K}_0) + \frac{\tau(K_0)(\varphi^i(\mathbf{K}_0) + b\mathbf{C}\cdot\mathbf{K}_0)}{b\bar{D}(\mathbf{K}_0) - 2\tau(\mathbf{K}_0)}\right)$$

$$= -\frac{(b\bar{D}(\mathbf{0}) + \sqrt{\Delta(\mathbf{K}_0)})\varphi^i(\mathbf{K}_0) + b(b\bar{D}(\mathbf{0}) - \sqrt{\Delta(\mathbf{K}_0)})\mathbf{C}\cdot\mathbf{K}_0}{2\sqrt{\Delta(\mathbf{K}_0)}}$$

Finally, we obtain a closed form expression of the welfare loss:

$$H(\mathbf{0}) - H(\mathbf{K}_0) = R_0(\mathbf{K}_0) + \mathbf{K}_0^T \cdot \left(\frac{b}{2}\mathbf{C}\cdot\mathbf{K}_0 - \mathbf{V}(\mathbf{0})\right) + \frac{1}{2b}\tau^2(\mathbf{K}_0).$$

Of course, the linear approximation is not exact, and the analyst must trade off simplicity against precision. The numerical analysis conducted for the UK shows that the actual marginal values and their linear approximation are reasonably close as long as the baseload technology produces every hour.

# 9 The Highly Visible Hand: Capacity Mechanism

It is not from the benevolence of the butcher, the brewer, or the baker that we expect our dinner, but from their regard to their own self-interest. We address ourselves not to their humanity but to their self-love, and never talk to them of our own necessities, but of their advantages.
—Adam Smith

## 9.1 Introduction

This last chapter discusses an important development in the structure of the electricity markets: over the past ten years, many jurisdictions have added a separate market for capacity to their energy market. The discussion of this important policy issue comes in the last chapter for two reasons. First, it uses many of the results and facts presented in previous chapters. Second, opinions on capacity mechanisms differ sharply among economists, regulators, and industry participants. This difference of opinion is really about the appropriate scope of government intervention in the operations of power markets and the costs and benefits of such intervention. This discussion is therefore a perfect coda to this text.

In most markets, the initial design included a wholesale energy spot market and forward energy markets. Policy makers expected that Adam Smith's "invisible hand" would guide producers' investment decisions and produce an adequate outcome.[1] This design was consistent with one essential policy objective for restructuring the power industry: push to the market investment decisions, that is, have firms and not bureaucrats decide which plants are built and where, and have investors and not rate payers bear risks associated with investment.

1. Exceptions include Spain, which adopted a regulated price for capacity, and Great Britain in the 1990s, which complemented the energy price with a capacity component. The markets in the Northeast United States inherited the short-term generation-adequacy mechanisms from their preceding power pools. They progressively morphed into long-term capacity mechanisms.

Over the years, policy makers and stakeholders have identified various imperfections with this energy-only market design, reviewed in section 9.2, which have lead to the development of capacity mechanisms. Two main imperfections stand out: first, some stakeholders are uncomfortable with extremely high power prices. Some, unconvinced of their economic necessity, object on general principles, while other believe it is impossible to protect customers from the exercise of producers' market power during on-peak hours, despite market monitors' significant increase in analytical skills and investigative authority, documented in chapter 3. This then leads regulators and SOs to cap wholesale spot prices, hence to create a separate capacity mechanism for investors to recover capacity costs.

Second, even in the absence of market power, there is no guarantee that installed capacity will meet the generation-adequacy criterion at all times. Capital-intensive industries are notorious for boom-and-bust cycles, during which periods of capacity expansions (the boom) lead to periods of overcapacity (the bust). Regulators and system operators set up capacity mechanisms to coordinate investment decisions, hence ensure the generation-adequacy criterion is met.

Section 9.2 examines these imperfections in detail and argues that they do not constitute a sound microeconomic justification for setting up additional markets, hence that rationale for capacity mechanisms lies in the realm of political economy, not in microeconomics.

Section 9.3 discusses experiences setting up capacity mechanisms in the United States and in Europe and draws lessons for future capacity market designs.

Finally, sections 9.4 and 9.5 present two slightly different analytical models of capacity mechanisms, each of them tailored to examine one specific failure of energy markets. The first model, presented in section 9.4, summarizes results presented in Léautier (2016). It supposes that competition in generation is imperfect, hence installed capacity is strategically lower than the optimum. Section 9.4 examines the impact of different capacity mechanisms that have been proposed: physical capacity certificates, financial (reliability) options, and coprocurement of operating reserves.

Section 9.5 draws on Lambin and Léautier (2018). It supposes that competition is perfect but that the installed capacity does not meet policy makers' generation adequacy criterion. Section 9.5 examines another capacity mechanism used in Europe, the creation of strategic generation reserves, and the impact of a capacity mechanism in one market on an adjacent market.

## 9.2 Rationale for Capacity Mechanisms

Policy makers in multiple countries have set up capacity mechanisms, that is, mechanisms aimed at ensuring that available generation capacity satisfies the generation-adequacy criterion.

Chapter 2 has shown that, left to its own devices, an electric power market converges in the long-term toward the optimal capacity. This market design is called the energy-only design. Why is there a need for an additional mechanism? Multiple justifications have been proposed over the years, that I have classified in four categories. I first present each argument, then its counterargument. A few sources are available for readers interested in more details—or other perspectives: the September 2008 issue of *Utilities Policy* is entirely dedicated to capacity markets and includes contributions from many academics working on this topic. A series of reports by Peter Cramton and his colleagues also makes a good read: Cramton and Stoft (2006) is the most detailed, Cramton and Ockenfels (2011) and Cramton, Ockenfels, and Stoft (2013) are the shortest ones. The material presented here borrows liberally from these sources, and others.

### 9.2.1 High Prices, Market Power, Price Caps, and Missing Money

#### 9.2.1.1 The argument

As shown in chapter 2, wholesale spot prices need to rise very high for a few hours to provide incentives for optimal investment. If demand were perfectly elastic, prices would have to rise, for example, to $1,000/MWh for 60 hours. If demand is not sufficiently elastic, administrative curtailment may occur. If the SO estimates the VoLL at $20,000/MWh, rationing would have to occur three hours per year on average.

Policy makers and citizens may object to seeing prices rise to that level. A first concern is that scarcity prices, which are much larger than variable production cost, are by definition the outcome of market power: generators, knowing that they are pivotal, bid a price significantly higher than their variable cost.

A second concern is that this pricing pattern is extremely risky: investors recover their cost of capital only if prices spike above variable cost, which should occur a few hours per year on average in a well-designed power system. Thus if Europe is blessed with a few consecutive mild winters (or if the United States encounters a few consecutive mild summers), supply is likely always to exceed demand, hence price will always remain close to variable cost: no recovery of the fixed capacity cost will occur. This risk has a cost: investors will require a high return to invest in generation assets, to compensate for this risk.

Price spikes also create a significant risk for consumers, especially non-price-reactive consumers, whose bills may jump unexpectedly. While large firms and high-income families may be able to absorb this volatility, it may be more challenging for small firms and low-income families.

The third concern is that, as is discussed in chapter 3, generators are tempted to increase the number of on-peak hours, for example, by withholding capacity in the spot market or strategically under investing.

To avoid these high prices, policy makers often impose price or offer caps, well below the prices required to cover the capacity costs. However, as discussed in section 3.4, imposing a price cap creates a missing-money problem: prices need to rise to very high levels for a few hours to cover capital costs. By construction, a price cap prevents prices from reaching these levels, hence leads to underinvestment compared with the required capacity. Thus a separate capacity mechanism is required to cover the capital cost of generation capacity, hence restore investment incentives.

#### 9.2.1.2 The counterargument

The first concern ignores the fact that scarcity prices are set by demand or by the market maker at the VoLL if curtailment is required. This may sound theoretical, but it has a clear practical implication: market makers can impose a cap on offer but allow prices to exceed the cap if demand is willing to pay more than the cap, or if curtailment is required. This practice is adopted for example by ISO New England.

The second concern starts from a correct premise: investors do require a premium to invest in risky assets. Specifically, creditors accept only a low debt level, so that operating profits are sufficient to cover mandatory debt and interest payments, even during mild winters. Since equity is more expensive than tax-advantaged debt, this increases producers' cost of capital. However, investors concerned about the volatility of their assets' profits can sell a fraction or the totality of their output forward in physical markets (presumably to retailers) or in financial markets. Baseload plants sell forward contracts, peaking units call options, and mid-merit plants a mix of these products. We observe robust forward markets' volume for near-dates (less than three to four years) and very little volume beyond four years.[2]

Thus capacity markets that guarantee a price for less than four years do not reduce the risk premium compared with existing forward markets. To provide a new risk reduction instrument, capacity markets would have to guarantee a price for more than four years. This is the case in New England and Great Britain, where new plants can get a longer price guarantee.

One is left wondering why existing producers do not enter into long-dated forward contracts, which last more than four years. To justify government intervention, the absence of long-dated forward markets would have to result from a market failure. However, this is unlikely. Generation-assets' owners and retailers today are financially sophisticated firms, fully aware of the existence of forward markets, and capable to build the team and the processes required to trade derivatives. Market makers are always eager to develop new markets, and financial intermediaries to enter into them. It is therefore plausible that the absence of long-dated forward products reflects investors' preferences, that is, that firms, pension funds, and individuals who finance electric generation assets do so precisely to get exposed to a potential increase in electricity prices, and that electricity retailers are unwilling to lock in a price.

2. This is true for electric power and most commodities.

So while it is true that high-risk increases the capital cost of generation assets, unless a market failure explains why investors do not use long-dated forward contracts to reduce the risk—hence the capital cost—this argument does not justify a new mechanism.

What about price spikes' risk to consumers? Most observers agree that large firms have the competencies to manage this risk, so they should not be a concern. While this may prove more difficult for small firms and residential users, they also can manage this risk contractually with their retailer. As discussed in chapter 5, there appears to be little value in exposing small residential users to spot prices. Thus their retailers may offer them fixed-price contracts for their entire consumption or for their baseline consumption, in which case consumers would be exposed to price spikes only on a small fraction of their demand.

The third concern is very valid. As discussed at length in chapters 3 and 7, market power is a reality in power markets, and left to their own devices traders will find and exploit all opportunities to increase their employers' profits, hence their bonus.

The counterargument, however, is that, as also discussed in chapter 3, market monitors are now well equipped to detect market power abuses. It is, of course, a never-ending race, but there is no structural reason that market monitors—hence consumers—should always lose it.

### 9.2.2 Generation Adequacy

#### 9.2.2.1 The argument

Even if generation were perfectly competitive, there is no guarantee that installed capacity would meet the generation-adequacy criterion. First, we know from experience that very few industries find themselves at their long-term equilibrium capacity. In fact, most industries are cyclical: overcapacity low-price periods alternate with undercapacity high-price ones. This boom-and-bust cycle can be observed in the oil, mining, and chemicals industries.

Second, even abstracting from the boom-and-bust cycle, there is no guarantee that the long-term equilibrium capacity meets the administrative generation-adequacy criterion. For example, equilibrium capacity could lead to 5 to 10 hours of curtailment per year on average, while the administrative criterion is only 3 hours.

#### 9.2.2.2 The counterargument

Again, the premise is absolutely correct: left to its own devices, the market is highly unlikely to meet the generation-adequacy criterion. Based on evidence from other industries and the limited history of market-driven investment in the power industry, I believe generators are more likely to overinvest than underinvest, but I cannot guarantee that the generation-adequacy criterion will always be met.

The counterargument is that an essential objective of the restructuring of the electric power industry in the 1990s was to "push to the market" decisions and risks associated with investment in power generation, that is, to have market forces, not bureaucrats, determine how much investment is required, and to have investors, not rate payers, bear the risks of excess capacity, construction cost overruns, and delays. As seen in chapter 2, the administrative reliability criterion was introduced historically because spot prices did not exist. Now that they do, a more economically sound approach is to set the price at the VoLL whenever curtailment is required. In other words, why adhere to a pre-spot-price methodology now that we have a spot price?

This logic has been embraced, for example, in New Zealand, which no longer has a generation-adequacy criterion. Policy makers, the electricity authority, and the SO carefully assess and communicate to industry participants future risk of rolling blackouts. However, no criterion is imposed.

### 9.2.3 RES Entry

#### 9.2.3.1 The argument

Entry by subsidized RES reduces the spot prices, hence the profitability of incumbent generation, which is needed to serve demand when the wind does not blow or the sun does not shine. In particular, flexible generation is required to compensate for short-term fluctuations in demand.

A more general argument is that, since the capital cost of one technology is covered by a subsidy outside of the spot market, the spot-market price no longer reflects the true opportunity cost of producing electricity. Therefore, other technologies also need to be subsidized.

#### 9.2.3.2 The counterargument

As shown in chapter 8, at the long-term equilibrium, the spot price finances exactly the optimal generation mix, given the mandated RES level. Unless demand increases significantly, which is unlikely in most OECD countries, owners of assets using incumbent technology will have to shut them down before the end of their economic life or, if they can, mothball them. It is unfortunate for them, but having less incumbent capacity present at the long-term equilibrium is precisely the objective of RES entry.

Thus the argument that, since RES are subsidized, incumbent technologies must also be subsidized to guarantee generation adequacy does not hold economically. Owners of incumbent technologies may find asymmetric subsidies unfair. In that case, they should push for reducing subsidies to RES, not jump on the subsidies bandwagon.

Finally, production from RES being variable in the short term, more flexible operating reserves are required to guarantee system reliability. This militates for better operating reserves markets, not capacity markets.

### 9.2.4 Poor Estimate of VoLL

#### 9.2.4.1 The argument

Another argument is that an estimate of the VoLL is required to balance the market in periods of scarcity and determine the level of capacity. Since the uncertainty on estimates of the VoLL is significant (estimates range from \$2,000 to \$200,000/MWh), an energy-only approach will be flawed. This point is best summarized in Cramton, Ockenfels, and Stoft (2013):

> The energy-only approach works because the market will build generators up to the point where an extra MW of generation makes revenues (VoLL x Duration) that exactly equals it costs....
>
> One problem is that it is difficult to estimate VoLL (Stoft 2002, Joskow 2007). The reason is that current markets have hardly any access to information concerning how consumers value reliability, because consumers take few market actions that are based on reliability considerations. This is obviously true for consumers who cannot be individually interrupted, because system operators typically have no control over the electricity flows that go to individual customers. The value of reliability may be revealed only for those (large) consumers, who do have real-time meters and can be interrupted, and if system operators are prepared to black them out based on the performance of their suppliers (see Chao and Wilson 1987, Joskow and Tirole 2006, 2007). But this is of little help since it is the average VoLL of those who cannot respond to price that is required for the energy-only market. Thus, the price-based approach to the adequacy problem ultimately depends on the quality of the regulator's estimate of VoLL.

#### 9.2.4.2 The counterargument

There is indeed very high uncertainty in current estimates of VoLL. As discussed in chapter 2, no single number can accurately represent the VoLL across a wide range of consumers, usages, and durations and circumstances of outages. However, as also discussed in chapter 2, a generation-adequacy criterion (e.g., 3 hours per year on average) is equivalent to a single VoLL (in this case $20,000/MWh). Therefore, if SOs and policy makers are comfortable selecting a generation-adequacy criterion, why should they not be comfortable selecting a single VoLL? The argument thus does not stand up to scrutiny.

### 9.2.5 (Almost) Every Market Has Set Up a Capacity Mechanism

#### 9.2.5.1 The argument

Capacity markets are now implemented in most electricity markets. Only Texas and Alberta in North America, and Australia and New Zealand do not have a capacity mechanism. In Europe, Great Britain has a centralized capacity market, and France a decentralized capacity mechanism.[3] Germany has decided not to adopt a capacity market but instead has opted for strategic reserves, similar to other continental Europe countries (e.g., Belgium, Scandinavia). More academic articles support capacity mechanisms and discuss improvements to their designs than advocate an energy-only design.

One final argument supporting capacity mechanisms is that all these people cannot be wrong.

3. Readers will appreciate the irony of liberal Great Britain having a more centralized capacity mechanism than dirigiste France.

#### 9.2.5.2 The counterargument

As the previous discussion has illustrated, the decision to set up a capacity mechanism owes more to political economy than to microeconomics. There is no right or wrong, simply different preferences. Let us review the incentives of the various stakeholders

**Elected officials** Mindful of the fate of Gray Davis, elected officials want to avoid blackouts, rolling or otherwise. Their preferred generation-adequacy criterion is "zero blackout, ever," irrespective of the cost. Similarly, elected officials dislike price spikes. Some may intellectually agree they are required to provide investment incentives, but they much prefer a spikeless electricity market. Elected officials also want to appear to be in charge. Generation adequacy, managed nationally, is probably one of the last visible issues controlled by energy ministers. Therefore, elected officials in most countries warmly welcome capacity mechanisms, which reduce the risk of having to explain rolling blackouts or power spikes. Box 9.1 presents a particularly telling example.

**Consumers** Consumers want to be protected. As discussed in chapter 4, they have shown little interest in switching suppliers. In some countries, they profoundly distrust the power industry.

In addition, most consumer associations have little trust in markets since the 2008 financial crisis and the 2000–2001 California electricity crisis (which is the only fact in the history of the power industry known to most consumer advocates). Consumer associations are therefore favorably inclined toward capacity mechanisms.

**Energy company managers** The fiduciary duty of energy company managers is to deliver to their investors the highest possible returns for the lowest possible risk. A capacity market appears to meet this objective. In Europe, existing producers are the loudest supporters of capacity mechanisms, arguing that entry of subsidized RES has caused them severe financial damage and that a capacity mechanism is required to right this wrong.

**Box 9.1**
"Keeping the lights on"

One afternoon in October 2015, Great Britain faced a high risk of power shortage. David Cameron, then the conservative British prime minister, immediately called a high-level ministerial reunion to discuss the possible electric power shortage Britain could face during the 2015/2016 winter, and vowed to "keep the lights on." Thus, a prime minister who has been comfortable reducing public spending on the beloved National Health Service, who has dared betting his country's future on the Brexit referendum, feels compelled to be seen participating in a technical discussion over possible power cuts affecting maybe 5 percent of British consumers for 5 hours during next winter. This illustrates elected officials' sensitivity to possible power cuts.

**Government, regulatory agencies, and system operator employees** To a large extent, most agencies are favorable to an increase in their purview, so welcome capacity markets as new markets to design, set-up, and monitor. Capacity mechanisms constantly need to be revised and improved, hence they are a steady source of employment in these agencies.

A less cynical view is that elected officials will blame these agencies in case of a rolling blackout. Since they face the risk, they need a tool to manage it.

**Consultants and academics** Consultants and academics want complex problems to study. Capacity mechanisms are difficult to structure, and multiple redesigns are required to correct unwanted side effects of the previous designs. This has proved highly profitable for economic consulting firms and their academic experts.

As discussed in the introduction, market design is an important and exciting field of economic research. Capacity mechanisms offer analytically complex yet highly practical problems to solve. They have contributed, for example, to important development in multi-unit auction theory.

These considerations contribute to academics' and consultants' broad support for capacity mechanisms.[4]

### 9.2.6 Synthesis

The previous discussion has shown that no market failure justifies the creation of capacity markets. High prices during a few hours do not constitute a market failure. On the contrary, they are perfectly legitimate in a well-functioning energy market. Administrative curtailment is not a market failure. It is expected in a well-functioning energy market until sufficient demand response makes it unnecessary.

As discussed at length in chapter 3, market power is a serious concern that should not be underestimated. However, market monitors have developed the skills to detect exercise of market power, and regulatory agencies have the authority to pursue firms and individuals.

Not meeting the administrative generation-adequacy standard is not a market failure, since the standard is a legacy from the previous industry structure, from which prices were absent. It is no longer required now that customers can—and do—adjust their demand to the wholesale spot price, as discussed in chapter 5.

In summary, the justifications for capacity markets boil down to recreating the previous industry features of (a) prices constant across states of the world and (b) an administratively determined generation-adequacy standard. It appears that stakeholders (policy makers, regulators, system operators, utilities, etc.) did not anticipate the implications of

4. I am absolutely not suggesting some academics have been recommending capacity mechanisms for personal gain, while being convinced they were not required. Rather, I am observing that, being convinced of the need for capacity mechanisms, many first-rate economists have actively promoted and designed them since the early 2000s.

the restructuring of the industry and are not ready to shift their mind-set. Failure to adapt is hardly a justification for government intervention.

I do not underestimate the challenge of explaining to voters that, in the new industry structure, market participants and not a governmental agency determines the number of hours of rolling blackout. While this may be possible if no blackout occurs, this is really a tough political sell in the aftermath of a blackout. All arguments supporting a capacity mechanism share the same premise: electricity is different. Because of its technical characteristics, but mostly because of its importance in our societies, generation adequacy cannot be left to market forces; it has to be determined and managed by governments.

We leave the realm of microeconomics and enter that of political economy. Policy makers face a choice between two imperfect solutions. If they believe (a) generation adequacy cannot be left to market and (b) they can design a market stable enough to provide incentives for long-term investment but flexible enough to accommodate new technologies, for example, RES, demand response, and energy storage, they should go the capacity market route. The next section discusses the best way to do this.

If on the other hand, they believe (a) their primary responsibility is to prevent exercise of market power and not interfere with investment decisions by private firms, (b) the cost of prescribing administratively the number of Gigawatts of installed generation capacity exceeds the cost of a few extra hours of rolling blackouts, and (c) they can explain to consumers—who are also voters—that installed generation capacity will no longer be determined by policy makers but by market forces, a capacity market is not required.

Setting up a capacity mechanism is a legitimate political choice, not a micro economic imperative. This decision boils down to a decision on the appropriate role for government in the economy and is not limited to the power industry. It has a philosophical dimension: Is government intervention legitimate? But also a practical one: Are the imperfections of government interventions less damaging than the market imperfections they correct?

## 9.3 Experience from Capacity Mechanisms

Different capacity mechanisms have been implemented in different countries. Convergence toward a robust design has been a long process.

### 9.3.1 The Basic Design

The standard capacity mechanism is a forward market for physical certificates. It is implemented in the ISOs on the East Coast of the United States and in Great Britain.

The basic design is quite simple: the SO runs a centralized forward market for installed capacity. To fix the ideas, suppose the SO wants to guarantee generation adequacy for year $T$. In year $(T-4)$ it first estimates the required generation capacity, using the technique described in chapter 2. Second, it runs an auction to purchase this required capacity. All winners in the auction are promised to receive in year $T$ the same market-clearing price for their capacity accepted in the auction. In exchange, this capacity has to be available when called upon by the SO during year $T$.

Owners of existing generation assets are allowed to participate in the auction, and so are developers of new assets, if they can demonstrate their assets will be online by year $T$. In New England, for example, new assets are allowed five years of price guarantee, that is, from years $T$ to $(T+4)$, to help asset developers secure financing, while existing assets receive only one year of price guarantee.

Retailers and consumers are also allowed to include demand-response bids into capacity markets. The demand they promise to shed is considered equivalent to production capacity.

The certificates auctioned in year $(T-4)$ are fully exchangeable until year $T$. Market participants can adjust their position as their circumstances evolve.

### 9.3.2 Implementation Challenges

While the basic design is simple, the devil, as usual, lies in the details of implementation. Below is a (nonexhaustive) list of issues to be solved. Each issue is conceptually simple, at least now that it has been extensively discussed. The challenges for the SO are to (a) build a consensus among stakeholders with potentially diverging interests on a practical approach to address it and (b) hire the resources and develop the process to implement this approach.

#### 9.3.2.1 How is availability defined?

Availability is not as simple as it may sound. All power plants need to be maintained, and none is available every hour of every year. It is therefore important to define precisely what availability means. Does a plant have to be available whenever called upon by the SO? Or is availability defined only for the high-demand periods (winter in Europe, summer in the United States)? During the availability period, does a plant have to be available 100 percent of the hours, 99 percent or another fraction?

#### 9.3.2.2 What fraction of plants' nameplate capacity is eligible for capacity credits?

A related issue is this: Are plants allowed to bid 100 percent of their nameplate capacity into the capacity market? Should historical capacity factors be taken into account? Which fraction of RES nameplate capacity is eligible? Since RES production is variable, many incumbents advocate for a small fraction. Unsurprisingly, RES developers advocate for a high fraction.

#### 9.3.2.3 How do we enforce the availability target?

If power producers face capped wholesale prices and cover a significant portion of their capacity cost through the capacity market, they have weak incentives to maximize their availability in spot markets. Under the early capacity market design in the United States, actual availability in the spot market was much lower than contracted availability, estimated using historical availability. Therefore, contrary to what may have been anticipated, setting up a capacity market does not automatically increase generation capacity available in the spot market. This finding is illustrated in the model presented in section 9.4.

The SO must design an enforcement mechanism. The most natural mechanism is to have a producer pay a very high price for the Megawatt-hours it was supposed to deliver when requested by the SO but did not. However, there is a risk that producers, expecting few requests from the SO, do not maintain their asset, that is, they find that the cost of maintenance exceeds the expected cost of not responding to a request from the SO. The SO would then find out too late that the capacity is not ready to produce. To solve this problem, the SO requests power from generators even when not required, to test a producer's readiness, and monitors that existing generation assets providing certificates are still operational, and that planned capacity having received certificates has indeed be installed.

#### 9.3.2.4 How is the demand curve for capacity defined?

Initial capacity markets in the Northeast of the United States, called installed-capacity (ICAP) markets, grew from the obligation power pools imposed on load-serving entities to demonstrate that they held generation certificates in excess of their demand. Cramton and Stoft (2006, page 31) describes the ICAP design: "All load is required to buy its proportional share of capacity credits so that the total capacity purchased equals the adequacy target. All installed capacity is allowed to sell capacity credits equal to its nameplate capacity. This is enforced by charging load something like \$150,000/MW-year of shortfall below its capacity requirement."

There was a concern that this market design led to unstable capacity prices, oscillating between zero and the penalty price. If installed capacity falls below the generation adequacy target for a given period, at least one LSE will miss its target and pay the fine. The value of a marginal Megawatt is the penalty (\$150,000/MW-year in the above example), hence the price for all Megawatts is the penalty. If installed capacity exceeds the generation adequacy target for a given period, the value of the marginal Megawatt is zero, hence the price for all Megawatts is zero. Capacity prices in the ICAP market thus exhibit the same pattern as electricity prices in the energy market: very high if demand exceeds capacity and rationing is required, very low otherwise.

To mitigate this instability, SOs have developed downward-sloping demand curves. This requires administrative determination of a set of parameters: maximum and possibly minimum price in the capacity market shape of the slope around the generation adequacy target.

#### 9.3.2.5 How to treat demand response?

As discussed in chapter 5, demand response is essential to improving the efficiency of the power industry. If the wholesale spot price is capped, RTP is not possible. Demand response has to find its value through the capacity market. Experience in the United States suggests this is challenging. As chapter 5 notes, measuring actual demand response is difficult: since baseline consumption is by definition unobservable, consumers (or their demand-response operator) have strong incentives to overstate their baseline to increase their compensation. Predicting future demand response three years in advance is even more challenging.

#### 9.3.2.6 How to treat adjacent markets?

Generation adequacy is the responsibility of the SOs: regional ISOs in the United States, national TSOs in Europe. Each jurisdiction decides whether or not to set up a capacity mechanism, and the SO decides on the specific rules. The treatment of adjacent markets raises a number of issues. Can market (or country A) set up a capacity mechanism if adjacent country B does not? The simple economic model presented in section 9.5 proves that the answer is a qualified yes: market A would have to have a mechanism to curtail exports in periods of stress. Since Megawatt-hours follow the laws of physics, this would require reducing available export capacity on the transmission lines.

If market A sets up a capacity mechanism, and country B does not, does country B free ride on country A's capacity? Should country B compensate country A? Section 9.5 proves that in some instances at least, the opposite occurs: consumers in market B experience higher expected loss of load when market A sets up a capacity market. The intuition is that generation capacity attracted to market A by the capacity remuneration crowds out capacity in market B.

Thus, capacity markets' architects need to at the minimum to coordinate with their neighbors.

### 9.3.3 Alternative Designs

#### 9.3.3.1 Financial commitment

The design with financial commitments is identical, except that participants sell a financial option instead of a physical commitment. The option is designed as follows: the SO sets a strike price. If the wholesale spot market exceeds the strike price, each auction winner pays the SO the wholesale spot price minus the strike price for its share of auctioned volume times the demand. For example, suppose the strike price is €100/MWh, the wholesale spot price €500/MWh, auctioned volume is 100 GW, and actual demand is 90 GW. For every hour at which the wholesale spot price is €500/MWh, a producer that has won 10 percent of the auctioned volume will pay

$$(500 - 100) \times 10\% \times 90{,}000 = 400 \times 9{,}000 = €3.6\,\text{million}.$$

This approach is preferable to the physical capacity certificates. The reason is that since producers are liable when prices are high, they have strong incentives to avoid or limit price spikes, thus they maximize their availability when the system is tight. This result is reminiscent of the Allaz and Villa (1993) result discussed in chapter 4: a forward commitment increases ex post competition. In this case, the forward commitment is organized centrally.

Rigorous analysis presented later in this chapter proves that reliability options alone are not sufficient to induce optimal investment. The SO must again impose a "no short sale" rule, that is, monitor that producers to make sure they do not sell more option than they have installed capacity.

#### 9.3.3.2 Decentralized market

Another possibility is to have decentralized certificates markets. This was the initial ICAP design in the United States, discussed above. It is still in force in the New York ISO and has been recently proposed in France.

This approach gives market participants more choice in procuring capacity. However, it is unclear that it guarantees generation adequacy: anticipating a loss of market share, or lower demand, retailers may underprocure. For that reason, the French SO has reserved the right to run a centralized auction to purchase sufficient generation capacity to meet the adequacy target.

#### 9.3.3.3 Strategic reserves

Another possibility, used in many European countries (e.g., Scandinavia, Belgium, Germany), is to constitute strategic reserves. The SO in effect purchases the capacity of generation assets it deems essential for the operation of the system.

This mechanism is reminiscent of the Reliability Must Run plants created by the California ISO in the late 1990s. There, the SO was purchasing energy to manage transmission constraints or ensure voltage support. Here, the SO purchases capacity to meet the generation adequacy target.

One argument in favor of strategic reserves is that they cost less than setting up a capacity market, since only a few plants and not the entire fleet are paid. However, this argument is mistaken. The intuition, developed by Lambin and Léautier (2018) and reproduced in section 9.5, is straightforward: all mechanisms that achieve the same generation-adequacy target lead to the same installed-generation capacity, hence generate the same total cost.

Some factors may weaken this result, for example, investors' risk aversion. Also, a strategic reserve alters the generation mix, which may modify the total cost. However, the intuition is robust: we should not expect generation adequacy to be cheaper using one approach or the other.

### 9.3.4 Market Power in Capacity Markets

As discussed previously, one important justification for capacity markets is exercise of market power in the energy market, which leads to the imposition of price caps and missing money. Does market power occur in capacity markets?

The original hope was that market power would be more manageable in capacity markets since they were long dated. If new producers are allowed to enter they will challenge whatever market power incumbents tried to exercise.

Alas, this hope turned out to be misplaced. When few bidders are competing, market power also occurs in capacity markets, although it may be less visible than a $10,000/MWh spike in the spot price or a rolling blackout owing to strategic output reduction. The Colombia capacity markets, where market participants have bid up the auction price is a classic, example, recounted by Harbord and Pagnozzi (2014). Another example, a group of researchers has conducted an experimental analysis of capacity markets, and conclude that "capacity markets are less competitive than predicted" (Le Coq, Orzen, and Schwenen 2017).

How big is the impact of market power in capacity markets? Analysis presented in section 9.4 offers a preliminary estimate. The premise of section 9.4 is that producers strategically underinvest, that is, they prefer aggregate capacity $K^C$ lower than the capacity required by the SO, denoted $K^*$. This situation is illustrated in figure 9.1. Consider the aggregate profit from a marginal MW of installed capacity, represented by the left curve in figure 9.1. Under certain regularity conditions, it is decreasing and crosses zero when aggregate capacity is equal to $K^C$: below $K^C$, firms find it profitable to invest, which is to say that the value of a marginal MW is positive. Firms do not willingly invest past $K^C$, since the value of a marginal MW would be negative. Since $K^* > K^C$, the value of a marginal MW of installed capacity around $K^*$ is negative: the left curve is indeed below zero for $K = K^*$. To induce producers to increase installed capacity to $K^*$, the SO must provide financial incentives. Specifically, the value of the marginal MW including the remuneration in the capacity market, which is the right curve in figure 9.1, must be larger than or equal to zero for $K = K^*$.

Total profit is the integral of marginal profit, that is, the surface under the marginal profit curve. Total profit absent a capacity mechanism and with a capacity mechanism are presented respectively on the left and right panels of figure 9.1. One verifies graphically that including a capacity mechanism significantly increases total profits.

The incremental aggregate profit from capacity payment in figure 9.1 is very large. The illustrative model presented in chapters 2 and 3 enables us to compute an estimate. I provide the average value over all relevant values of the price cap $\bar{p}$, and for the proportion of price-reactive customers $\alpha = 5\%$, which appears appropriate for most markets. For price elasticity of demand $\eta = -0.01$, the incremental profit from capacity markets is around €5,100/per MW per year, approximately 10 percent of investment cost; for $\eta = -0.1$, it is around €8,400/per MW per year, approximately 16 percent of investment cost. These estimates illustrate that the cost of the capacity market to consumers is not trivial.

This analysis has a disturbing implication: if the SO considers setting up a capacity market, he increases the incentives for producers to exercise their market power. Suppose

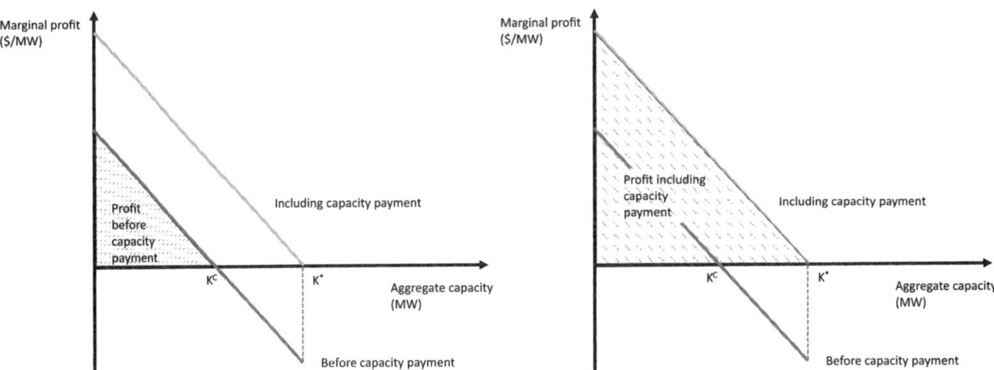


**Figure 9.1**
Marginal and aggregate profit before and including capacity payments. Marginal profit before capacity payment (left line) is null when aggregate capacity is $K^C$, hence negative when aggregate capacity is $K^*$. Marginal profit including capacity payment (right line) is null when aggregate capacity is $K^* > K^C$, hence positive when aggregate capacity is $K^C$. The left panel presents aggregate profit before capacity payment. The right panel presents aggregate profit including capacity payment.

policy makers are concerned about market power. If they set up a capacity mechanism, they promise producers higher profits than the imperfectly competitive level. Producers then have strong incentives to appear imperfectly competitive to induce policy makers to set up a capacity market.

### 9.3.5 The Process of Designing Capacity Mechanisms

As the list of issues above suggests, designing capacity mechanisms is challenging. The rules must be extremely detailed. Significant value is at stake for industry participants, hence the consultation and rule-making processes are long and arduous.

In the United States, the process has lasted more than ten years. Starting from the ICAP markets in the Northeast, different versions have been proposed, each improving on the previous one. Optimists see the glass half full: by now, we have ironed out all initial defects and converged towards a stable design. Pessimists see the glass half empty: we are constantly discovering unanticipated side effects that must be corrected. Furthermore, since technology and policy objectives are constantly evolving, we must constantly adapt capacity market designs.

The recent experience in Great Britain seems to prove the pessimists right. Great Britain adopted the ISO New England's forward-capacity market design, arguably one of the most advanced. However, the fall 2015 auction resulted in diesel generators winning a significant share of the credits: British consumers found themselves providing £850 million in subsidies to highly polluting diesel generators. The political backlash that followed led to a change in the capacity market rules to guarantee that the share won by diesel generators in the future would be lower. Market designers, however, must tread carefully, since capacity markets must be technology neutral. So, if diesel generation is the most economic solution, this is the outcome that a fair market will produce. Still in Great Britain, in the summer of 2016, it emerges that the SO had not been able to secure enough demand response through the capacity market. Again, rules must be adapted.

### 9.3.6 New Technologies in Capacity Mechanisms

One important consequence of capacity market is that the wholesale spot prices no longer fully cover the cost of producing electricity. Consider first that policy makers are concerned about extremely high on-peak prices, and impose a cap on wholesale prices. Then, expected operating profits under these capped prices fall short of the capacity cost of a marginal generation unit: this is the "missing money" problem. The capacity cost of power generation units is recovered through revenues from both spot and capacity markets.

Consider now that policy makers are concerned that equilibrium-installed capacity does not meet the generation-adequacy criterion, and set up a capacity market to achieve their target. If they are correct, additional generation capacity (compared to the equilibrium) is installed, which reduces spot-market prices. Revenues from the capacity market are required to compensate for that shortfall.

In both cases, the wholesale spot prices do not fully reflect the value users put on a marginal Megawatt-hour.

Therefore, investors cannot use their anticipation of the wholesale spot price to build business plans for new technologies. Instead, they will have to negotiate with policy makers and SOs the conditions of their participation into capacity markets, or the subsidies they should receive. Experience shows that the outcome of these negotiations is highly uncertain, and highly political. Incumbents will find excellent arguments to slow down or reduce new technologies' entry. The resulting regulatory environment will thus be permanently changing to adapt to new circumstances. Investors will face significant regulatory risks, which will certainly deter investment. They will devote more effort to convincing policy makers of the merits of their innovation than to actually developing said innovation. Government, and not competition, will determine the technology mix, which appears to contradict the objective of the restructuring of the power industry.

Consider, for example, demand response. An industrial user who invests in the ability to reduce its demand for a few hours for a few days is no longer able to receive the "true" value of the Megawatt-hours he decides not to consume. His demand-response capacity must then be valued in the capacity market, which creates measurement challenges as previously discussed.

Energy storage is another example. Most observers argue that storage will become more economical in the next few years. A storage unit arbitrages between on-peak and off-peak prices: it buys at low prices and resells at high prices. Some units perform daily arbitrage, that is, buy at night and resell the following afternoon, others perform weekly or monthly arbitrage, some hydro reservoirs perform annual arbitrage. If the on-peak prices are lower than the value of the marginal Megawatt-hour, the market value of storage units will be lower than their "true" economic value, hence underinvestment in storage units occurs. To avoid this, storage units will have to participate in capacity mechanisms, that is, new rules to enable their participation will have to be devised (see, e.g., Waterson 2017).

Designing new rules is not as simple as it may seem. First is the technical challenge. For example, how can we predict the volume of demand response an industrial site will be able to deliver three years from now? Some firms probably do not know what their industrial production will be three years from now. Similarly, how can we predict the availability of batteries? While the technology is being developed, very little information is available. Using financial and not physical commitments solves partially this technical challenge.

Second is the political economy challenge. New entrants will want highly welcoming rules, tailored to their technical specificities to increase their profits. Incumbents will, of course, have a different view. Policy makers will be torn between facilitating the deployment of new technologies (Harbord and Pagnozzi 2014) and responding to the arguments of existing firms. The example of Colombian wind farms illustrates the challenge.

The slippery slope is to go from RES subsidies and capacity markets to central planning of the generation mix, that is, having the government—through the SO or other entities—decide how much total capacity is required and organize a tender for different types of generation technologies. To a certain extent, this seems to be the route the British government is following in the new energy market design it enacted in 2013 (see, e.g., the 2013 Energy Act and the Electricity Market reform, Department of Energy and Climate Change 2015, October 2015 update).

New technologies are expected to profoundly transform the electricity industry. This transformation is essential, as it is a critical part of the meeting the climate change challenge. If the market design constitutes an obstacle to investment, new technologies' entry may be delayed, or may never occur, which would be immensely counterproductive.

### 9.3.7 Implications for Future Capacity Mechanism Design

Analytics as well as international experience suggest that a centralized financial-commitments market constitutes the best capacity mechanism.

Unfortunately, selecting a design is only one part of setting up a capacity mechanism, as a large number of detailed implementation rules must be designed and choices made. As is often the case, the devil lies in the details.

Market designers face a challenging trade-off: on the one hand, rules must be constant across time to elicit investment, since regulatory risk deters investment more effectively than price risk. On the other hand, flexibility is required to accommodate changes, in particular, new technologies that are rapidly coming into the market.

## 9.4 A Model of Capacity Mechanisms and Market Power

### 9.4.1 Academic Literature on Capacity Mechanisms

A rich academic literature describes and analyzes different capacity mechanisms. Stoft (2002) discusses average VoLL pricing, Hogan (2005) proposes an energy cum operating reserves market, and Cramton and Stoft (2006, 2008) and Cramton and Ockenfels (2011) propose a financial reliability options mechanism.[5] Joskow and Tirole (2007) show that a capacity market and a price cap do not restore the first best with more than two states of the world. Chao and Wilson (2005) examine the impact of options on spot-market equilibrium, investment, and net surplus.

### 9.4.2 Physical Capacity Certificates

#### 9.4.2.1 Set-up

The set-up is that of section 3.7: all customers are assumed to respond to the spot price, $N$ producers compete imperfectly facing inverse demand $P(Q, \theta)$, and the SO imposes cap $\bar{p}$ on the wholesale spot price. As discussed in chapter 3 and illustrated in figure 3.3, for a given price cap $\bar{p}$, the long-term equilibrium capacity $K^C(\bar{p})$ is lower than the optimal capacity $K^*(\bar{p})$ for two reasons: producers strategically underinvest to increase the probability of the on-peak period, and when the price cap is binding, the value received by producer (equal to the cap $\bar{p}$) is lower than the value of the marginal Megawatt-hour (equal to the VoLL $v(\bar{p}, \theta)$).

Suppose the SO sets up a market to procure from producers a volume of physical capacity certificates equal to or larger than its target capacity $K^T > K^C$. The target capacity can be the economic optimum, that is, $K^T = K^*(\bar{p})$, or not. To simplify the notation and analysis, operating reserves are ignored, since their inclusion does not modify the economic insights, as was proved in section 2.6. All units (old and new) receive the same compensation in the physical certificates markets. $\phi^n$ and $\Phi = \sum_{m=1}^{N} \phi^m$ are respectively the certificates sold by producer $n$ and the aggregate volume of certificates sold.

This section provides the main results and their intuition. Interested readers can find the proofs in Léautier (2016), from which this section borrows heavily.

The timing is as follows:

1. The SO designs the rules of the energy and capacity markets. All parameters are set.
2. Each producer builds capacity $k^n$.
3. Each producer sells physical capacity certificates $\phi^n$ to the SO, according to the rules set up previously.

5. Strictly speaking, these options ensure resource adequacy, not reliability. Nevertheless, I use the term "reliability options" as it was the term used in the original Cramton and Stoft articles.

4. The spot markets are played. In each state of the world $\theta$, producers compete à la Cournot facing inverse demand $P(Q,\theta)$, given their installed capacity $k^n$ and their physical certificates commitment $\phi^n$. The SO pays the physical certificates to the producers and passes the cost of purchase to customers.

To simplify the analysis, this pass-through is assumed not to distort consumption decisions in the spot market, for example, the pass-through is proportional to the peak consumption.

Stages 2 and 3 can be inverted or simultaneous: generators first build the plants, then sell physical capacity certificates, or build and sell simultaneously.

As discussed in section 9.3, SOs offer a "smoothed" (inverse) demand curve:

$$H(\Phi)=\begin{cases} r & \text{if} \quad \Phi\leq K^T \\ h(\Phi) & \text{if} \quad K^T<\Phi<K^T+\Delta\bar{K} \\ 0 & \text{if} \quad \Phi\geq K^T+\Delta\bar{K} \end{cases}$$

where (a) $r$, the capital cost of capacity, is the maximum price the SO is offering for capacity, (b) $\Delta\bar{K}>0$ is an arbitrary capacity increment, and (c) $h(.)$ is such that $H(.)$ is $C^2$, except maybe at $K^T$ and $K^T+\Delta\bar{K}$, $h'(\Phi)<0$, $2h'(\Phi)+\phi h''(\Phi)<0$ for all $\phi$, and

$$|h'(K^T)|\geq\frac{Nr}{K^T}. \tag{9.1}$$

Condition (9.1) simplifies the exposition but is not essential. It is met in practice. For example, Cramton and Ockenfels (2011) suggest a linear form for $h(.)$ with $\frac{\Delta\bar{K}}{K^T}=4\%$. Condition (9.1) is then equivalent to $N\frac{\Delta\bar{K}}{K^T}\leq 1$ and holds as long as fewer than twenty-five producers compete.

The three-stage game (stages 2, 3, and 4) is solved below, assuming the target $K^T$ and the function $H(.)$ are given, and the SO uses $r$ as the reserve price. Determination of the optimal stage 1 parameters is discussed later.

#### 9.4.2.2 Main Result

**Proposition 9.1.** The SO must impose and monitor that the installed capacity exceeds the capacity certificates sold by each generator: $k^n\geq\phi^n$. Then (a) producers sell as many credits as they install capacity, and (b) $K^T$ is the unique symmetric equilibrium investment. A capacity market increases producer's profit and net surplus if $K^T\leq K^*$.

The first observation is that existence of a physical capacity certificates market alone does not alter investment incentives. Suppose first the SO imposes no condition on certificates sales. Denote $\Pi^n(k^n,K,\bar{p})$ producer $n$'s expected profit when her capacity is $k^n$, aggregate capacity is $K$, a price cap $\bar{p}$ is imposed. Producer $n$'s expected profit, including revenues from the capacity market, is $\Pi^n_{CM}(k^n,\phi^n,K,\Phi,\bar{p})=\Pi^n(k^n,K,\bar{p})+\phi^n H(\Phi)$. The total derivative of profit with respect to $k^n$, denoted $\frac{d\Pi^n}{dk^n}$ with a slight abuse, is

$\frac{d\Pi^n}{dk^n}(k^n, K, \bar{p}) = \frac{\partial\Pi^n}{\partial k^n}(k^n, K, \bar{p}) + \frac{\partial\Pi^n}{\partial k^n}(k^n, K, \bar{p})$. We have

$$\frac{d\Pi^n_{CM}}{dk^n}(k^n, \phi^n, K, \Phi, \bar{p}) = \frac{d\Pi^n}{dk^n}(k^n, K, \bar{p}):$$

a capacity market alone does not modify investment incentives, $K(\bar{p})$ remains the installed capacity. This observation was articulated very early on (e.g., Wolak 2004).

In the model, producers exercise market power by reducing capacity ex ante, and not by withholding output on-peak. The SO must therefore ensure that producers cannot sell more certificates than their installed capacity, that is, the SO must impose $k^n \geq \phi^n$.

When she does, producers sell exactly as many certificates as they have installed capacity since incremental capacity is unprofitable unless it collects capacity markets revenues. Then, since $k^n = \phi^n$ at the equilibrium, producer $n$'s program is

$$\max_{k^n} \Pi^n_{CM}(k^n, K, \bar{p}) = \Pi^n(k^n, K, \bar{p}) + k^n H(K).$$

Léautier (2016) proves $k^n = \frac{K^T}{N}$ is the unique symmetric equilibrium. Producers' profit is also increased, as illustrated in figure 9.1.

The net surplus $W^C(K, \bar{p})$ is independent of the transfers between consumers and producers. As proved in section 3.7, $W^C(K, \bar{p})$ is increasing for $K \leq K^*(\bar{p})$, thus increasing aggregate capacity from $K^C(\bar{p})$ to $K^T \in [K^C(\bar{p}), K^*(\bar{p})]$ increases net surplus. If $K^T > K^*(\bar{p})$, the net surplus may be reduced. This point is illustrated in section 9.5.

#### 9.4.2.3 Robustness

Is there an optimal structure to the physical certificates market? Since the expected net surplus $W^C(K, \bar{p})$ is independent of the transfers between consumers and producers, all sets of parameters that lead to the same equilibrium capacity yield the same net surplus. For example, a reserve price $p^K$ lower than $r$ but high enough that $k^n = \frac{K^T}{N}$ remains the unique equilibrium would lead to the same equilibrium, would reduce transfers from consumers to producers, hence increase consumers' surplus and reduce producers' profits.

Actual inverse demand curves for certificates are slightly different from the function $H(.)$. In particular, the target $K^T$ lies in the downward sloping portion, and not at the upper corner. Proving formally existence of a unique equilibrium is more difficult, but the intuition is unchanged.

If Condition (9.1) is not met, the aggregate capacity at the unique symmetric equilibrium is $K^C_{CM} \in (K^*, K^* + \Delta\bar{K}]$. Welfare increases if and only if $\Delta\bar{K}$ is small enough that $W(K^* + \Delta\bar{K}) \geq W(K^C)$.

### 9.4.3 Financial Reliability Options

Financial contracts constitute another mechanism used in power markets. This section examines financial reliability options, proposed by Oren (2005), Cramton and Stoft (2006, 2008), and, more recently, Cramton and Ockenfels (2011). Options and not forward contracts are the financial instruments analyzed here, since they are in general preferable (Chao and Wilson 2005, which examines a slightly different option design). These options constitute an insurance against spot energy prices higher than a pre-agreed strike price $\bar{p}^S$, sold by producers to customers. If the spot price $p(\theta)$ is lower than $\bar{p}^S$, producer $n$ does not make any payment. If $p(\theta) > \bar{p}^S$, producer $n$ pays $(p(\theta) - \bar{p}^S)$ times a fraction of the realized demand equal to his fraction of the total options sale.

The SO does not impose a cap on wholesale prices, and runs an auction for financial reliability options. $\phi^n$ and $\Phi = \sum_{m=1}^{N} \phi^m$ are respectively the options sold by producer $n$ and the aggregate volume of options sold. To limit the potential exercise of market power, Cramton and Ockenfels (2011) propose the SO requires all available generation capacity must be committed forward through option sales: $\phi^n \geq k^n$. This limits the timings considered: capacity $k^n$ is committed before reliability options $\phi^n$ are sold.

The timing and notation are identical to the capacity market case, except that the subscript $_{RO}$ is added when appropriate. A very simple auction set-up is assumed, similar to the one suggested by Cramton and Stoft (2008): the SO determines the volume she desires to purchase, denoted $K^T$, sets the capital cost of capacity $r$ as the reserve price for the auction, and proposes a downward-sloping inverse demand curve for options. To simplify the exposition, assume that $H_{RO}(\Phi) = H(\Phi)$.

When the spot price exceeds the strike price, consumers receive rebate $\max(p(\theta) - \bar{p}^S, 0)$ per unit of energy purchased. We assume they behave perfectly rationally, that is, they use $\bar{p}^S$ as the effective price guiding their consumption decision. Then, actual demand does not depend on the spot price, which leads to rationing. The SO then sets the spot price at the VoLL: $p(\theta) = v(\bar{p}^S, \theta)$. This point is extremely important: the price protection (or the option) has the same economic impact on price-reactive consumers as a cap on the wholesale spot price.[6]

Chao and Wilson (2005) examine a slightly different market structure: they consider physical options paired (or not) with a complementary price insurance, and compute the linear supply-function equilibrium for options forward sales and power spot sales. Their findings are aligned with those presented below.

6. Léautier (2016) proves the results also hold if consumers do not behave perfectly rationally and use $P(K, \theta)$ as the price guiding their consumption.

### 9.4.4 Expected Profits with Financial Reliability Options

The producers' profit function is characterized below:

**Lemma 9.1.** The expected profit of producer $n$ is

$$\Pi^n_{RO}(k^n, K, \phi^n, \Phi, \bar{p}^S) = \phi^n H(\Phi) + \Pi^n(k^n, K, \bar{p}^S) + \left(k^n - \frac{\phi^n}{\Phi}K\right)\int_{\hat{\theta}_0(K,\bar{p}^S)}^{+\infty}(v(\bar{p}^S,\theta) - \bar{p}^S)f(\theta)d\theta. \tag{9.2}$$

Producer $n$ receives the revenues from options sale $\phi^n H(\Phi)$, plus profits from the energy market. As discussed in chapter 3, the imperfectly competitive spot price may reach the cap after or before generation produces at capacity. Suppose first the cap starts binding on-peak, that is, $\hat{\theta}^C_N(K,c) \le \hat{\theta}_0(K,\bar{p}^S)$. Off-peak, producers receive Cournot operating profit as discussed in section 3.5. On-peak, as long as the wholesale spot price is lower than the strike price $\bar{p}^S$, operating profit is $(P(K,\theta) - c)$.

When the wholesale spot price reaches the strike price $\bar{p}^S$, consumers effectively pay $\bar{p}^S$ per unit, hence demand no longer reacts to price. The SO must curtail a fraction of customers, and set the price at the VoLL. Producers' operating profit is $(v(\bar{p}^S,\theta) - c)$. In addition, producers also pay back the difference $(v(\bar{p}^S,\theta) - \bar{p}^S)$ between the spot and the strike prices times their fraction $\frac{\phi^n}{\Phi}$ of the total demand. Since $\hat{\theta}_0(K,\bar{p}^S) \ge \hat{\theta}^C_N(K,c)$, total demand is equal to total capacity $K$ and the payment is proportional to $\frac{\phi^n}{\Phi}K$.

The profit realized in states higher than $\hat{\theta}_0(K,\bar{p}^S)$ is

$$\begin{aligned}
\pi^n_{RO}(k^n, K, \phi^n, \Phi, \bar{p}^S, \theta) &= k^n(v(\bar{p}^S,\theta) - c) - \frac{\phi^n}{\Phi}K(v(\bar{p}^S,\theta) - \bar{p}^S) \\
&= k^n(\bar{p}^S - c) + \left(k^n - \frac{\phi^n}{\Phi}K\right)(v(\bar{p}^S,\theta) - \bar{p}^S) \\
&= k^n\left(\left(1 - \frac{\phi^n}{\Phi}\frac{K}{k^n}\right)v(\bar{p}^S,\theta) + \frac{\phi^n}{\Phi}\frac{K}{k^n}\bar{p}^S - c\right).
\end{aligned}$$

The second line in the above expression shows that expected profit can be expressed as $\Pi^n(k^n, K, \bar{p}^S)$, the Cournot profit for price cap $\bar{p}^S$, plus an additional term as indicated on equation (9.2).

The third line in the above expression shows that producers face a weighted average of the spot price and the option price, hence are less sensitive to an increase in spot price. As seen in chapter 3, producers competing à la Cournot and holding forward contracts face lower incentives to exert market power in the spot market.

Equation (9.2) also obtains if $\hat{\theta}^C_N(K,c) > \hat{\theta}_0(K,\bar{p}^S)$.

### 9.4.5 Equilibrium Capacity with Financial Reliability Options

We now establish existence and properties of a symmetric equilibrium. The details of the proofs are presented in Léautier (2016).

**Proposition 9.2.** Suppose the SO imposes $\phi^n \geq k^n$ and chooses strike price $\bar{p}^S$ such that

$$\int_{\hat{\theta}_0(K,\bar{p}^S)}^{+\infty} (v(\bar{p}^S,\theta) - \bar{p}^S) f(\theta) d\theta \leq r. \tag{9.3}$$

In equilibrium, each producer sells $\phi^n = \frac{K^T}{N}$ options. Reliability options reduce but do not eliminate the underinvestment problem. $K_{RO}^C(\bar{p}^S)$, the unique symmetric equilibrium of the options and investment game, verifies

$$K^C(\bar{p}^S) \leq K_{RO}^C(\bar{p}^S) < K^*(\bar{p}^S),$$

with equality occurring when $N = 1$.

As usual, we solve the game by backward induction. Start first with the second stage, reliability options. We have

$$\frac{\partial \Pi_{RO}^n}{\partial \phi^n}(k, K, \phi^n, \Phi, \bar{p}^S) = H(\Phi) + +\phi^n H'(\Phi)$$
$$- \frac{\Phi - \phi^n}{\Phi^2} K \int_{\hat{\theta}_0(K,\bar{p}^S)}^{+\infty} (v(\bar{p}^S,\theta) - \bar{p}^S) f(\theta) d\theta.$$

Increasing the volume of options sold has two effects on profits: first, an effect on the revenues from the options sale, which is independent of installed capacity. Second, it increases the on-peak contingent liability, hence reduces expected profits. This second effect depends only on aggregate capacity, not on individual capacity choice.

Suppose $(N-1)$ firms choose $\phi^m = \frac{K^T}{N}$, and firm $n$ deviates. Consider first an upward deviation, $\phi^n > \frac{K^T}{N}$. As was the case with physical certificates, condition (9.1) guarantees that the first effect is negative. Then, since the second effect is also negative, an upward deviation is not profitable.

Consider now a downward deviation, $\phi^n < \frac{K^T}{N}$. The first effect is $H(\Phi) + \phi^n H'(\Phi) = r > 0$: Reducing volume of options sold reduces firm $n$'s profits. If we assume aggregate capacity $K \geq K^C(\bar{p}^S)$, we prove that $\frac{\partial \Pi_{RO}^n}{\partial \phi^n}(k, K, \phi^n, \Phi)(K, \phi^n, \Phi) > 0$ for all $\phi^n \in \left[k^n, \frac{K^T}{N}\right]$: a downward deviation is not profitable. $\phi^n = \frac{K^T}{N}$ constitutes a symmetric equilibrium, which is unique. Condition (9.3) simplifies the exposition, as it guarantees that $\Phi = K^T$ is the unique equilibrium of the options market, however it is not essential.

We then move on to the first stage of the game:

$$\Pi_{RO}^{n}\left(k^{n}, K, \frac{K^{T}}{N}, K^{T}, \bar{p}^{S}\right) = \frac{K^{T}}{N} H(K^{T}) + \Pi^{n}(k^{n}, K, \bar{p}^{S}) + \left(k^{n} - \frac{K}{N}\right) \int_{\hat{\theta}_{0}(K, \bar{p}^{S})}^{+\infty} (v(\bar{p}^{S}, \theta) - \bar{p}^{S}) f(\theta) d\theta.$$

Equilibrium capacity is independent of the target capacity $K^{T}$, since only the ratio $\frac{\phi^{n}}{\Phi} = \frac{1}{N}$ appears in the expression. Using techniques presented in chapter 3, Léautier (2016) proves the equilibrium exists and is unique. The marginal value of capacity at the symmetric equilibrium is characterized by

$$\Psi_{RO}(K_{RO}^{C}(\bar{p}^{S}), \bar{p}^{S}) = \Psi_{N}^{C}(K_{RO}^{C}(\bar{p}^{S}), \bar{p}^{S}) + \left(1 - \frac{1}{N}\right) \int_{\hat{\theta}_{0}(K_{RO}^{C}(\bar{p}^{S}), \bar{p}^{S})}^{+\infty} (v(\bar{p}^{S}, \theta) - \bar{p}^{S}) f(\theta) d\theta = 0,$$

where $\Psi_{N}^{C}(K, \bar{p}^{S})$ is the marginal value of capacity at the symmetric equilibrium of a Cournot game, given by Proposition 3.3. Thus,

$$\Psi_{N}^{C}(K_{RO}^{C}(\bar{p}^{S}), \bar{p}^{S}) = -\left(1 - \frac{1}{N}\right) \int_{\hat{\theta}_{0}(K_{RO}^{C}(\bar{p}^{S}), \bar{p}^{S})}^{+\infty} (v(\bar{p}^{S}, \theta) - \bar{p}^{S}) f(\theta) d\theta \leq 0 = \Psi_{N}^{C}(K^{C}(\bar{p}^{S}), \bar{p}^{S}).$$

Since $\Psi_{N}^{C}(K, \bar{p})$ is decreasing in its first argument, $K_{RO}^{C}(\bar{p}^{S}) \geq K^{C}(\bar{p}^{S})$, with equality only if $N = 1$.

Finally, Léautier (2016) proves that $\Psi_{RO}(K^{*}(\bar{p}^{S}), \bar{p}^{S}) < 0 = \Psi_{RO}(K_{RO}^{C}(\bar{p}^{S}), \bar{p}^{S}) \Leftrightarrow K_{RO}^{C}(\bar{p}^{S}) < K^{*}(\bar{p}^{S})$ since $\Psi_{RO}(K, \bar{p})$ is decreasing in its first argument.

For $N > 1$, reliability options curb the exercise of market power: the resulting installed capacity is higher than the Cournot capacity. Thus, they are more effective than physical certificates alone, that have no impact on installed capacity without the "no short sale" obligation.

However, reliability options are not sufficient to completely eliminate market power and restore optimal investment incentives. This result may appear surprising, since reliability options impose a penalty of $(v(\bar{p}^{S}, \theta) - \bar{p}^{S})$ on each unit a producer is short energy. However, a closer examination of the mechanism reveals that, at the symmetric equilibrium, this penalty represents only $\frac{N-1}{N}(v(\bar{p}^{S}, \theta) - \bar{p}^{S})$, which is not sufficient to fully compensate for the "missing money" $(v(\bar{p}^{S}, \theta) - \bar{p}^{S})$.

This results mirrors Allaz and Vila's (1993) analysis of the interaction between spot and forward markets: assuming Cournot competition in both, they show that introducing forward markets reduces but does not eliminate market power and has no impact on a monopoly ($N = 1$).

Finally, observe that generators sell more certificates than installed capacity in equilibrium: $\theta^n = \frac{K^T}{N} \geq \frac{K^C_{RO}}{N}$.

### 9.4.6 Equivalence between Physical Certificates and Financial Reliability Options When "No-Short-Sale" Conditions Are Added

If the SO cannot impose a no-short-sale condition, the previous result proves that financial reliability options yield higher investment. Which one should the SO choose if she can impose a no-short-sale condition? Proposition 9.3 below shows that both mechanisms are equivalent, if the technical parameters are equivalent:

**Proposition 9.3.** Suppose (a) the SO imposes and monitors that the installed capacity equals the options sold by each generator: $k^n = \phi^n$, (b) the wholesale price cap in the capacity market is set equal to the strike price of the reliability option ($\bar{p} = \bar{p}^S$) and satisfies Condition (9.3), and (c) the demand functions for reliability options and for capacity credits are identical and satisfy Condition (9.1). Then, financial reliability options yield the same equilibrium as a capacity market with a no-short-sale condition.

***Proof.*** Since the SO imposes $\phi^n = k^n$, equation (9.2) yields

$$\Pi^n_{RO}(k^n, \mathbf{k}_{-n}) = \Pi^n(k^n, k_1\bar{p}^s) + k^n H(K).$$

If $\bar{p}^S = \bar{p}$ then, with the no-short-sale conditions, $\Pi^n_{RO} = \Pi^n_{CM}$. Thus the equilibria are identical.

As mentioned earlier, since producers sell exactly as many options as their installed capacity, the profit net of the payment on the option is equivalent to a cap on prices. Therefore, if the technical parameters are identical, both approaches are equivalent.

### 9.4.7 Coprocurement of Energy and Operating Reserves

Suppose the SO coprocures energy and operating reserves as described in section 2.6, and that $\overline{v}$, the best estimate of the VoLL, which acts as a price cap, is high enough that it is reached on-peak. Applying the analysis of Cournot equilibrium developed in chapter 3, we find that the marginal value of capacity for a producer at the symmetric equilibrium is

$$\bar{\Psi}^{OR}(K, \bar{v}, c) = \int_{\hat{\theta}^{OR}(K,c,N)}^{\hat{\theta}_0^{OR}(K,\overline{v})} \left( P\left(\frac{K}{1+h(\theta)}, \theta\right) + \frac{1}{N}\frac{K}{1+h(\theta)}\partial_Q P\left(\frac{K}{1+h(\theta)}, \theta\right) - c\right) f(\theta)d\theta$$
$$+ \int_{\hat{\theta}_0^{OR}(K,\overline{v})}^{+\infty} \frac{\overline{v} - c}{1+h(\theta)} f(\theta)d\theta - r,$$

and there exists a unique symmetric equilibrium for which each generator invests $\frac{K_{OR}^{C}}{N}$ defined by

$$\bar{\Psi}_{OR}(K_{OR}^{C}(\bar{v}), \bar{v}, c) = 0. \tag{9.4}$$

**Proposition 9.4.** Suppose the SO runs an energy-cum-operating-reserves market and imposes a price cap set at the best estimate of the VoLL $\overline{v}$. The problem is isomorphic to standard peak-load pricing. $K^{ORC}(\bar{v}) < K^{OR*}(\bar{v})$ unless (a) generation is perfectly competitive ($N \to +\infty$), and (b) the price cap is never binding ($\hat{\theta}_0^{OR}(K, \overline{v}) \to +\infty$).

***Proof.*** The result follows immediately from equations (9.4) and (2.15).

Including an operating reserve market leads to the same investment incentives as average VoLL pricing. This result is surprising: one would have expected the operating reserves market to alleviate the missing-money problem, since (a) all producing units receive a higher price and (b) units providing capacity but not energy are remunerated.

However, these two effects are already included in the determination of the socially and privately optimal capacities $K^{OR*}(\bar{v})$ and $K^{ORC}(\bar{v})$. Then, units providing reserve capacity receive the same profit $(w(\theta) - c)$ as units producing electricity, to avoid arbitrage between markets. No additional profit is generated. The operating reserves market remunerates reserves, which are needed, not capacity investment.

## 9.5 Capacity Mechanisms and Generation Adequacy

The analysis presented in this section borrows heavily from Lambin and Léautier (2018). The analytical description of the industry is different from the remainder of this text. Since we focus on generation adequacy, we assume that demand does not respond to price, while competition is perfect. Specifically, we assume that inelastic load $l$ is distributed on $[0, +\infty]$ according to cumulative distribution function $F(.)$ and probability distribution function $f(.) = F'(.)$. Denote $G(.) = 1 - F(.)$. The value of energy consumed is $V$, the VoLL is $v \geq V$, both expressed in €/MWh. Since all Megawatt-hours consumed are valued at $V$, if rationing is anticipated and proportional, $v = V$, as shown in chapter 2. If rationing is not anticipated, we may have $v > V$.

We consider a single production technology with variable cost $c$ and capacity cost $r$ per unit, both expressed in €/MWh. Off-peak, price is the variable cost of production $c$. On-peak, rationing occurs and the SO sets the price at the VoLL $v$.

This simplest set-up enables us to (a) characterize the net surplus loss from reliability higher than the economic optimum, (b) prove the equivalence between a capacity market and operating reserves, and (c) examine the impact of introducing a capacity market in one market on an adjacent market.

### 9.5.1 Single Market Analysis

#### 9.5.1.1 Equilibrium

The equilibrium-installed capacity $K^*$ is such that the marginal operating profit when rationing occurs is exactly equal to the marginal capacity cost, hence is uniquely defined by

$$(v-c)G(K^*)=r.$$

The probability of rationing at equilibrium is

$$Pr(l>K^*)=G(K^*)=\frac{r}{v-c}.$$

The net surplus is

$$\begin{aligned} W^* &= \int_0^{K^*}(V-c)lf(l)dl+\int_{K^*}^{+\infty}(v-c)K^*f(l)dl-rK^* \\ &= \int_0^{K^*}(V-c)lf(l)dl. \end{aligned}$$

Since we have assumed constant returns to scale, producers' profit is equal to zero. The net surplus is the surplus from consumers. The latter derive no surplus on-peak, since they purchase Megawatt-hours valued at $v$. The net surplus is thus the consumers' surplus off-peak.

Finally, unserved load at the equilibrium is

$$\mathcal{L}^*=\mathcal{L}(K^*)=\int_{K^*}^{+\infty}(l-K^*)f(l)dl=\int_{K^*}^{+\infty}(1-F(l))dl.$$

#### 9.5.1.2 Capacity market

Policy makers may find that the probability of rationing or the unserved load are too high. Given our assumptions, the equilibrium maximizes net surplus. Therefore, requiring a lower probability of rationing or unserved load reduces net surplus. In this example, we suppose policy makers require unserved load does not exceed a target $\mathcal{L}^T<\mathcal{L}^*$. The analysis would be similar if policy makers had imposed a target on the probability of rationing.

To achieve this objective, the SO sets up a capacity mechanism, that is, a payment $m$ to induce investment $K^T=\mathcal{L}^{-1}(\mathcal{L}^T)$. In equilibrium, the payment $m$ is such that expected operating profit (including $m$) is equal the capacity cost:

$$m+(v-c)G(K^T)=r.$$

The probability of rationing is

$$Pr(l>K^T)=G(K^T)=\frac{r-m}{v-c}.$$

Observing that $r=(v-c)G(K^*)$, the payment $m$ can be expressed as

$$m=(v-c)(G(K^*)-G(K^T)):$$

the revenue from the capacity market exactly compensates the true missing money, that is, the revenue lost from setting a capacity target $K^T$ higher than optimal.

Since producers' profit is by construction zero, the net surplus is simply consumers' surplus. Consumers receive $(V-c)$ per unit until $l=K^T$, and pay $mK^T$:

$$\begin{aligned} W^T &= \int_0^{K^T} (V-c)lf(l)dl - mK^T \\ &= W^* + \int_{K^*}^{K^T} [(V-c)l-(v-c)K^T]f(l)dl \\ &= W^* - \int_{K^*}^{K^T} [(v-V)l+(v-c)(K^T-l)]f(l)dl < W^*. \end{aligned}$$

As expected, introducing a capacity market reduces the net surplus: the gain from less frequent rationing $(V-c)\int_{K^*}^{K^T} lf(l)dl$ is lower than the additional capacity cost $mK^T$, for two reasons. First, consumers value energy at $V$ when they are not rationed, while producers receive the VoLL $v \geq V$. This inefficient consumption arises because demand does not respond to price. Second, consumers pay the payment $m$ on all units of capacity, while they receive surplus only on volume $l \leq K^T$.

#### 9.5.1.3 Strategic reserves

Another approach to meet the unserved load target is for the SO to procure strategic reserves. Different mechanisms are implemented. The simplest one is the SO purchases a physical "swing" option from producers, that is, she purchases the right to take delivery of any volume of energy up to her target capacity $K^T$ and pay the agreed-upon strike price, denoted $s \in [c, v]$. Denoting $K$ the equilibrium capacity, that is, excluding strategic reserves, the SO purchases $(K^T-K)$ options to achieve generation adequacy.

The market rules are otherwise unchanged, in particular, $p(l)=c$ for $l \leq K$, and $p(l)=v$ for $l>K$. Therefore, equilibrium capacity is unchanged: $K=K^*$.

As long as $l \leq K^* \Leftrightarrow p(l)=c$, the SO does not exercise her option. When $l>K^* \Leftrightarrow p(l)=v$, the SO exercises her option: she purchases quantity $(l-K^*)$as long as $l \leq K^T$, then $(K^T-K^*)$ for $l \geq K^T$ at price $s$ and resells it at price $v$.[7] Since competition is perfect, producers sell that option at price $\mu(s)$ such that they realize no economic profit:

$$\mu(s)(K^T-K^*)+(s-c)[\int_{K^*}^{K^T}(l-K^*)f(l)dl+(K^T-K^*)G(K^T)]=r(K^T-K^*).$$

7. The SO executes the transaction even if $s=v$, despite capturing no economic gain.

The SO profit from the options is independent of the strike price and negative:

$$\begin{aligned}
\Pi^{SO} &= (v-s)\left[\int_{K^*}^{K^T}(l-K^*)f(l)dl + (K^T-K^*)G(K^T)\right] - \mu(s)(K^T-K^*).\\
&= (v-c)\left[\int_{K^*}^{K^T}(l-K^*)f(l)dl + (K^T-K^*)G(K^T)\right] - r(K^T-K^*)\\
&= (v-c)\int_{K^*}^{K^T}(l-K^T)f(l)dl + (K^T-K^*)[(v-c)(G(K^*)\\
&\quad - G(K^T)+G(K^T)) - r]\\
&= (v-c)\int_{K^*}^{K^T}(l-K^T)f(l)dl + (K^T-K^*)[(v-c)G(K^*) - r]\\
&= (v-c)\int_{K^*}^{K^T}(l-K^T)f(l)dl < 0.
\end{aligned}$$

The sale of energy does not cover the options' purchase price since the latter must cover the cost of the entire committed capacity, while the SO consumes only a portion of the committed capacity. The SO must then levy a charge on consumers to cover the residual cost of the options.

Since generation capacity is $K^T$, identical to the capacity market case, the net surplus is also identical:

$$W = \int_0^{K^T}(V-c)lf(l)dl + K^T[(v-c)G(K^T)-r] = \int_0^{K^T}(V-c)lf(l)dl - mK^T = W^T.$$

### 9.5.2 Multiple Markets Analysis

Consider two identical markets, indexed by $i = 1, 2$. Suppose first they are both energy-only. Since markets are identical, prices are identical for every state of the world, and there are no power flows between them. Each one behaves as an isolated energy-only market: equilibrium-installed capacity is $K^*$, net surplus is $W^*$, and unserved demand is $\mathcal{L}^*$.

#### 9.5.2.1 Capacity mechanism in one market, no export reduction

Suppose now market 2 sets up a capacity mechanism to achieve generation-capacity target $K^T$. As in chapter 6, denote $Q_i^s$ , $Q_i^d$ , $K_i$ and $p_i$ the production, consumption, installed capacity, and prices in market $i$. Assume that transmission capacity is never binding, and that $K_1 \leq K_2$. The dispatch is as follows:

- For $l \leq K_1$, demand in each market is served domestically.

- For $K_1 < l \leq \frac{K_1+K_2}{2}$, producers in market 2 serve demand in market 1: $Q_1^d = Q_2^d = l$, $Q_1^s = K_1$, $Q_2^s = 2l - K_1$, $p_1 = p_2 = c$, and $z = l - K_1$.
- For $\frac{K_1+K_2}{2} < l$, producers in market 2 also produce at capacity: $Q_1^s = K_1$ and $Q_2^s = K_2$.

Suppose first the SO in market 2 cannot reduce exports to market 1 to avoid rationing in her own market. Curtailment occurs in both markets: $Q_1^d = Q_2^d = \frac{K_1+K_2}{2}$, $p_1 = p_2 = v$.

The free-entry condition for generation in market 1 is

$$(v-c)G\left(\frac{K_1+K_2}{2}\right) = r \Leftrightarrow \frac{K_1+K_2}{2} = K^* \Leftrightarrow K_1 = 2K^* - K_2,$$

while the free-entry condition for generation in market 2 is

$$m + (v-c)G\left(\frac{K_1+K_2}{2}\right) = r \Leftrightarrow m = 0.$$

Thus if market 2's SO cannot reduce exports to avoid rationing in her own market, she cannot use a capacity mechanism to achieve her generation-capacity target. Conversely, if an SO wants to achieve a generation-capacity target, she must have the right to reduce exports to avoid rationing in her own market.

This result makes intuitive sense: both markets are perfectly connected by an interconnection; in reality, they are a single electricity market. For the SO of one portion of this single market to implement a generation-adequacy standard, it must rely on an administrative rule to "trap" the Megawatt-hours in her market in time of crisis. Lambin and Léautier (2018) show that this result extends to any distribution of loads (i.e., if markets are no longer identical) and if transmission constraints are present on-peak.

#### 9.5.2.2 Capacity mechanism in one market, selected export reduction

Suppose now market 2's SO is allowed to reduce exports to avoid curtailing her market. Such a rule is not necessarily compatible with the rules governing the European common market. Export reduction requires the cooperation of market 1's SO, and may be difficult to implement in practice. As before, suppose $K_1 \leq K_2$. For $l \leq \frac{K_1+K_2}{2}$, production, consumption, and prices are identical to the previous situation.

For $\frac{K_1+K_2}{2} < l \leq K_2$, both producers produce at capacity: $Q_1^s = K_1$, $Q_2^s = K_2$. Market 2's SO reduces exports to prevent domestic load shedding: $Q_2^d = l$. Exports are $z = K_2 - l$, hence consumption in market 1 is $Q_1^d = K_1 + K_2 - l$. Since rationing occurs in market 1, $p_1 = v$. Rationing does not occur in market 2. However, if market 2's SO were to set $p_2 = c$, producers in market 2 would strongly object to reduced exports. Thus, $p_1 = p_2 = v$.

For $l > K_2$, producers in market 2 no longer export: $Q_1^s = Q_1^d = K_1$, $Q_2^s = Q_2^d = K_2$, and $p_1 = p_2 = v$.

Curtailment in market 2 occurs only when $l > K_2$, thus the SO can achieve its reliability target and set $K_2 = K^T$. The free-entry condition for generation in market 1 is

$$(v-c)G\left(\frac{K_1+K^T}{2}\right)=r \Leftrightarrow \frac{K_1+K^T}{2}=K^* \Leftrightarrow K_1=2K^*-K^T<K^*.$$

Equilibrium capacity in market 1 is determined by the capacity target in market 2; the higher the target in market 2, the lower the installed capacity in market 1. Capacity mechanism in market 2 crowds out investment in market 1.

The free-entry condition for generation in market 2 is

$$m+(v-c)G\left(\frac{K_1+K^T}{2}\right)=r \Leftrightarrow m=0.$$

Market 2's consumers enjoy higher generation adequacy without paying any capacity payment! Market 2's net surplus is therefore unchanged compared with the energy-only market and higher than isolated capacity market:

$$W_2=(v-c)\int_0^{K^*} lf(l)dl = W^*.$$

Thus consumers in market 2 achieve the optimal net surplus corresponding to capacity $K^*$ while the unserved load is $K^T < K^*$.

Market 1's net surplus is also the energy-only one

$$W_1=(v-c)\int_0^{K^*} lf(l)dl = W^*.$$

Consumers in market 1 also receive optimal net surplus $W^*$. However, unserved demand is higher in market 1

$$\begin{aligned}\mathcal{L}_1 &= \int_{K^*}^{K^T}(l-(K_1+K^T-l))f(l)dl + \int_{K^T}^{+\infty}(l-K_1)f(l)dl\\ &= 2\int_{K^*}^{+\infty}(l-K^*)f(l)dl + \int_{K^T}^{+\infty}[l-K_1-(2l-K_1-K^T)]f(l)dl\\ &= 2\mathcal{L}(K^*)-\mathcal{L}(K^T) > \mathcal{L}(K^*),\end{aligned}$$

and increases with $K^T$

$$\frac{\partial \mathcal{L}_1}{\partial K^T}=(1-F(K^T))>0.$$

Total unserved demand is constant

$$\mathcal{L}_1 + \mathcal{L}(K^T) = 2\mathcal{L}^*.$$

Introduction of the capacity mechanism in market 2 leaves prices in each market unchanged, however through an increase in their unserved demand it harms consumers in market 1: equilibrium capacity is reduced ($K_1 < K^*$), hence unserved demand increases ($\mathcal{L}_1 > \mathcal{L}^*$). Therefore, consumers in market 1 "finance" market 2's reduction in unserved demand.

# 10 Concluding Observations

This book's main message is that, twenty-five years after the beginning of the restructuring of the power industry, the economics of electricity markets are by and large well understood. The building blocks are in place: economic models grounded in robust theory have been developed and are tested in a variety of countries and circumstances. This does not mean that we know everything about power markets. On the contrary, section 10.1 reviews a selection of important issues for which additional research is required. As final words, I take a step back from the analysis to reflect in section 10.2 on three broad policy issues that will shape power markets for the years to come.

## 10.1 Economic Research Questions

The list of open research questions related to the economics of power markets is very long. I choose three that have the highest relevance to policy makers: redesigning access and usage rate for the transmission and distribution networks, empirically understanding customers' behavior and preferences, and continuing to improve market design. Solving these problems will require truly interdisciplinary teams.

### 10.1.1 Network Access and Usage Rates Design

The evolution of network rates is a pressing and challenging issue in most developed countries. Historically, users paid a unit price to access and use the transmission and distribution networks equal to the revenues allowed by the regulator divided by the number of Megawatt-hours transiting on the grid (the variable part of the tariff), or by peak demand on the grid measured in Megawatts (the fixed part of the tariff), or by a combination of the two. Since users were almost undifferentiated, this was considered an acceptable approximation. The split between fixed and variable parts varied across countries and was mostly the result of history.

Consumers today are much more differentiated. For example, a consumer who has a solar panel on his roof purchases less energy from the market and at times may resell unused energy to the market. A consumer who has a solar panel and a storage unit may almost never purchase energy from the market. Yet both retain a connection to the grid, which they use as an insurance against days without sun.

As discussed in chapter 6, redesigning the network rate structure is urgent: in many OECD countries, users increasingly opt for decentralized generation. This reduces the number of Megawatt-hours transiting on the grid. Since network costs are essentially fixed, the usage rate per Megawatt-hour charged to remaining customers has to increase, which induces them to opt for decentralized generation. Many utilities in the United States fear that the cycle will continue indefinitely, hence their grid costs will no longer be recoverable.

The challenge is to charge users for the true cost they impose on the network. It has two components: designing the rate and implementing it. The first issue is economic, the second political. Both are intimately linked.

Most practitioners advocate rebalancing the rate structure to increase the fixed part and reduce the variable part of the tariff. This may solve the utilities problem in the short term, but I do not believe it will be sufficient: changing the structure of the distribution rate creates winners and losers. The latter will be vocally against any change and likely to block it.

A more fruitful approach is to leverage the insights of cooperative game theory to design a fair, hence politically acceptable, rate structure. This issue has been extensively studied in a somewhat esoteric branch of the academic literature, where it is known as the cost-allocation problem (Young 1994). It involves sophisticated and highly abstract mathematics. Applying the insights of cooperative game theory to a power or distribution network requires a robust understanding of the cost structure of these networks, in particular, the true cost one user impose on the network, which varies with the user's profile and location. To my best knowledge, no utility has this information today.

This makes for a hard problem to solve. However, this added complexity is everything but superfluous: cooperative game theory is precisely the approach required to design a tariff structure that is politically implementable. Nicolas Astier, a graduate student at the Toulouse School of Economics, working with Michel Lebreton, who teaches at the school and is among the happy few specialists of cooperative game theories, has started this arduous journey (Astier 2017, chap. 2). I am confident he will make great strides in solving that problem and that other researchers will follow in his footpath.

### 10.1.2 Empirical Analysis

While the simple theory underlying power markets is well understood, significant scientific work remains to be conducted to measure empirically the impact of different market structures and policies.

Maybe the most important issue is customers' behavior. Since retail prices have been constant since the beginning of the power industry, little is actually known about (future) consumers' response to price changes. A few experiments have been conducted: Aubin et al. (1995) use data from the critical-peak-pricing retail rate structure implemented in

France in the 1990s, Reiss and White (2005) use data from California residential customers to estimate demand for different usages (e.g., heating, cooling, lighting), Jessoe and Rapson (2014) run a field experiment to estimate how providing information affects consumers behavior, an approach also pursued using a different data set by Kahn and Wolak (2013). Both latter studies confirm that information matters (Jessoe and Rapson's title is self-explanatory: "Knowledge Is [Less] Power"). Still, much remains to be learned.

This issue matters tremendously. Demand response is essential to the decarbonization of electric power generation. It is therefore essential that policies and market rules facilitate and encourage demand response. Grounding these policies and rules in sound empirical analysis will significantly enhance their quality.

### 10.1.3 Further Market Design

Throughout this book, I have made two simplifying assumptions. Relaxing these is fertile ground for further research.

First, I have not opened the black box of price formation, that is, I have assumed that the system and market operators have a mechanism that produces a unique price. A very exciting field of research examines the available price formation mechanisms and looks for the best one. For example, how should day-ahead auctions be structured? Should producers bid an energy price only? Should they bid an entire supply curve? Should they bid their start-up and shut-down costs as well? How should day-ahead and real-time markets be coordinated to limit the exercise of market power?

Technically, this is the field of multi-unit, multi-attribute auctions, which produces some of the finest economic theory research. It is complemented by cutting-edge empirical and operations research work. The power industry presents a specific market design challenge: start-up and shut-down costs link real-time markets together. Which plants are running at this hour is largely determined by which plants were running one hour ago. Technically, this creates nonconvex cost functions.

A number of researchers are exploring these issues. David Salant (2014), who visits the Toulouse School of Economics, has applied his vast knowledge of auction theory to design auctions in power markets. Mar Reguant, who currently teaches at Northwestern University, has studied the impact of inter market linkages on bidding strategies using data from the Spanish market (Reguant 2014) and the arbitrage between day-ahead and real-time markets (Reguant and Ito 2016). Estelle Cantillon, who teaches at the Université Libre de Bruxelles and visits Toulouse, is examining whether pre dispatch information sharing facilitates market power, using data from New Zealand. This research will lead to more robust market designs.

Second, I have limited the results to the long-term equilibrium, while at the same time recognizing that no industry is ever at the long-term equilibrium. The implicit underlying assumption is that an out-of-equilibrium economic system will find its way to the equilibrium. Real markets are characterized by frictions, which may preclude the system from reaching its long-term equilibrium. For example, incumbent generators may keep assets operating as long as their revenue covers their cash cost, thus precluding entry by other more efficient technologies. Out-of-equilibrium analysis is an important field of research, which will also improve the quality of decision making.

### 10.1.4 Interdisciplinary Research

One exciting development is the emergence of truly interdisciplinary research teams, combining, for example, economists, engineers, sociologists, and operations research scientists, to tackle these issues. Cross-disciplinary teams have been heralded for years, yet academic researchers have remained specialized. I believe this is slowly evolving, for the better. For example, economists have teamed up with biologists and climate scientists to tackle climate change issues; Richard Green, professor of economics at the Business School of Imperial College, works closely with his colleagues in the electrical engineering department. We can hope this will become the norm.

## 10.2 Policy Choices

Over the next decade, societies and elected officials who represent them will make a large number of choices directly impacting electricity markets. I briefly discuss three of them: our collective response to the global-warming challenge, the electrification path in developing countries, and the architecture of electricity markets.

### 10.2.1 Response to global warming

Our most important policy choice is our collective determination in facing the global-warming challenge. The significant decrease in the cost of installing RES has translated into an equally significant reduction in the cost of decarbonizing electricity generation. In parallel, the cost of storing electricity is fast decreasing, and new reactor designs are likely to bring down the cost of nuclear energy. The building blocks are all present: in a few years, we will be able to produce electricity without emitting $CO_2$ at a cost comparable to burning fossil fuel. This will then make it possible to decarbonize transport and heating, by replacing fossil fuels with electricity as their energy source.

Being close to having the technology to produce low-$CO_2$ electricity is necessary to prevent global warming, but far from sufficient.

A team of scientists has compiled a list of more than one hundred solutions to stop climate change and evaluated each of them: their cost and their potential reduction in atmospheric tons of $CO_2$-equivalent. This effort, called Project Drawdown, is summarized by Paul Hawken (2017). The team finds that energy alone (mostly wind and solar electric power generation) contributes 23 percent to the total reduction in atmospheric $CO_2$-equivalent. Adding transport, buildings, and cities (mostly electrification of these) contributes another 10 percent of the total reduction in atmospheric $CO_2$-equivalent. By developing the technology to produce low $CO_2$ electricity, we have about one third of the solution. We still have two thirds to go.

Project Drawdown also includes food (30%), land use (14%), materials (11%), and women and girl education, mostly family planning (12%). Included in the solutions examined in Project Drawdown are also reverse $CO_2$ emissions, that is, activities that pump $CO_2$ out of the atmosphere, either plants or mechanical solutions, such as carbon capture and storage.[1]

The breadth of the Project Drawdown list underscores the need for sound and predictable policies, taken by individual governments with international coordination, to spur the required investment. Will our species rise to the challenge? If so, at what speed? Climate-change policies will have tremendous impact on electricity markets. As discussed in chapter 8, here lies a first major divergence between micro-economics, which advocates for a high carbon price, and political economy.

### 10.2.2 Electrification Path in Developing Countries

The second policy choice is the path to electrification that developing countries choose. We all know that more than 1 billion people still have no access to electricity, and we all hope the electricity fairy will reach them by midcentury. The electrification challenge is likely to be an urban one, for two reasons. First, urbanization is progressing inexorably and rapidly in developing countries: the vast majority of humanity will soon live in cities. Second, recent academic analysis (Lee, Miguel, and Wolfram 2016) suggests that the cost of rural electrification in eastern Africa exceeds potential consumers' willingness to pay. The electricity fairy is important, but not of the utmost importance for those leaving in rural poverty. Access to clean water and sanitation appears higher on the priority list. Should this finding prove robust, rural electrification on a large scale will not be funded in poor countries.

In telecommunications, Africa has leapfrogged the fixed network and adopted directly mobile technology. Will the same happen for electricity? Will developing countries move directly to decentralized solutions using minigrids, decentralized production, and storage? Or will they follow the tried-and-true route of centralized production?

### 10.2.3 Architecture of Electricity Markets

Another (and final in this list) policy choice is the market architecture adopted (mostly) in OECD countries. Broadly speaking, countries can choose to rely on wholesale spot-market prices to guide decisions or to rely on administered forward markets, such as capacity markets and a variety of subsidies.

1. The magazine *The Economist* put reversing $CO_2$ emissions on its cover the exact week I am writing up this chapter. What a striking coincidence!

The economists' prescriptions are clear and presented throughout this text. The wholesale market is an energy-only market (chapter 2), with geographically differentiated locational marginal prices to capture the impact of congestion (chapter 6). The spot market, used to make consumption and production decisions, is complemented by forward markets.

No structural intervention is required in the retail sector, as long as switching to a different supplier is easy and information about potential savings from competition is shared widely (chapter 4). Legitimate social concerns are addressed through transfers. Large users opt for real-time pricing, and retailers offer critical-peak pricing to commercial and small industrial users, and possibly residential users (chapter 5). Retail and wholesale electricity markets operate under the watchful gaze of an analytically strong and legally empowered market monitor to prevent exercise of market power (chapters 3, 4, and 7).

Finally, $CO_2$ is priced, either through a market or a tax. Residual market subsidies for RES may persist for a while, under the form of tenders (chapter 8). The bulk of government support to the development of new RES is directed toward R&D.

Of course, policy makers never fully embrace economists' prescriptions. One reason is that there are as many prescriptions as they are economists (if not more). The other is that policy makers incorporate legitimate political concerns in electricity markets' design. Five issues stand out. First, policy makers may object to very high on-peak prices, hence impose a cap an wholesale market prices. Second, they may consider they are responsible to guarantee the reliable provision of electricity, hence they (or their agent) may maintain the historical administrative reliability criterion and design capacity mechanisms (chapter 9) to guarantee that criterion is met.

Third, they may also decide to promote one (or more) technologies and design support mechanisms. As a result, the wholesale spot price no longer represents the opportunity cost of power during on-peak hours, when that information is the most useful. Innovative technologies, such as demand response and storage, have to be remunerated through the capacity mechanism.

Fourth, most policy makers prefer to encourage consumers to resell their electricity to the wholesale market when price is high (Peak-time resell), than to expose these consumers to spot price (real-time pricing), as discussed in chapter 5.

Finally, policy makers in most of Europe object to spatial differentiation of electricity prices within their country, that is, locational marginal pricing, as they feel it would threaten hard-won national unity.

While these choices are legitimate, they run counter to the underlying economics, hence require complex market rules. Furthermore, it is unclear that the resulting market will satisfy these objectives.

Policy makers must find reasonably efficient compromises between economics and political realities.

Unfortunately, there is no silver bullet, that is, no one has proposed a market architecture that is consistent with microeconomics and policy makers' preferences, and efficiently ushers in the energy transition.

Policy makers must therefore find the most efficient compromises between economic and political realities.

It is my hope that readers of this text will design such compromises. Failing to do so would dramatically increase the cost of the energy transition and potentially put the entire transition in jeopardy. This would be a tragedy.

# Afterword

The power industry is going through a period of fundamental transformation that is exciting and compelling. It is vital that public and private decision makers understand and have a full overview of the economic principles that shape the industry. These principles are clearly set out in Thomas-Olivier Léautier's book.

## Historical Background: Economic Research at EDF

Électricité de France (EDF) has always invested in economic research into the power industry. In the 1950s, EDF, under the initiative of Marcel Boiteux, developed the principles of marginal-cost pricing that gave rise to the peak pricing tariffs Effacement Jours de Pointe (EJP) and Tempo, which were deployed in the 1990s. During the 1960s, EDF established a General Economic Research department (EEG), tasked with analyzing the economics of the power industry. That a strong proportion of the group's executives came out of the EEG bears witness to its importance.

In the 1990s, the EEG opened up to the outside world, and EDF began a partnership with what would become the Toulouse School of Economics (TSE). EDF thus contributed to the research on regulation carried out by Jean-Jacques Laffont and Jean Tirole, which helped the latter earn the Nobel Prize in Economic Sciences.

## EDF's Contribution to the Book

From 2006 to 2016, Thomas-Olivier Léautier coordinated TSE's research partnership with EDF. Some chapters in this book are the result of scientific enquiries that were spawned by discussions between TSE researchers and EDF practitioners. In addition, Léautier has taught electricity market economics to a few hundred executives at EDF. This text is, in part, the product of these interactions. We at EDF are proud to have contributed to the development of this book through our joint research and executive education programs. We remain convinced that knowledge is a public good and that producing and disseminating knowledge is an important part of our mission.

## A Few Words about the Book

This is the first book to collate the main economic results that structure power markets. It covers most important topics, from price formation on wholesale markets to transmission pricing, demand response, and the impact of renewable energy producers on wholesale markets. This is a significant achievement and a welcome addition to our bookshelves. In addition, Léautier highlights the difficulties in reconciling microeconomic results and public expectations of the power industry.

## Conclusion

Humankind faces the enormous challenge of reversing $CO_2$ emissions by the end of the century to limit the impact of climate change. Rising to this challenge requires increasing the share of electricity in the energy mix and the share of low-$CO_2$ technologies in the power-generation mix. Over the coming years, billions will be invested in developing low-carbon electricity generation as well as demand response technologies and storage. A solid understanding of the economics underlying power markets is required to design policies that will foster investment and to make the best investment choices. This book offers a significant contribution by setting out the economics with great clarity. EDF is proud to have played a part and warmly congratulates Thomas-Olivier for this tour de force.

Jean-Bernard Lévy
Chairman and Chief Executive Officer, EDF

# Bibliography


Aghion, Philippe, and Patrick Bolton. 1987. "Contracts as a Barrier to Entry." *American Economic Review* 77 (3): 388–401.

Aid, Rene, A. Chemla, A. Porchet, and N. Touzi. 2001. "Hedging and Vertical Integration in Electricity Markets." *Management Science* 57 (8): 1438–1452.

Allaz, B., and J-L. Vila. 1990. "Cournot Competition, Futures Markets, and Efficiency." *Journal of Economic Theory* 59:1–16.

Allcott, Hunt. 2011. "Rethinking Real Time Electricity Pricing." *Resource and Energy Economics* 33 (4): 820–842.

Allcott, Hunt. 2012. "The Smart Grid, Entry, and Imperfect Competition in Electricity Markets." NBER Working Paper No. 18071, May.

Ariely, Dan. 2008. *Predictably Irrational: The Hidden Forces That Shape Our Decisions*. New York: HarperCollins.

Astier, Nicolas. 2017. "Essays on the Economics of Modern Electricity Markets." PhD diss., Toulouse School of Economics.

Astier, Nicolas, and Thomas-Olivier Léautier. 2016. "Demand Response: Smart Market Designs for Smart Consumers." Working paper, Toulouse School of Economics.

Aubin, Christophe, Denis Fougere, Emmanuel Husson, and Marc Ivaldi. 1995. "Real-Time Pricing of Electricity for Residential Customers: Econometric Analysis of an Experiment." *Journal of Applied Econometrics* 10: S171–S191.

Baron, David P., and Roger B. Myerson. 1982. "Regulating a Monopolist with Unknown Costs." *Econometrica* 50 (4): 911–930.

Bell Keith, Richard Green, Ivana Kockar, Graham Ault, and Jim McDonald. 2011. "Academic Review of Transmission Charging Arrangements." Report produced on behalf of the Gas and Electricity Markets Authority, Project TransmiT.

Benabou, Roland, and Jean Tirole. 2016. "Bonus Culture: Competitive Pay, Screening, and Multitasking." *Journal of Political Economy* 124 (2): 305–370.

Bertrand, Joseph. 1883. "Théorie mathematique de la richesse sociale." *Journal des Savants*, 499–508.

Boiteux, M. 1949. "La Tarification des Demandes en Pointe: Application de la Theorie de la Vente au Cout Marginal." *Revue Generale de l'Electricite* 58: 321–340.

Borenstein, S. 2005. "The Long-Run Efficiency of Real-Time Pricing." *Energy Journal* 26 (3): 93–116.

Borenstein, S., J. Bushnell, and S. Stoft. 2000. "The Competitive Effect of Transmission Capacity in a Deregulated Electricity Industry." *RAND Journal of Economics* 31 (2): 294–325.

Borenstein, S., J. Bushnell, and F. Wolak. 2002. "Measuring Market Inefficiencies in California's Restructured Wholesale Electricity Market." *American Economic Review* 92 (5): 1376–1405.

Borenstein, S., and S. Holland. 2005. "On the Efficiency of Competitive Electricity Markets with Time-Invariant Retail Prices." *RAND Journal of Economics* 36 (3): 469–493.

Boyer, Marcel, Michel Moreaux, and Michel Truchon. 2014. "Partage des coûts et tarification des infrastructures." Technical Report 2006MO-01, Centre interuniversitaire de recherche en analyse des organisations (CIRANO).

British Petroleum. 2017. "BP Energy Outlook."

Brunekreeft, Gert, Karsten Neuhoff, and David Newbery. 2005. "Electricity Transmission: An Overview of the Current Debate." *Utilities Policy* 13: 73–93.

Bushnell, J., and S. Stoft. 1996. "Electric Grid Investment under a Contract-Network Regime." *Journal of Regulatory Economics* 10: 61–79.

Bushnell, Jim, Erin Mansur, and Celeste Saravia. 2008. "Vertical Arrangements, Market Structure, and Competition." *American Economic Review* 98 (1): 237–266.

Chao, H. P., and Robert Wilson. 2005. "Resource Adequacy and Market Power Mitigation via Option Contracts." Technical Report 1010712, Electric Power Research Institute. http://stoft.com/metaPage/lib/Chao-Wilson-2003-04-resource-adequacy-options.pdf.

Cournot, Augustin. 1838. *Recherches sur les Principes Mathematiques de la Theorie des Richesses.* Paris: Hachette.

Crampes, Claude, and Matthias Laffont. 2016. "Retail Price Regulation in the British Energy Industry." Working Paper, Toulouse School of Economics.

Crampes, Claude, and Thomas-Olivier Léautier. 2015. "Demand Response in Adjustment Markets for Electricity." *Journal of Regulatory Economics* 48 (2): 169–193.

Cramton, P., and A. Ockenfels. 2011. "Economics and Design of Capacity Markets for the power sector." Mimeo, University of Maryland. http://www.cramton.umd.edu/papers2010-2014/cramton-ockenfels-economics-and-design-of-capacity-markets.pdf.

Cramton, P., and A. Ockenfels. 2016. *Interdisziplinäre Aspekte der Energiewirtschaft. Energie in Naturwissenschaft, Technik, Wirtschaft und Gesellschaft*, chapter Economics and Design of Capacity Markets for the Power Sector. Springer Vieweg, Wiesbaden, June. http://www.cramton.umd.edu/papers2010-2014/cramton-ockenfels-economics-and-design-of-capacity-markets.pdf.

Cramton, P., A. Ockenfels, and S. Stoft. 2013. "Capacity Markets Fundamentals." *Economics of Energy and Environmental Policy* 2 (2): 27–46.

Cramton, P., and S. Stoft. 2006. "The Convergence of Market Designs for Adequate Generation Capacity." White Paper for the California Electricity Oversight Board.

Cramton, P., and S. Stoft. 2008. "Forward Reliability Markets: Less Risk, Less Market Power, More Efficiency." *Utilities Policy* 16 (3): 194–201.

Cramton, Peter, and Jeffrey Lien. 2000. "Value of lost load." Mimeo, University of Maryland.

Deller, David, Monica Giulietti, Joo Young Jeon, Graham Loomes, Ana Moniche, and Catherine Waddams. 2014. "Who Switched at "the Big Switch" and Why?" Technical report, ESRC Centre for Competition Policy, University of East Anglia. http://competitionpolicy.ac.uk/documents/8158338/8194340/Big+Switch+-+Results.pdf/2e01588d-6564-4e28-b06d-233eaad389c4.

Department of Energy and Climate Change. 2011. "Review of the Generation Costs and Deployment Potential of Renewable Electricity Technologies in the UK." Technical report, October.

Department of Energy and Climate Change. 2013. "Electricity Generation Costs." Technical report, July.

Department of Energy and Climate Change. 2013. "UK Greenhouse Gas Emissions, Final Figures." Technical report, February.

Department of Energy and Climate Change. 2015. "Electricity Market Reform: 2015 Update." Technical report, October.

Edgeworth, Francis. 1897. La teoria pura del monopolio. *Giornale degli Economisti*, 40: 13–31.

Edgeworth, Francis. 1925. *Papers Related to Political Economy*, volume 1, chapter "The pure theory of monopoly." London: Macmillan.

European Commission. 2014. "Guidelines on State Aid for Environmental Protection and Energy 2014-2020." *Official Journal of the European Union.*

Faruqui, A., D. Harris, and R. Hledik. 2009. "Unlocking the USD 53 Billion Savings from Smart Meters in the EU." The Brattle Group Discussion Paper, October.

Federal Energy Regulatory Commission. 2013. "JP Morgan Unit Agree to $410 Million in Penalties, Disgorgement to Ratepayers." July.

Frankfurt School-UNEP. 2017. "Global Trends in Renewable Energy Investment 2017." Technical report, Frankfurt School- UNEP Collaborating Centre for Climate & Sustainable Energy Finance.

Friedman, Milton. 1970. "The Social Responsibility of Business Is to Increase Its Profits." *The New York Times Magazine*, 32(13): 122–126, September.

Fabra, N., N.-H. von der Fehr, and M.-A. de Frutos. 2011. "Market Design and Investment Incentives." *Economic Journal* 121: 1340–1360.

Fabra, Natalia, Nils-Henrik von der Fehr, and David Harbord. 2006. "Designing Electricity Auctions." *RAND Journal of Economics* 37 (1): 23–46.

Garcia, Alfredo, and Zhijian Shen. 2010. "Equilibrium Capacity Expansion under Stochastic Demand Growth." *Operations Research* 58 (1): 30–42.

Gibney, Alex. 2005. *Enron: The Smartest Guys in the Room.* Distributed by Magnolia Pictures.

Gilbert, R., K. Neuhoff, and D. Newbery. 2002. "Allocating Transmission to Mitigate Market Power in Electricity Networks." Cambridge-MIT Electricity Project, Working Paper CMI EP07.

Gilbert, Richard, Karsten Neuhoff, and David Newbery. 2004. "Allocating Transmission to Mitigate Market Power in Electricity Networks." *RAND Journal of Economics* 35 (4): 691–709.

Gollier, Christian, and Jean Tirole. 2015. "Effective Institutions against Climate Change." *Economics of Energy and Environmental Policy*, April.

Gollier, Christian, and Jean Tirole. 2017. *Global Carbon Pricing: The Path to Climate Cooperation*, chapter, Effective institutions against climate change. Cambridge, Mass.: MIT Press.

Gowrisankaran, G., S. S. Reynolds, and M. Samano. 2015. "Intermittency and the Value of Renewable Energy." *Journal of Political Economy*.

Green, R. 2007. "Nodal Pricing of Electricity: How Much Does It Cost to Get It Wrong?" *Energy Journal* 31:125–149.

Green, R., and T.-O. Léautier. 2016. "Getting Paid to Stay Idle: Comparing Renewable Energy Support Mechanisms." Mimeo, Toulouse School of Economics.

Green, R., and N. Vasilakos. 2011. "The Economics of Offshore Wind." *Energy Policy* 39 (2): 496–502.

Green, R. J., and N. Vasilakos. 2012. "Storing Wind for a Rainy Day: What Kind of Electricity Does Denmark Export?" *The Energy Journal* 33(3): 1–22.

Green, Richard. 1996. "Increasing Competition in the British Electricity Spot Market." *Journal of Industrial Economics* 44 (2): 205–216.

Green, Richard, and Thomas-Olivier Léautier. 2017. "Do Costs Fall Faster than Revenues? Renewable Electricity Subsidies Dynamics." https://www.tse-fr.eu/fr/publications/do-costs-fall-faster-revenues-dynamics-renewables-entry-electricity-markets.

Green, Richard, and David Newbery. 1992. "Competition in the British Electricity Spot Market." *Journal of Political Economy* 100: 929–953.

Grodecka, Anna, and Karlygash Kuralbayeva. 2015. "The Price vs. Quantity Debate: Climate Policy and the Role of Business Cycles." OxCarre Research Paper 137. https://www.economics.ox.ac.uk/materials/papers/13830/paper137.pdf.

Hansen, Jean-Pierre, and Jacques Percebois. 2012. *Transition(s) electrique(s): ce que l'Europe et les marches n'ont pas su vous dire*. Paris: Odile Jacob.

Harbord, David, and Marco Pagnozzi. 2014. "Britain's Electricity Capacity Auctions: Lessons from Colombia and New England." *Electricity Journal* 27(5): 54–62.

Hart, Oliver, and Jean Tirole. 1990. "Vertical Integration and Market Foreclosure." *Brookings Papers on Economic Activity: Microeconomics*, 205–286.

Hawken, Paul, ed. 2017. *Drawdown: The Most Comprehensive Plan Ever Proposed to Reverse Global Warming*. New York: Penguin Books.

Hecht, Gabrielle, 2009. *The Radiance of France: Nuclear Power and National Identity after World War II*. Cambridge, Mass.: MIT Press.

Hicks, John Richard. 1935. "Annual Survey of Economic Theory: The Theory of Monopoly." *Econometrica*, 3(1): 8, January.

Hogan, W. 1992. "Contract Networks for Electric Power Transmission." *Journal of Regulatory Economics* 4(3): 211–242.

Hogan, W. 2005. "On an "Energy Only" Electricity Market Design for Resource Adequacy." Mimeo, Center for Business and Government, John F. Kennedy School of Government, Harvard University.

Hogan, W., J. Rosellon, and I. Vogelsang. 2010. "Towards a Combined Merchant-Regulatory Mechanism for Electricity Transmission Expansion." *Journal of Regulatory Economics* 38 (2): 113–143.

Holland, Stephen, and Erin Mansur. 2006. "The Short-Run Effects of Time-Varying Prices in Competitive Electricity Markets." *Energy Journal* 27 (4): 127–155.

Holmberg, P., and A. D. Philpott. 2015. "On Supply-Function Equilibria in Constrained Transmission Networks." Mimeo, Research Institute of Industrial Economics (IFN).

Holmberg, Par, and David Newbery. 2010. "The Supply Function Equilibrium and Its Policy Implications for Wholesale Electricity Auctions." *Utilities Policy* 18 (4): 209–226.

Holmberg, Par, David Newbery, and Daniel Ralph. 2018. "Supply Function Equilibria: Step Functions and Continuous Representations." *Journal of Economic Theory* 148 (4): 1509–1551.

Hotelling, H. 1929. "Stability in Competition." *Economic Journal* 39: 41–57.

Intergovernmental Panel On Climate Change. 2014. Fifth Assessment Report. November.

International Energy Agency. 2010. *Projected Cost of Generating Electricity*. OECD/IEA. IRENA (International Renewable Energy Agency).

International Energy Agency. 2015. "Special Report on Energy and Climate Change."

IRENA. 2016. "The Power to Change: Solar and Wind Cost Reduction Potential to 2025." Technical report, IRENA, June.

IRENA. 2017. "Electricity Storage and Renewables: Costs and Markets to 2030." Technical report, IRENA, June.

Jessoe, K., and D. Rapson. 2014. "Knowledge Is (Less) Power: Experimental Evidence from Residential Energy Use." *American Economic Review* 104 (4): 1417–1438.

Jin, Y., and P. Jorion. 2006. "Firm Value and Hedging: Evidence from U.S. Oil and Gas Producers." *Journal of Finance* 61: 893–919.

Joskow, P., and E. Kahn. 2002. "A Quantitative Analysis of Pricing Behavior in California's Wholesale Electricity Market during Summer 2000: The Final Word." CMI Working Paper 02.

Joskow, P., and R. Schmalensee. 1983. *Markets for Power: An Analysis of Electric Utility Deregulation*. Cambridge, Mass.: MIT Press.

Joskow, P., and J. Tirole. 2000. "Transmission Rights and Market Power on Electric Power Networks." *Rand Journal of Economics* 31 (3): 450–487.

Joskow, P., and J. Tirole. 2005. "Merchant Transmission Investment." *Journal of Industrial Economics* 53 (2): 233–264.

Joskow, P., and J. Tirole. 2007. "Reliability and Competitive Electricity Markets." *RAND Journal of Economics*, 38 (1): 60–84.

Joskow, P. L. 2007. "Competitive Electricity Markets and Investment in New Generating Capacity." In *The New Energy Paradigm*, edited by D. Helm. Oxford: Oxford University Press.

Joung, Manho, Ross Baldick, and You Seok Son. 2008. "The Competitive Effects of Qwnership of Financial Transmission Rights in a Deregulated Electricity Industry." *Energy Journal* 29 (2): 165–184.

Kahn, M., and F. Wolak. 2013. "Using Information to Improve the Effectiveness of Nonlinear Pricing: Evidence from a Field Experiment." Mimeo, Department of Economics Stanford University.

Klemperer, Paul D., and Margaret A. Meyer. 1989. "Supply Function Equilibria in Oligopoly under Uncertainty." *Econometrica* 57 (6): 1243–1277.

Kreps, D., and J. Scheinkman. 1983. "Quantity Precommitment and Bertrand Competition Yield Cournot Outcomes." *Bell Journal of Economics* 14: 326–337.

Laffont, J.-J., and J. Tirole. 1993. *A Theory of Incentives in Procurement and Regulation*. Cambridge, Mass.: MIT Press.

Lambin, X., and T.-O. Léautier. 2018. "Capacity Mechanisms across Borders: Who Is Free Riding?" Working paper, Toulouse School of Economics.

Léautier, T.-O. 2000. "Regulation of an Electric Power Transmission Company." *Energy Journal* 21 (4): 61–92.

Léautier, T.-O. 2001. "Transmission Constraints and Imperfect Markets for Power." *Journal of Regulatory Economics* 19 (1): 27–54.

Léautier, T.-O. 2013. "Marcel Boiteux Meets Fred Schweppe: Nodal Peak Load Pricing." Mimeo, Toulouse School of Economics.

Léautier, T.-O. 2014a. "Is Mandating Smart Meters Smart?" *Energy Journal* 35 (4): 135–158.

Léautier, T.-O. 2014b. "Transmission Constraints and Strategic Underinvestment in Electric Power Generation." Working paper, Toulouse School of Economics. https://www.tse-fr.eu/fr/publications/transmission-constraints-and-strategic-underinvestment-electric-power-generation.

Léautier, T.-O. 2016. "The Invisible Hand: Ensuring Optimal Investment in Electric Power Generation." *Energy Journal* 37 (2): 89–109.

Léautier, T.-O. 2018. "On the Long-Term Impact of Exogenous and Endogenous Price Caps: Investment, Uncertainty, Imperfect Competition, and Rationing." Mimeo, Toulouse School of Economics.

Léautier, Thomas-Olivier, and Benoit Peluchon. 2016. "Capacity Mechanisms and Asset Beta." Mimeo, Toulouse School of Economics.

Leautier, T.-O., and J.-C. Rochet. 2014. "On the Strategic Value of Risk Management." *International Journal of Industrial Organization* 37: 153–169, November.

Léautier, Thomas-Olivier, and Veronique Thelen. 2009. "Optimal Expansion of the Power Transmission Grid: Why Not?" *Journal of Regulatory Economics* 36: 127–153.

LeCoq, Chloe, Henrik Orzen, and Sebastian Schwenen. 2017. "Pricing and Capacity Provision in Electricity Markets: An Experimental Study." *Journal of Regulatory Economics* 51 (2): 122–158.

Lee, Kenneth, Edward Miguel, and Catherine Wolfram. 2016. "Experimental Evidence on the Demand for and Costs of Rural Electrification." Working Paper 22292, National Bureau of Economic Research.

Lewis, Michael. 2010. *The Big Short: Inside the Doomsday Machine*. New York: W. W. Norton.

Lijesen, M. 2001. "The Real-Time Price Elasticity of Electricity." *Energy Economics* 29: 249–258.

MacKay, David J. C. 2008. *Sustainable Energy – without the Hot Air*. UIT Cambridge, Cambridge, England.

Mahenc P., and F. Salanie. 2004. "Softening Competition through Forward Trading." *Journal of Economics Theory* 116 (2): 282–293.

Markowitz, H. 1952. "Portfolio Selection." *Journal of Finance* 7: 77–91.

McLean, Bethany, and Peter Elkind. 2003. *The Smartest Guys in the Room: The Amazing Rise and Scandalous Fail of Enron*. New York: Portfolio Trade.

Meyer, Warren. 2016. "Coyote's Bipartisan Climate Plant: A Climate Skeptic Calls for a Carbon Tax." Coyote Blog, March.

Modigliani, F., and M. Miller. 1958. "The Cost of Capital: Corporate Finance and the Theory of Investment." *American Economic Review* 48(3): 261–297.

Modigliani, F., and M. Miller. 1963. "Corporate Income Taxes and the Cost of Capital: A Correction." *American Economic Review* 53(3): 433–492.

Murphy F., and Y. Smeers. 2005. "Generation Capacity Expansion in Imperfectly Competitive Restructured Electricity Markets." *Operations Research* 53 (4): 646–661.

NASA. 2018. Vital Signs. https://climate.nasa.gov/vital-signs/carbon-dioxide/.

Neuhoff, K., J. Barquin, J. W. Bialek, R. Boyd, C. J. Dent, F. Echavarren, T. Grau, C. von Hirschhausen, B. F. Hobbs, F. Kunz, C. Nabe, G. Papaefthymiou, C.h Weber, and H. Weigt. 2013. "Renewable Electric Energy Integration: Quantifying the Value of Design of Markets for International Transmission Capacity." *Energy Economics* 40: 760–772.

Newbery, David. 2008. "Analytic Solutions for Supply Function Equilibria: Uniqueness and Stability." Energy Policy Research Group, University of Cambridge Working Paper 0824.

Newbery, David. 2011. "High Level Principles for Huiding GB Transmission Charging and Some of the Practical Problems of Transition to an Enduring Regime." Report produced on behalf of the Gas and Electricity Markets Authority, Project TransmiT.

Nocke, Volker, and Patrick Rey. 2016. "Exclusive Dealing and Vertical Integration in Interlocking Relationships." Working paper, Toulouse School of Economics.

Nordhaus, William. 1994. *Managing the Global Commons: The Economics of Climate Change*. Cambridge, Mass.: MIT Press.

Nordhaus, William D. 2014. "The Perils of the Learning Model for Modeling Endogenous Technological Change." *The Energy Journal* 35 (1): 1–13.

Oren, Shmuel. 2005. "Ensuring Generation Adequacy in Competitive Electricity Markets." In James Griffin and Steven Puller, editors, *Electricity Deregulation: Choices and Challenges*, 388–415. Chicago: University of Chicago Press.

Ovaere, Martin. 2017. "Economics of Power Transmission Reliability." PhD diss. KU Leuven, Belgium.

Patrick, R., and F. Wolak. 2001. "Estimating the Customer-Level Demand for Electricity under Real-Time Market Prices." NBER working paper 8213, April.

Perez-Arriaga, J. I., F. J. Rubio, and J. F. Puerta Gutierrez. 1995. "Marginal Pricing of Transmission Services: An Analysis of Cost Recovery." *IEEE Transactions on Power Systems* 10: 546–553.

Philpott, Andy, and Geoffrey Pritchard. 2004. "Financial Transmission Rights in Convex Pool Markets." *Operations Research Letters* 32 (2): 109–113.

Proust, Marcel. 1919. *Du côté de chez Swann*. Paris: Gallimard.

Proust, Marcel. 2003. *In Search of Lost Time*. New York: Modern Library.

Reguant, Mar. 2014. "Complementary Bidding Mechanisms and Start-up Costs in Electricity Markets." *Review of Economic Studies* 81 (4): 1708–1742.

Reguant, Mar, and Koichiro Ito. 2016. "Sequential Markets, Market Power, and Arbitrage." *American Economic Review* 106 (7): 1921–1957.

Reiss, Peter, and Matthew White. 2005. "Household Electricity Demand, Revisited." *Review of Economic Studies* 72 (3): 853–883.

Rey, Patrick, and Jean Tirole. 2007. "A Primer on Foreclosure." In *Handbook of Industrial Organization,* edited by Mark Armstrong and Robert H. Porter, 3: 2145–2220. New York: Elsevier.

Rifkin, Jeremy. 2011. *The Third Industrial Revolution*. New York: Palgrave MacMillan.

Rosellon, J., and T. Kristiansen, ed. 2013. *Financial Transmission Rights: Analysis, Experiences, and Prospects*. London: Springer.

Salant, David. 2014. *A Primer on Auction Design, Management, and Strategy*. Cambridge., Mass.: MIT Press.

Sappington, David E. M., and David S. Sibley. 1988. "Regulating with Cost Information: The Incremental Surplus Subsidy Scheme." *International Economic Review* 29 (2): 297–306.

Schlöndorff, Volker. 1984. *Swan in love*. Distributed by Orion Classics.

Schweppe, F., M. Caramanis, R. Tabors, and R. Bohn. 1988. *Spot Pricing of Electricity*. Norwell, Mass.: Kluwer.

Smith, A. 1776. *An Inquiry into the Nature and Causes of the Wealth of Nations*. London: W. Strahan and T. Cadell.

Staffell, Ian, and R. J. Green. 2014. "How Does Wind-Farm Performance Decline with Age?" *Renewable Energy* 66: 775–786.

Stern, Nicholas. 2007. *The economics of climate change: the Stern review*. Cambridge: Cambridge University Press.

Stigler, George J. 1971. "The Theory of Economic Regulation." *Bell Journal of Economics and Management Science* 2 (1): 3–21.

Stoft, S. 2002. *Power System Economics*. Wiley-Interscience.

Thisse, Jacques-Francois, and Xavier Vives. 1988. "On the Strategic Choice of Spatial Price Policy." *American Economic Review* 78 (1): 122–137.

Tirole, Jean. 1988. *The Theory of Industrial Organization*. Cambridge, Mass.: MIT Press.

Tirole, Jean. 2006. *The Theory of Corporate Finance*. Princeton, NJ: Princeton University Press.

Tirole, Jean. 2017. *Economics for the Common Good*. Princeton, NJ: Princeton University Press.

US Energy Information Administration. 2016. "International Energy Outlook."

Vogelsang, I., and J. Finsinger. 1979. "A Regulatory Adjustment Process for Optimal Pricing by Multiproduct Monopoly Firms." *Bell Journal of Economics* 10: 157–161.

Waterson, Michael. 2017. "The Characteristics of Electricity Storage, Renewables and Markets." *Energy Policy* 104: 446–473.

Weitzman, Martin. 1974. "Price versus Quantities." *Review of Economics Studies* 45 (2): 229–238.

Wiser, Ryan, Andrew Mills, Joachim Seel, Todd Levin, and Audun Botterud. 2017. "Impacts of Variable Renewable Energy on Bulk Power System Assets, Pricing, and Costs." Technical report, Lawrence Berkeley National Laboratory, November.

Wolak, Frank A. 2003. Measuring unilateral market power in wholesale electricity markets: The California market, 1998–2000. *American Economic Review* 93(2).

Wolak, Frank. 2004. "What's Wrong with Capacity Markets?" http://stoft.com/metaPage/lib/WolaK-2004-06-contract-adequacy.pdf. Mimeo, Department of Economics, Stanford University.

Wolak, Frank. 2013. Economic and political constraints on the demand-side of electricity industry restructuring processes. *Review of Economics and Institutions*, 4(1):1–42.

Yergin, Daniel. 1990. *The Prize: The Epic Quest for Oil, Money, and Power*. New York: Simon and Schuster.

Yergin, Daniel, and Joseph Stanislaw. 1998. *The Commanding Heights: The Battle for the World Economy*. New York: Free Press.

Young, H. P. 1994. "Cost Allocation." *In Handbook of Game Theory*, with economic applications, Volume 2 chapter 34 2: 1195–1235. Amsterdam: Elsevier Science.

Zottl., G. 2011. "On Optimal Scarcity Prices." *International Journal of Industrial Organization* 29 (5): 589–605.

# Index

Note: Footnote information is indicated with n and note number following the page number.